Springer Series on Atomic, Optical, and Plasma Physics

Volume 101

The Springer Series on Atomic, Optical, and Plasma Physics covers in a comprehensive manner theory and experiment in the entire field of atoms and molecules and their interaction with electromagnetic radiation. Books in the series provide a rich source of new ideas and techniques with wide applications in fields such as chemistry, materials science, astrophysics, surface science, plasma technology, advanced optics, aeronomy, and engineering. Laser physics is a particular connecting theme that has provided much of the continuing impetus for new developments in the field, such as quantum computation and Bose-Einstein condensation. The purpose of the series is to cover the gap between standard undergraduate textbooks and the research literature with emphasis on the fundamental ideas, methods, techniques, and results in the field.

More information about this series at http://www.springer.com/series/411

Victor Kolikov · Alexander Bogomaz
Alexander Budin

Powerful Pulsed Plasma Generators

Research and Application

Springer

Victor Kolikov
Institute for Electrophysics
and Power Energetics
Russian Academy of Sciences
St. Petersburg, Russia

Alexander Budin
Institute for Electrophysics
and Power Energetics
Russian Academy of Sciences
St. Petersburg, Russia

Alexander Bogomaz
Institute for Electrophysics
and Power Energetics
Russian Academy of Sciences
St. Petersburg, Russia

ISSN 1615-5653 ISSN 2197-6791 (electronic)
Springer Series on Atomic, Optical, and Plasma Physics
ISBN 978-3-030-07000-7 ISBN 978-3-319-95249-9 (eBook)
https://doi.org/10.1007/978-3-319-95249-9

Softcover re-print of the Hardcover 1st edition 2018

Printed on acid-free paper

This Springer imprint is published by the registered company Springer Nature Switzerland AG
The registered company address is: Gewerbestrasse 11, 6330 Cham, Switzerland

Foreword

The book contains a lot of the unique experimental data and results of theoretical estimations in the field of powerful pulsed electric discharges in superdense gases. The main motivation for the authors of the book to publish it is the unflagging interest of researchers and practitioners to the electrophysical methods of ultrahigh energy gas plasma obtaining and use of such plasma in physical experiments. At the authors' participation, many studies have been carried out to identify the possibilities for using the low-temperature plasma as a source of intense X-ray and other types of radiation. It has been experimentally established that an increase in the initial gas pressure to hundreds of megapascals is increased the stability of arc burning, efficiency of energy transfer from the arc to the gas, and growth of the enthalpy of the plasma. The obtained data are the basis for the development of electrophysical devices with a high energy density of the gas plasma. These include high-intensity sources of visible, UV, and X-ray radiation for laser pumping, generators of high-enthalpy plasma jets, and plasmachemical reactors. One among the main results of the studies is that the increase in particle concentration in the arc at growth of temperature causes intense X-ray radiation. Besides, it was found that a shock wave reflected from the wall of the electrodischarge chamber compresses the arc and increases the energy density of the plasma and its temperature. Based on the original designs of electric discharge chambers, the highly efficient pulsed plasma generators had been developed those withstand the repeated impact of ultrahigh energy plasma. The book summarizes the results of researches published in articles and reports presented at conferences. The book includes chapters devoted to a detailed description of the results of investigations of both pulsed electric arcs and the design of pulsed plasma generators.

I am sure that this book will be useful both for professionals in this field, and for students of relevant specialties.

St. Petersburg, Russia — Prof. G. A. Shneerson

Preface

This book is based on the results of the experimental and theoretical studies of the electrical mega-ampere pulsed arcs and mechanisms of energy transfer from arc to hydrogen, helium, and air under initial pressure up to 360 MPa, and current up to 2 MA. The principal motive for carrying out of these researches was the pressing need of transition from already mastered in a science and technics range of energy of 10^6–10^7 J to higher values. Such problem, in turn, has demanded the increase in the discharge current amplitude till 1–10 MA and more at time length of the first half-period of the discharge current $\geq 10^{-4}$ s.

It was found experimentally that increasing to hundreds of MPa of the initial gas pressure leads to arc stability, high efficiency of the energy transfer from arc to gas, and plasma enthalpy growth. In the book, the revealed features of such discharges caused in high density of energy in the arc channel and a high pressure of surrounding gas are described. These features are absent or weakly are present in the discharges in less dense environment and with smaller density of plasma energy.

Obtained data are the basis for the development of electrophysics devices with high energy density. Those include high-intensity sources of visible, UV, and X-rays irradiation for the lasers pumping, generators of high-enthalpy plasma jets, and plasmachemical reactors.

The interest of researchers in these studies is explained by achieving the extreme plasma parameters and increasing the energy density in the arc. For these purposes, the X- and Z-pinches, plasma focuses, and multiwire assemblies are used. Furthermore, in these complex electrical devices with a rate of current rise $\geq 10^{14}$ A/s, plasma exists less than one microsecond in the volume not more than few cubic millimeters.

High-density plasma of more volume with the electron concentration of 10^{19}–10^{21} cm^{-3} and temperature of 10^5–10^6 K can be produced by means of Z-pinches devices those have low MHD instabilities. To supply such arcs, it can use the capacitive storages with a lower rate of current rise, i.e., cheaper than mentioned above.

It was found that the growth of the particle concentration in arc at temperature increasing generates the intense X-ray irradiation. The cumulation of the shock wave reflected from the cylindrical wall of the Z-pinch discharge chamber compresses the arc and increases its energy density and temperature.

High-power thermal flows to the electrodes produce such kind of erosions as ejection of surface layer of molten metal from the electrodes and electrode jets. These jets form the turbulent zones and greatly impact the properties of arc, and the heat and mass transfer to the gas. All these are contrasted sharply against the arcs with lower current density, in which the jets do not play a significant role.

Vapor and particles of electrodes' metal under the high concentration change the arc properties. This applies in particular to the sharp growth of the near-electrode voltage drops, which has not still a convincing explanation. The results of our study show that one reason for this phenomenon is the electrode jets.

Under the initial gas pressure of tens and hundreds of MPa, the radiative characteristics of the arc are changing. These changes are caused by growth of temperature and concentration of particles in the arc. As a result, it is changing the transparency of the arc itself, and the transparency of the transition layer between the arc and the surrounding gas. Under these conditions, the effect of radiation imprisonment is observed, and arc temperature is increased. In its turn, growth of arc temperature and concentration of the radiative ions cause the arc compression and increase of energy density in arc due to excess of radiation power above the electric power.

In the book, the description of experimental stands with the maximal storage energy up to 9 MJ and current amplitude up to 10 MA and designs of pulse generators of plasma take the important place, due to which there was the possibility of carrying out the researches in a unique range of parameters. Experiments were realized in hydrogen, helium, and air at initial pressure from 0.1 up to 360 MPa and current amplitude of 0.2–2.0 MA at current rise rate of 10^8–10^{10} A/s. It is necessary to note that by virtue of features of the discharge initiation at high pressures, for this purpose, the metal wire was used. Thereof, and also because of erosion of electrodes in rather "slow" discharges, the channel plasma consisted of ions of metal. In faster discharges for erosion-resistant electrodes (or other ways of initiation), the structure of plasma was defined by surrounding gas.

In our opinion from the results of researches described in the book, the greatest scientific and practical interest is represented with the following:

- The developed designs of plasma generators that allow experiments to be performed at an initial gas molecular concentration of $>10^{22}$ cm^{-3} and currents as high as 10 MA.
- Mode of programmed energy input in an arc which allows to increase essentially an energy input at constant current amplitude.
- Occurrence of a mode of turbulent heat exchange between an arc and surrounding gas due to which efficiency of gas heating can exceed 90%.
- Formation of near-electrode zones with is abnormal voltage drop (>1 kV), caused by intensive electrode jets.

- At high density of working gas, the contribution of shock waves to gas heating considerably grows.
- The mode of the discharge channel contraction which was observed at current amplitude of 0.5–1.5 MA and gas pressure of 5–30 MPa and arose at excess of power of radiative losses over the Joule heating power enclosed in an arc is discovered and investigated.

The main contents of the book are presented in papers, in conferences, and in authors' dissertations. The book may be of interest for both undergraduate and graduate students in related disciplines, and for professionals employed in other fields of studies who would want to further explore this subject in depth.

St. Petersburg, Russia

Victor Kolikov
Alexander Bogomaz
Alexander Budin

Acknowledgements

We would like to thank Yu. Ardentov, L. Bukhteev, Yu. Gusev, M. Heyfits, A. Koval', V. Mamutov, M. Meshkov, N. Solomakhov, V. Tanevsky, and A. Tkachenko for contribution in the research, discussion, and interpretation; former co-workers: D. Andreev, V. Borodin, A. Glukhov, A. Kulishevich, A. Kuprin, B. Levchenko, S. Losev, I. Makarevich, A. Radiushin, A. Savvateev, N. Shirokov, V. Svetova, A. Voronov, S. Zakharenkov, V. Zhuravlev, and M. Znesin for their contribution; co-workers: G. Antonov, A. Leks, V. Leontiev, M. Pinchuk, and A. Pozubenkov for their collaboration. Special thanks go to our mentor Ph. Rutberg for leadership and inestimable help.

St. Petersburg, Russia

Victor Kolikov
Alexander Bogomaz
Alexander Budin

Contents

Abbreviations

AEDC	Arnold Engineering Development Center (USA)
CPS-10	Capacitive power supply 10 kV
CPS-25	Capacitive power supply 25 kV
MEG	Magnetic explosive generators
MHD	Magnetohydrodynamic (generator, instability)
MWC	Pseudoalloy (50% Mo, 45% W, 5–10% Cu)
NRL	Naval Research Laboratory (USA)
OHHM	Oxygen–hydrogen–helium mixture
PG	Plasma generators
PPG	Powerful plasma generators
SXR	Soft X-Ray radiation
WNC	Pseudoalloy (90% W, 1–5% Nil, 5–10% Cu)
WNF	Pseudoalloy (80% W, 10% Ni, 10% Fe)

Symbols

a_o	Speed of sound in cold gas
b	Thickness of skin layer
C	Heat capacity
C_d	Distributed electric capacity
C_s	Capacity of capacitor storage
c	Electric capacity
D	Dissociation energy of hydrogen molecule
d	Debye length
d_a	Arc diameter
E	Electric field intensity
e	Charge of electron
f_d	Frequency band of voltage divider
G_a	Conductivity of arc plasma
g	Acceleration of gravity
I	Current
I_a	Root-mean-square current
I_H	Ionization potential of hydrogen
j	Current density
k	Boltzmann constant
k_{ar}	Irradiation absorption coefficient
k_{bd}	Breakdown coefficient
k_{op}	Ejection coefficient
k_{td}	Turbulent diffusion coefficient
L_a	Arc inductance
L_{dc}	Inductance of discharge circuit
l_a	Arc length
l_d	Voltage divider length
l_{il}	Interelectrode length
l_T	Scale of turbulence
$\ln \Lambda$	Coulomb logarithm

M_1	Ratio of shock wave velocity to speed of sound
m_{am}	Atomic weight of metal
m_m	Weight of vapor metal
m_s	Weight of measuring shunt element
N_L	Loschmidt number
n_a	Concentration of atoms
n_{dp}	Concentration of perturbing particles
n_e	Concentration of electrons
n_{Ha}	Concentration of hydrogen atoms
n_{Hm}	Concentration of hydrogen molecules
n_i	Concentration of ions
n_m	Concentration of metal atoms
P_g	Gas-dynamic pressure
P_m	Magnetic pressure
Q	Electric charge
R_a	Shunt active resistance AC
R_c	Shunt resistance DC
R_{dc}	Active resistance of discharge circuit
R_e	Erosion durability
R_f	Final shunt resistance
R_{hv}	Resistance of voltage divider high-voltage arm
R_i	Initial shunt resistance
r_a	Arc radius
r_i	Average distance between ions
r_1	Outer radius of voltage divider resistance
r_2	Inner radius of divider housing
S_a	Surface area of arc
T	Period of oscillations
T_a	Arc temperature
T_{as}	Surface arc temperature
T_i	Initial gas temperature
T_{mt}	Melting temperature
T_w	Wall temperature of discharge chamber
t_{cur}	Time of current flow
t_p	Discharge pulse duration
t_{rd}	Reaction time of voltage divider
U_a	Voltage drop across arc
U_i	Charge voltage
U_{ne}	Near-electrode voltage drop
u	Atomic mass unit
V_{am}	Maximal volume of arc
V_{ch}	Volume of discharge chamber
v_a	Velocity of arc movement
v_{as}	Rate of arc expansion
v_p	Average thermal velocity of particles

v_{pp}	Velocity of plasma piston
v_T	Velocity of inhomogeneities along temperature gradient direction
W_{ac}	Energy stored in capacitor supply
W_{in}	Energy input
Z	Atomic number
Z_{dc}	Characteristic impedance of discharge circuit
α	Temperature coefficient
α_{Hd}	Hydrogen dissociation degree
α_{Hi}	Hydrogen ionization degree in arc
α_i	Ionization degree
Γ_D	Debye coupling parameter
γ	Adiabatic index
δ	Damping decrement
Δc_V	Difference in van der Waals potential constants of initial and final states
ΔL_a	Change of arc inductance
Δl	Difference in lengths of basic and additional pressure sensors' channels
Δt	Time delay
η	Dynamic viscosity coefficient
λ_t	Turbulent heat transfer coefficient
μ	Magnetic permeability of shunt metal
ρ_g	Gas density
σ	Stefan–Boltzmann constant
ω	Circular frequency of current

Chapter 1
Introduction

Abstract Till now, the study of low-temperature dense gas plasma generated by powerful pulsed electrical arcs is active enough. These researches were begun in the middle of XX century and were inspired mainly by the needs in high-enthalpy plasmas for aerospace and other applications. Currently, physics of high-power pulsed electrical discharges in gases under high and ultrahigh pressure is an essential part of high-energy density physics and physics of extreme states of matter. Relevance and importance of the performing research are determined by the needs of modern science and technology in increasing the energy of plasma above the achieved now of 10^5 to 10^7 J. Solution of this problem involves the implementation of technical tools and scientific methods to perform the experiments with current ≥ 10 MA and discharge pulse duration $\geq 10^{-4}$ s.

The book presents the results of development of plasma generators' designs and study of powerful pulsed electrical arcs in gases at high and ultrahigh pressure. Discharge current in the arcs is $\geq 10^5$ to 10^6 A, initial pressure of the gas is 0.1–200 MPa and duration of the first half-cycle of current is 2–500 μs.

Pulsed arcs of such parameters in comparison with the arcs with lower currents and initial pressures are differed, primarily, by the high energy and greater stability. High temperatures and concentrations enable to use such arcs as sources of UV radiation and soft X-rays [1–9], in various electrophysical installations [10–16], in pulsed plasma generators [17, 18], in high-current switching systems [19, 20], etc.

Pulsed arcs at current of 10^5 A and atmospheric pressure for relatively long time are used in the above electrophysics devices and much experimental and theoretical studies are devoted to them. Since the processes in such arcs are very diverse because of the wide range of parameters, there is no need to examine them all in detail. At the same time, the most important results of these studies will be presented in the review of the works and some of them will be used at further discussion.

The review [21] presents the main directions of researches of high-pressure arcs and shows the main methods of diagnostics of the dense gas-discharge plasma. It is indicated that at increasing of pressure, the dominant component of the energy

V. Kolikov et al., *Powerful Pulsed Plasma Generators*,
Springer Series on Atomic, Optical, and Plasma Physics 101,
https://doi.org/10.1007/978-3-319-95249-9_1

balance of the arc becomes the radiation. Reference [4] was one of the first works to show experimentally that in mega-ampere arcs, the electrical energy is efficiently converted into light energy.

The works [22–25] present the results of the study, the conditions of equilibrium and stability of the high-current self-constricted arc in plasma with varying degrees of transparency. It is indicated that the arc in dense, low-temperature, optically opaque plasma is more stable than in the high-temperature plasma. In the transparent plasma, the main cause of instabilities is the overheating caused by the slow rate of heat flow connected with the radiation, which is unable to compensate the increase of the temperature fluctuations because of Joule heating [23].

Besides, in plasma at increasing density due to nonideality, the speed of sound c is decreased because $c = \sqrt{P/\rho}$, and pressure P is raised slower than density ρ. For example, plasma of iron [26] under 7 eV and increasing of the ions concentration from 10^{17} to 10^{21} cm^{-3}, the speed of sound decreases by $\sim$4 times. Since for the most dangerous long-wave instability in the form of necks, the increment γ is proportional to the speed of sound [27]. At the increasing of density, the time of rise of this instability is increased.

Since the speed of sound is also included in increments of the flexural and shorter wave instabilities, higher particle density in the arc must increase the arc stability. This is true in relationship to the metal plasma, which in our case is formed at the explosion of the initiating wire and electrode erosion. To hydrogen filling the discharge chamber in the most experiments, this effect is less pronounced, as in the case of metal plasma, where nonideality at increasing of plasma density is appeared earlier.

Along with an increase of density in the arc, its stability is increased at increasing of the density around the pinch. At number of experimental and theoretical studies it was shown that in the arc with a voltage of about megavolt, the appearance of the plasma around the denser central part of the pinch, so-called corona, it led to increase its stability. In this case, the main part of the current flows through the outer rarefied and relatively hot part of the pinch—small fraction of its mass [28–31]. The work [32] by the confirmation of this structure was the absence of radiation collapse, which would appear under uniform distribution of the mass along the radius. Continuous plasma generation around an artificial neck during the current rise increases the time of its steady state [30]. Perhaps the increase of the steady state is connected with the constant rise rate of current, on this possibility was pointed out in the work [33].

The data of the work [34] indicate that the Z-pinch in a dense gas is a new type of discharge. In such pinches, the plasma is heated by the ohmic processes, and the magnetic field only keeps the plasma from expanding. Plasma of the dense Z-pinches has no thermal contact with the wall. Involvement of the cold gas into the arc may increase power of the X-ray by increasing of the gas density in the arc. Pinch surrounded by the dense neutral gas is more stable in relationship to the magnetohydrodynamic instabilities than the arc plasma in the vacuum [35, 36].

The idea thermal insulation of hot plasma was proposed in the work [37]. At high temperatures, when the gas is ionized, its thermal conductivity can be greatly reduced by a magnetic field. Such plasma is not has a number of instabilities, which are inherent to vacuum plasma. Besides, this plasma is protected from wall impurities by thick layer of gas at high pressure. In the work [38], the dynamics of the high-temperature pinch with condensed dispersed phase was studied and it was concluded that high pressure and dust phase makes the discharge more stable.

The works [39–45] are presented the results of study of pulsed ultrahigh pressure arcs in hydrogen, helium, and argon at pressure of up to 100 MPa. It was found the formation of a high-temperature area because of the radiation "imprisonment" from the central part of the arc. The estimation the diffusion time of photons from the central area to the periphery of the arc channel was performed. It was pointed out that if the time is longer than the characteristic time of energy input, the axial area of the arc is additionally heated. It is noted that such arcs are opaque to visible light, so at their research is necessary to use soft X-ray radiation that and has been done by the authors. The phenomenon of radiation "imprisonment" in the pulsed high-current arc in hydrogen is indicated also in the work [46]. Radiation "imprisonment" is appeared because of the strong absorption of the radiation by the multiply charged ions. As a result of these factors action, the maximum temperature T in the arc is $\sim 2 \times 10^5$ K and concentration of ions $n_i \sim 10^{18}$ cm^{-3}.

Similar results are also presented in the works [47, 48] where in a capillary discharge at increasing capillary length up to 30–50 mm, a sharp increase in temperature on the arc axis to $(1.5–2.0) \times 10^5$ K took place. This phenomenon is explained the radiation shielding by the vaporized near-wall layer of capillary material. In the standard reference source developed by the authors, at a current of 10 kA and capillary length of 10 mm, the brightness temperature is equal to the gas kinetic 3.9×10^4 K [49]. These studies show that the temperature of the central zone can be increased by reducing the thermal conductivity of the peripheral parts of the arc by increasing the density of the gas.

The important role in the mega-ampere arcs plays the erosion of the electrodes. In these arcs, the electric power is 10^8 to 10^{10} W, main part of which is directed onto the electrodes. Erosion of the electrodes may reach 10^{-3} to 10^{-2} g/C [50, 51], what corresponds to the loss of mass of the electrode per discharge of few grams. This leads to the appearance significant mass of metal in the arc channel that changes the characteristics of the discharge and increases the erosion of the electrodes.

The electrode material is eroded in the form of the individual particles, drops, and vapor consisting of neutral atoms and ions of the electrode metal [52, 53]. The share of these fractions depends on the specific conditions of the discharge: discharge current, current density, duration of discharge, initial pressure, material, and geometry of the electrodes.

For arcs in air at pressure ≤ 0.1 MPa and the current of several kA, these questions are presented in detail in the work [52]. In many papers relating to electrode erosion at current $\geq 10^5$ A, it is indicated that the main erosion is in the liquid phase [20, 51, 54, 55]. In our view, as will be shown below, this is true only at average surface current density $j \leq 10^5$ A/cm^2. In these studies for solving the task on determining the thickness of the electrode melting, the volt equivalent U_e is assumed equal to 10–20 V [20, 55, 56]. It corresponds to the actual heat flow onto the electrode $q = jU_e \sim 10^6$ W/cm^2 and is close to the near-electrode voltage drop $U_0 \sim$ 10–20 V [53, 57].

At mega-ampere discharges in gases of high and ultrahigh pressure, the arc is compressed due to thermal contraction and pinch effect. In this case, the average current density j on the electrode can be up to 10^6 to 10^7 A/cm^2, volt equivalent U_e up to hundred volts [58]. At this conditions heat flow onto the electrode is $q \sim 10^9$ W/cm^2. At such energy densities at the electrode is possible an emergence of new types of erosion.

Estimations from the works [51, 54] show that the main contribution to the magnitude of q is given with radiation. Therefore, such impact onto the electrode can be compared with the laser radiation. The density of the heat flow onto the electrodes under above conditions is comparable with laser radiation of power flow density 10^7 to 10^9 W/cm^2. The effect of laser radiation on the various metals was studied in detail in the works [59–61]. In this case as under the impact of laser radiation can be formed erosion jets those change the properties of the arcs.

Cathode and anode jets in arcs with a current of up to several tens of kA have long been known and studied in detail [19, 62]. Speeds of the cathode and anode jets, concentration of charged particles at the base of the jets, electrode consumption have been determined. Reviews on the results of these researches are presented in the works [63, 64]. In number of studies was investigated the interaction between the cathode and anode jets with each other and opposite electrode [65, 66]. The discharge current in these studies does not usually exceed 10 kA.

In mega-ampere arcs at high-energy input and high pressure of gas, the electrode jets can impact on the properties of the arc and heat transfer to the gas. As noted, value of total near-electrode voltage drops at discharge current of $\sim$60 kA is 500 V [44], and for mega-ampere arcs, this value can be up to $\sim$1 kV [67].

At the study of impact of electrode jets onto the structure of the arc under current $\sim$10 kA, it was seen the formation of luminous shock-compressed areas, which in dependence of the initial conditions are formed in the middle of the discharge gap or in near electrode regions [68]. If the electrode jet reaches the opposite electrode at a supersonic speed, the electrode destruction occurs by melting the metal and then blowing out it by jet.

Destruction of the electrodes may be caused by the jet of plasma from the region of arc contraction [69]. The jet can reach the opposite electrode at supersonic speed only at asymmetric input of the electrode metal into the interelectrode gap. Otherwise, the interaction of jets leads to the formation of the shock-compressed plasma area in the electrode gap [70]. Shock-compressed areas arising at the interaction of the jets can be observed in the visible range of the spectrum, and in the soft X-rays [71]. The formation of the shock-compressed areas are accompanied by the growth and instability of the voltage drop across the arc.

In the area of the jets collision, the high energy is outputted. At collision of opposing deuterium plasma flows with energies up to 600 J at speed of 3.5×10^5 m/s were recorded the neutron emission and soft X-rays [72]. At collision of the plasma flow with the target at speed of 2×10^6 m/s was obtained completely ionized aluminum plasma with electron temperature of 30 keV [73]. Electrons with temperature of 150–300 eV can be obtained in the neck of the cathode jet at current $I \sim 10^3$ to 10^4 A and rise rate $dI/dt \sim 10^9$ to 10^{10} A/s [74].

As noted in the work [67], quasi-stationary arc with current density of 10^5 to 10^7 A/cm^2 has a number of specific properties. One of these properties is a sharp increase in voltage drops near the electrodes, which reaches of several hundred volts. Such growth can be caused by electrode jets, which can cause the turbulence in light gas, which improves heat transfer from the arc to the gas.

In the work [75], it is shown that the cause of abnormally high rate of the gas cooling after the extinction of the current is the turbulent hydrodynamic processes. Based on experimental data, the estimation of the coefficient of turbulent thermal conductivity was performed, which by two orders of magnitude is higher than conventional thermal conductivity at temperature when the gas electric strength is restored. It is assumed that in our experiments, the high thermal conductivity is also due to the electrode jets. One of the tasks was to clarify the mechanism of formation of such mode.

To achieve the high-energy density in the arc, it is necessary to increase the current and pressure. This leads to that the main component of the balance of the energy outputted from the arc is the radiation. Estimation of the energy balance of the arc in hydrogen at pressure of 10 MPa and current up to 10^4 A showed that almost all energy is outputted from the arc by the radiation [76]. This also applies to different configurations of the pinches with mega-ampere current (plasma focus, Z-pinches, X-pinches). Recently interest grows to such discharges due to the fact that they allow obtaining the high-temperature plasma with high density [7–9, 38, 77–79].

Generation of X-rays with low-energy output was carried out in the work [80], where was used the capacitive power supply with energy of few kilojoules. To generate the radiation with energy of hundreds kilojoules and above, the systems with complex high-voltage equipment are required.

In some cases, for example, in the final stages of convergence of the plasma sheath at the point in the "plasma focus" and at the explosion of deuterated polyethylene filaments were obtained the longer steady state of plasma ~200 ns. It should be borne in mind that the duration of the steady-state of plasma is determined by its temperature.

Applications of such discharges are the generation of soft X-ray photons with energy of ~1 keV, and usage them as high-brightness light sources. Generating the hard X-ray in such discharges occurs within a few nanoseconds due to the short lifetime of the radiation sources, known as "hot spots."

For most practical applications, the volume generation of soft X-ray is needed, when the main share of the arc energy is radiated. This raises the problem to obtain more stable state of the arc. Progress in this direction is connected with the search for the new geometries, for example, Z-pinch as a "book" [81].

To stabilize the Z-pinch in the work [82] was used the current sheath with high speed of convergence to the point. In this case, the mode is obtained in which are formed not the individual “hot spots,” but the continuous arc between the electrodes. Its time of existence on an order of magnitude is more than that of “hot spots.”

In the work [83] was investigated the possibility of transition from the channel mode to the mode of the micropinch or “hot spot” in the installation of the “plasma focus.” There were presented the criteria for such transition performance. The increase in steady-state time of Z-pinches would give more energy output from the arc.

Particular importance in this connection has the pinches in the gas of high density [34]. Interest in such pinches is associated with their possible use as a compact source of soft X-rays. The size of the arc in these pinches can be smaller than in conventional arcs, but emission characteristics are higher due to the ability to achieve higher temperature and higher concentration of radiating atoms and ions.

These arcs have a number of features. Since the arcs in the dense gas are more resistant to the MHD instabilities than the arc in a vacuum [35], so for their feeding, you can use capacitor banks with greater discharge duration. In the work [84] is pointed out that during the flow of mega-ampere current more than critical current Pisa–Braginskii through the copper wire with a specific weight of 100 μg/cm, the modes are possible, when time of radiation instability occurrence, which associated with a change in the transparency of the arc, is ~ 1.0 μs. It is noted the complexity of the dynamics of the homogeneous along the axis Z-pinch due to the instability of the steady state occurring during the transition from the volume radiation to the surface radiation.

In general, the possibility to obtain the steady Z-pinch, taking into account the radiation and heat conductivity is considered in the work [85]. The dependence of the steady state of the arc on two dimensionless parameters was researched. First of them characterizes the influence of the external pressure, the second δ characterizing the effect of heat conductivity, and is equal to the ratio of the cooling time due to the loss of energy in the radiation to the time of equalization of the temperature by the thermal conductivity. It is shown that there are δ values, at which the steady state is absent, and the value, at which the only one steady state is possible. In addition, there are values of δ, for which there are two steady states.

Availability of the current of Pisa–Braginskii is a special case of a more general solution of the problem of the steady state of the arcs. Contraction of the high-current arc due to the achievement of the critical current of Pisa–Braginskii I_{cr} [86–88], since the discovery of this phenomenon and to date has been investigated theoretically and described in large number of the works. Already in the first papers [86–89] contraction is associated with excess of radiation losses over the energy inputted in arc that occurs when $I > I_{cr}$. Later, this phenomenon became known as the radiative contraction [90, 91], or the radiation collapse [92–94], what means the nonstop contraction of the pinch under the conditions mentioned above.

In fact, the radiative collapse is continued as long as the energy outputted by radiation, and the energy inputted in the arc is not equalized. Model of the radiative collapse allowed to explain many phenomena in the high-current arcs contracting under the action of its own magnetic field. For example, it allows to describe the formation of the "hot spots" [91], development of constriction in X-pinch [95], dynamics of the plasma focus [96], final stage of which is determined by heavy impurities [97].

It has been shown the possibility of the arc contraction modes while increasing the temperature and density of the particles in the arc [90, 92]. Modeling of the Z-pinch dynamics at radiative contraction has shown that the radiation increases the hold parameter in high-temperature plasma in Z-pinch constrictions, and at discharges in the deuterium, it could lead to increase of the neutrons output [98]. Most of these researches are related to vacuum arcs and arcs under the gas pressure, which is significantly less than atmospheric. Under such conditions, the arcing is occurred in the metal plasma produced by the explosion of the initiating wire and by evaporation of the electrodes and at current of I_{cr} is 100–200 kA. According to results of the work [91],

$$I_{cr} = 0.27\sqrt{\ln \Lambda / K} \text{ (MA)}, \tag{1.1}$$

where $\ln \Lambda$ is Coulomb logarithm, and K is ratio of total power of volume radiation to power of volume bremsstrahlung radiation.

At high emissivity ability of metal due to bar-shaped radiation, when $K \sim 100$, I_{cr} is substantially lower than for the arcs in hydrogen. As described in the work [89], the reduction in the radiation intensity increases I_{cr}. Consequently, I_{cr} is not constant, as at increasing of the current, the radiation in lines is decreased, and therefore, K and I_{cr} are increased [94].

Value of K may vary and for other reasons, for example, due to decrease of transparency of the arc [93]. As shown in the work [89], the velocity of the arc moving at radiative contraction depends on the radiation losses. Therefore, changing the radiative characteristics of the arc, it is possible to change the speed of its contraction in one direction or another. Since, the increase in the concentration of xenon atoms in the deuterium plasma results in growth in the increment of azimuthal radiative instability [99]. And vice versa, complete ionization of the neon reduces the rate of the radiation losses and prevents the development of the radiative collapse [100].

High initial pressure of hydrogen and helium contributes to the emergence of acoustic and shock waves. The emergence of the cylindrical acoustic oscillations at pulsed discharges known for relatively long time and is described in the numerous papers. For example, in one of these works, acoustic oscillations were observed in pulsed tube lamps that have been used to determine the average gas temperature [101].

It is noted that the type of oscillations depends on the method of initiation of the arc. At more powerful discharges at discharge current of 100–600 kA and its rise rate of 10^{11} to 10^{12} A/s, the shock waves are formed, which are focused on axis of the arc [102–106]. For example, in the work [107] was recorded the shock wave, which is passed through the core and is reflected from the arc axis.

At increasing of initial gas pressure, the shock wave amplitude is increased, and at increasing of the gas initial density in which the discharge is occurred, the energy input into the arc is increased. In the work [108], it was showed that into the initiating wire at the stage of its heating in the water it is possible to input two times more energy than in the air.

The study of electrical explosion of conductors in the zone of converging shock wave [104] showed that the increase of the initial gas pressure increases the energy in the metal wire before the explosion. In the work [109], it is noted that the heating of dense plasma on the mutual axis of the plasma jets generates sharp peak of temperature, what can produce the short-wave X-ray radiation. At the initial gas pressure of 10–15 MPa, amplitude of the shock waves can reach the several hundred MPa. Reflection of these waves from the cylindrical wall of the discharge chamber may significantly change properties of the arc. Besides in the arc are generated oscillations associated with the interaction of the magnetic and gas-kinetic pressures.

Let us consider the arc in which the kinetic pressure P_g was placed in equilibrium by the magnetic pressure P_m. We assume that under the action of external disturbance, the arc is contracted and the change of pressure in the arc is adiabatic. That disturbance can be an acoustic wave generated by breakdown of the inter-electrode gap and converging in the center of the discharge chamber after reflection from the cylindrical wall.

At the adiabatic process,

$$P_g V^{\gamma} = \text{Const} \tag{1.2}$$

where V is volume of the cylindrical arc, and γ is adiabatic index.

Since $V \sim r^2, P_g \sim I/r^2\gamma$ and $P_m \sim I^2/r^2$. If the arc is contracted, since the $\gamma > 1$ and current I is constant, then $P_{g,}$ there will be more than P_m, and the arc should expand, and vice versa. This results in the arc oscillations, whose frequency is determined by the speed of sound, radius of the arc and current.

Theoretically, such oscillations were considered in the works [110–113]. In the work [110], it is indicated that the interaction of the plasma pressure acting outward and magnetic force $J \times B$ acting inside causes radial oscillations. Solution of MHD-equations describes the adiabatic radial oscillations of the ideal conducting Z-pinch of DC. If $I \geq I_B$ (I_B is current at which the magnetic pressure is balanced by gas-kinetic pressure), then period of oscillations corresponds to the time of few runs of the acoustic waves from the center to the wall of the discharge chamber.

The sustainability analysis shows that the oscillation period may be shorter than the time constant of the constriction instability that allowing to observe such oscillations at the experiment. In the work [111], the radial MHD-oscillations of the plasma cylinder in the adiabatic approximation were considered. The solution for the frequency of the nonlinear radial oscillations, in which the initial current distribution can be set arbitrarily, was obtained. In the particular case, where the current density is the same over the cross section of the arc, oscillations of the arc boundary occur with a frequency corresponding to the period T

$$T = \frac{56r^2}{I}\sqrt{\frac{mn}{\gamma - 1}}, \tag{1.3}$$

where r is radius of the arc, I is current, m is mass of the metal atom in the arc, and n is concentration of the metal in the arc.

In the works [112, 113] found solution, describing the pinch oscillations during the output of energy from the arc by radiation. The case of the collapsing pinch, when at high current the outputted energy due to radiation higher than the Joule heat input, was considered. As an example, the model of Pisa–Braginskii, when energy output is carried out by bremsstrahlung radiation accompanying electron-ion collisions, was used. It has been shown that, of all dissipative processes the main is output of energy by radiation. If the outputted energy is much less than inputted energy, the pinch performs large number of oscillations near the Bennett-type equilibria in, which the plasma pressure is approximately balanced against the magnetic forces. At increasing of the energy outputted by the radiation, the period, amplitude and number of oscillations are reduced.

References

1. A.F. Aleksandrov, A.A. Rukhadze, Sov. J. Adv. Phys. Sci. **112**, 193 (1974). (in Russian)
2. A.F. Aleksandrov, A.A. Rukhadze, *Physics of High-Current Electrodischarge Light Sources* (Moskow, 1976) (in Russian)
3. A.F. Aleksandrov, O.V. Karpov, G.D. Petrov, O.I. Surov, A.T. Savichev, I.B. Timifeev, Sov. J. High Temp. **16**, 1134 (1978). (in Russian)
4. B.P. Gitrman, D.I. Zenkov, A.I. Pavlovskii, N.N. Petrov, E.N. Smirnov, G.M. Spirov, Sov. J. Tech. Phys. **52**, 1983 (1982). (in Russian)
5. P.L. Usenko, A.N. Subbotin, V.V. Gaganov, Instr. Exp. Tech. **45**, 376 (2002)
6. V.I. Zaitsev, G.S. Volkov, Instr. Exp. Tech. **47**, 498 (2004)
7. K.M. Matzen, Phys. Plasmas **4**, 1519 (1997)
8. R.W. Spielman, C. Deeney, G.A. Chandler, Phys. Plasmas **5**, 2105 (1998)
9. T.A. Shelkovenko, S.A. Pikuz, R.D. McBride, D.B. Sinars, P.F. Knapp, G. Wilhelm, D.A. Hammer, N.Y. Orlov, Plasma Phys. Rep. **36**, 50 (2010)
10. J.G. Kelley, M.A. Levine, A.L. Besse, A. Tatarian, Phys. Fluids, **12**, I-76 (1969)
11. R. Dannenberg, Rocket Techn. and Astron. **10**, 172 (1972). (in Russian)
12. J.V. Makarov, E.K. Chekalin, *Physical Processes in Electromagnetic Shock Tubes* (Moskow, 1968) (in Russian)
13. G.C. Vlases, J. Fluid Mech. **16**, 82 (1963)
14. J.C. Camm, P.H. Rose, Phys. Fluids **6**, 663 (1963)
15. PhG Rutberg, G.A. Shvetsov, A.F. Savvateev, IEEE Trans. Magn. **39**, 29 (2003)
16. K. Lekont, *Physics of Hypervelocity Processes* (Moskow, 1971) (in Russian)
17. I.A. Glebov, F.G. Rutberg, *High-Power Plasma Generators* (Moscow, 1985) (in Russian)
18. I.N. Romanenko, *Pulse Arcs in Gases* (Cheboksarui, 1976) (in Russian)
19. O.B. Bron, L.K. Sushkov, *Plasma Flows in an Electric Arch of switching off Devices* (Leningrad, 1975) (in Russian)

20. D.F. Alferov, V.A. Nevrovskii, V.A. Sidorov, High Temp. **40**, 15 (2002)
21. E.I. Asinovskiy, V.A. Zeigarnik, Sov. J. High Temp. **12**, 1278 (1974). (in Russian)
22. A.A. Rukhadze, S.A. Triger, Sov. J. Appl. Mech. Tech. Phys. **3**, 11 (1968). (in Russian)
23. V.B. Rozanov, A.A. Rukhadze, S.A. Triger, Sov. J. Appl. Mech. Tech. Phys. **5**, 18 (1968). (in Russian)
24. A.A. Rukhadze, S.A. Triger, Sov. J. Appl. Mech. Tech. Phys. **4**, 14 (1969). (in Russian)
25. A.F. Aleksandrov, E.P. Kaminskaja, A.A. Rukhadze, Sov. J. Appl. Mech. Tech. Phys. **1**, 33 (1971). (in Russian)
26. B.V. Zamyshljaev, E.L. Stupitskiy, A.G. Guz', V.G. Zhukov, *Composition and Thermodynamic Functions of Plasma, Handbook* (Moscow, 1984) (in Russian)
27. Ju.S. Protasov, S.N. Chuvashev, *Physical electronics of gas-discharge devices* (Moscow, 1993) (in Russian)
28. P. Kubeš, J. Kravárik, O. Renner, E. Krouský, YuL Bakshaev, P.I. Blinov, A.S. Chernenko, E.M. Gordeev, S.A. Dan'ko, V.D. Korolev, A.Y. Shashkov, Plasma Phys. Rep. **28**, 296 (2002)
29. Y.L. Bakshaev, P.I. Blinov, V.V. Vikhrev, E.M. Gordeev, S.A. Dan'ko, V.D. Korolev, E.A. Smirnova, V.I. Tumanov, A.S. Chernenko, A.Y. Shashkov, S.F. Medovshchikov, S.L. Nedoseev, Plasma Phys. Rep. **27**, 1039 (2001)
30. Y.L. Bakshaev, P.I. Blinov, V.V. Vikhrev, S.A. Dan'ko, V.D. Korolev, B.R. Meshcherov, S.L. Nedoseev, E.A. Smirnova, G.I. Ustroev, A.S. Chernenko, A.Y. Shashkov, Plasma Phys. Rep. **32**, 7 (2006)
31. N.A. Bobrova, T.L. Raznikova, P.V. Sasorov, Sov. J. Plasma Phys. Rep. **14**, 1053 (1988). (in Russian)
32. L.E. Aranchuk, S.L. Bogolubskiy, G.S. Volkov, V.D. Korolev, J.V. Koba, V.I. Liksonov, A.A. Lukin, L.B. Nikandrov, O.V. Telkovskaja, M.V. Tulupov, A.S. Chernenko, V.Ja. Tsarfin, V.V. Jankov, Sov. J. Plasma Phys. Rep. **12**, 1324 (1986) (in Russian)
33. A.E. Robson, Nucl. Fusion **28**, 2171 (1988)
34. N.V. Ardeljan, A.S. Kovalev, A.M. Popov, A.T. Rakhimov, V.A. Feoktistov, Sov. J. Plasma Phys. Rep. **11**, 1167 (1985). (in Russian)
35. W.M. Manheimer, M. Lampe, J.P. Boris, Phys. Fluids **16**, 1126 (1973)
36. C.C. Fälthammer, Phys. Fluids **4**, 1145 (1961)
37. H. Alfven, E. Smars, Nature. Lett. Editors **4753**, 801 (1960)
38. V.P. Vinogradova, M.A. Karakin, V.I. Krauz, A.N. Mokeev, V.V. Myalton, V.P. Smirnov, EYu. Khautiev, V.E. Fortov, Plasma Phys. Rep. **32**, 642 (2006)
39. J.P. Knjazev, E.S. Borovik, R.V. Mitin, V.I. Petrenko, Sov. J. Tech. Phys. **XXXVII**, 523 (1967) (in Russian)
40. V.I. Petrenko, R.V. Mitin, Ju.P. Knjazev, A.V. Zvjagintsev, Sov. J. Tech. Phys. **XXXIX**, 1827 (1969) (in Russian)
41. E.S. Borovik, V.P. Kantsedal, J.P. Knjazev, R.V. Mitin, V.I. Petrenko, Sov. J. Tech. Phys. **XXXVII**, 703 (1967) (in Russian)
42. E.S. Borovik, V.I. Petrenko, R.V. Mitin, V.P. Kantsedal, J.P. Knjazev, Sov. J. Tech. Phys. **XXXIX**, 1416 (1969) (in Russian)
43. V.I. Petrenko, R.V. Mitin, Sov. J. Tech. Phys. **XLV**, 1225 (1975) (in Russian)
44. R.V. Mitin, *Properties of Low-Temperature Plasmas and Diagnostics Methods* (Novosibirsk, 1977) (in Russian)
45. R.V. Mitin, V.I. Petrenko, Y.L. Evetskiy, Sov. J. High Temp. **11**, 1147 (1973). (in Russian)
46. S.G. Alihanov, I.K. Konkambaev, B.S. Estrin, Sov. J. High Temp. **5**, 410 (1967). (in Russian)
47. N.N. Ogurtsova, I.V. Podmyshenskiy, V.M. Shelemina, High Temp. **6**, 48 (1968)
48. N.N. Ogurtsova, I.V. Podmyshenskiy, V.L. Smirnov, Sov. J. High Temp. **14**, 1 (1976). (in Russian)
49. M.I. Demidova, N.N. Ogurtsova, I.V. Podmyshenskiy, V.M. Shelemina, J. Appl. Spectrosc. **23**, 1545 (1975)

50. PhG Rutberg, A.V. Budin, V.A. Kolikov, B.P. Levchenko, V.V. Leontiev, I.P. Makarevich, N.A. Shirokov, High Temp. **32**, 589 (1994)
51. I.V. Tsvetkov, Izvestia RAS **58**, 156 (1994). (in Russian)
52. G.A. Mesyats, *Ectons in the Vacuum Discharge: Breakdown, a Spark, an Arch* (Moscow, 2000) (in Russian)
53. G.A. Mesyats, S.A. Barengol'ts, Adv. Phys. Sci. **45**, 1001 (2002)
54. I.V. Tsvetkov, Plasma Phys. Rep. **19**, 1281 (1993). (in Russian)
55. G.S. Belkin, V.J. Kiselev, Sov. J. Tech. Phys. **XXXVI**, 384 (1966) (in Russian)
56. B.A. Urjukov, *Experimental Researches of Plasma Generators* (Novosibirsk, 1977) (in Russian)
57. I.N. Romanenko, MEI Pap. **70**, 263 (1968). (in Russian)
58. V.P. Ignatko, G.M. Chernjavskiy, in *I All-Union Seminar on Dynamics of the High-Current Arc Discharge* (Novosibirsk, 1990) (in Russian)
59. S.I. Anisimov, Sov. J. High Temp. **6**, 116 (1968). (in Russian)
60. N.N. Ruikalin, A.A. Uglov, A.N. Kokora, *Laser Processing of Materials* (Moscow, 1975) (in Russian)
61. S.I. Anisimov, J.A. Imas, G.S. Romanov, J.V. Hoduiko, *Action of High-Power Radiation on Metals* (Moscow, 1970) (in Russian)
62. A.A. Pljutto, V.N. Rozkov, A.T. Kapin, Sov. J. Exp. Tech. Phys. **47**, 494 (1964). (in Russian)
63. G.A. Lujbimov, V.I. Rahovskiy, Sov. J. Adv. Phys. Sci. **125**, 665 (1978). (in Russian)
64. B. Juttner, IEEE Trans. Plasma Sci. **PS-15**, 474 (1987)
65. M.A. Sultanov, L.I. Kiselevskiy, Sov. J. High Temp. **4**, 375 (1966). (in Russian)
66. A.F. Bretskih, V.I. Suisun, J.D. Hromoy, in *VI All-Union Conference on Physics of Low-Temperature Plasmas* (Novosibirsk, 1989) (in Russian)
67. V.P. Ignatko, in *IV All-Union Conference on Physics of Low-Temperature Plasmas* (Kiev, 1975) (in Russian)
68. M.A. Sultanov, V.A. Ageev, Sov. J. Appl. Spectrosc. **XVIII**, 584 (1973) (in Russian)
69. A.N. Dolgov, G.K. Salakhutdinov, Plasma Phys. Rep. **29**, 757 (2003)
70. M.A. Sultanov, L.I. Kiselevskiy, Sov. J. High Temp. **4**, 40 (1966). (in Russian)
71. H. Chuaqui, M. Favre, R. Saavedra, E.S. Wyndham, L. Soto, P. Choi, C. Dumitrescu-Zoita, IEEE Trans. Plasma Sci. **26**, 1162 (1998)
72. G.N. Dudkin, B.A. Nechaev, V.N. Padalko, V.M. Bystritsky, V.V. Gerasimov, R.V. Kublikov, S.S. Parzhitsky, V.L. Stolupin, J. Vozniak, V.I., Veretel'Nik, E.G. Furman, Plasma Phys. Rep. **31**, 1039 (2005)
73. P.J. Turchi, N.F. Roderick, J.H. Degnan, M.H. Frese, D.J. Amdahl, IEEE Trans. Plasma Sci. **36**, 1 (2008)
74. V.I. Krasov, I.A. Krinberg, V.L. Papernyĭ, E.A. Zverev, Plasma Phys. Rep. **31**, 843 (2005)
75. E.A. A zizov, S.A. Kravchenko, S.G. Solodovnikov, Plasma Phys. Rep. **31**, 616 (2005)
76. A.T. Onufriev, V.G. Sevastjanenko, Sov. J. Appl. Mech. Tech. Phys. **2**, 17 (1968). (in Russian)
77. V.I. Krauz, M.G. Levashova, M.A. Karakin, V.S. Lisitsa, A.N. Mokeev, V.V. Myalton, V.P. Smirnov, O.N. Krokhin, V.Ya. Nikulin, A.V. Oginov, V.E. Fortov, Plasma Phys. Rep. **34**, 43 (2008)
78. S.S. Anan'ev, Y.L. Bakshaev, P.I. Blinov, V.A. Bryzgunov, V.V. Vikhrev, S.A. Dan'ko, A.A. Zelenin, E.D. Kazakov, Y.G. Kalinin, A.S. Kingsep, V.D. Korolev, E.A. Smirnova, G.I. Ustroev, A.S. Chernenko, V.A. Shchagin, Plasma Phys. Rep. **36**, 601 (2010)
79. A.A. Akunets, S.S. Anan'ev, Y.L. Bakshaev, P.I. Blinov, V.A. Bryzgunov, V.V. Vikhrev, I.V. Volobuev, S.A. Dan'ko, A.A. Zelenin, E.D. Kazakov, V.D. Korolev, B.R. Meshcherov, S.L. Nedoseev, E.A. Smirnova, G.I. Ustroev, A.S. Chernenko, V.A. Shchagin, V.G. Pimenov, Plasma Phys. Rep. **36**, 699 (2010)
80. C.R. Negus, N.J. Peacock, J. Phys. D Appl. Phys. **12**, 91 (1979)
81. V.D. Selemir, A.E. Dubinov, E.A. Rjaslov, V.I. Kargin, B.G. Ptitsin, Rus. J. Tech. Phys. **75**, 123 (2005). (in Russian)

82. W. Kies, G. Decker, U. Berntien, Y.V. Sidel'nikov, D.A. Glushkov, K.N. Koshelev, D.M. Simanovskii, S.V. Bobashev, Tech. Phys. Lett. **25**, 802, (1999)
83. R. Lebert, A. Engel, J. Applied Phys. **78**, 6414 (1995)
84. V.N. Dokuka, A.A. Samokhin, Plasma Phys. Rep. **15**, 267 (1989)
85. N.A. Bobrova, T.L. Raznikova, Sov. J. Plasma Phys. Rep. **13**, 92 (1987). (in Russian)
86. R.S. Pease, Proc. Phys. Society **70**, 11 (1957)
87. S.I. Braginskii, Sov. Phys. JETP **6**, 494 (1958)
88. S.I. Braginskii, *Physics of Plasma and Problem of Controlled Thermonuclear Reactions* (RAS USSR, 1958) (in Russian)
89. J.W. Shearer, Phys. Fluids **19**, 1426 (1976)
90. V.V. Vikhrev, JETP Lett. **27**, 95 (1978)
91. V.V. Vikhrev, V.V. Ivanov, K.N. Koshelev, Sov. J. Plasma Phys. Rep. **8**, 688 (1982)
92. S.I. Blinnikov, V.S. Imshennik, Sov. J. Plasma Phys. Rep. **8**, 110 (1982)
93. K.N. Koshelev, N.R. Pereira, J. Appl. Phys. **69**, R21 (1991)
94. E.V. Aglickiy, V.V. Vihrov, A.V. Gulov, *Multicharge Spectroscopy in Hightemperature Plasma* (Moscow, 1991) (in Russian)
95. G.V. Ivanenkov, S.Y. Gus'kov, W. Stepniewskiy, Plasma Phys. Rep. **34**(8), 675 (2008)
96. V.V. Vikhrev, K.G. Gureev, Sov. J. Tech. Phys. **23**, 1295 (1978)
97. K.N. Koshelev, V.I. Krauz, N.G. Reshetnjak, R.G. Salukvadze, J.V. Sidelnikov, E.J. Hautiev, Sov. J. Plasma Phys. Rep. **15**, 1068 (1989) (in Russian)
98. V.V. Vikhrev, Sov. J. Plasma Phys. Rep. **12**, 454 (1986). (in Russian)
99. V.S. Imshennik, V.V. Neudachin, Sov. J. Plasma Phys. Rep. **13**, 1226 (1987). (in Russian)
100. S.V Bobashev., D.M. Simanovskii, G. Decker, W. Kies, P. Röwekamp, Kh. Zol', U. Berntin, Rus. J. Tech. Phys. Lett. **23**, 316, (1997)
101. I.V. Demenik, V.E. Mnuskin, N.I. Sereda, D.P. Solovei, Sov. J. High Temp. **8**, 443 (1970). (in Russian)
102. S.G. Alihanov, E.L. Bojarintsev, V.A. Kornilov, T.S. Melnikova, Sov. J. High Temp. **5**, 970 (1967). (in Russian)
103. K.V. Dubovenko, Rus. J. Tech. Phys. **62**, 83 (1992). (in Russian)
104. A.G. Rousskikh, V.I. Oreshkin, AYu. Labetsky, S.A. Chaikovsky, A.V. Shishlov, Tech. Phys. **52**, 571 (2007)
105. V.S. Komelkov, J.V. Skvortsov, V.N. Tereshchenko, Sov. J. Tech. Phys. **XXXIII**, 719 (1963) (in Russian)
106. E.M. Barkhudarov, I.A. Kossyi, N.M. Tarasova, S.M. Temchin, M.I. Taktakishvili, T.S. Zhuravskaya, V.A. Levin, V.V. Markov, N.A. Popov, Plasma Phys. Rep. **35**, 924 (2009)
107. S.Y. Gus'kov, G.V. Ivanenkov, A.R. Mingaleev, S.A. Pikuz, V.B. Rozanov, T.A. Shelkovenko, V.V. Nikishin, V.F. Tishkin, W. Stepniewskiy, D.A. Hammer, Plasma Phys. Rep. **26**, 745 (2000)
108. A.E. Ter-Oganesyan, V.M. Romanova, A.R. Mingaleev, T.A. Shelkovenko, S.A. Pikuz, S.I. Tkachenko, Plasma Phys. Rep. **31**, 919 (2005)
109. G.A. Askar'yan, S.V. Bulanov, I.V. Sokolov, Plasma Phys. Rep. **25**, 549 (1999)
110. F.S. Felber, Phys. Fluids **25**, 643 (1982)
111. L.S. Solov'ev, Sov. J. Plasma Phys. Rep., **10**, 1045 (1984) (in Russian)
112. B.E. Mejerovitch, Sov. J. Plasma Phys. Rep. **11**, 1446 (1985). (in Russian)
113. B.E. Mejerovitch, *High Current Channel* (Moscow, 1999) (in Russian)

Chapter 2
Test Benches of the IEE RAS' Pulsed Plasma Generators

Abstract Two test benches have been designed to perform research with powerful pulsed plasma generators operating at the IEE RAS. Benches are equipped with capacitive power supplies for plasma generators and measuring hardware for registration of the parameters of pulsed arc in various gases of high density, and properties of the low-temperature plasma. Test benches differ in their energy performance and diagnostic capabilities and hereinafter called as "small" and "big" ones. "Small" test bench is equipped with a capacitive power supply CPS-10 based on the capacitor bank voltage of 10 kV, energy store 6 MJ, and a maximum discharge current of 2 MA. CPS-10 has option the programmable input energy in arc and is used, currently, for power supplying the powerful pulsed plasma generator PPG-3, diagnostic discharge chamber, and electrodischarge installation with adiabatic compression of gas. Powerful pulsed plasma generator PPG-3 is designed for large-scaled tests under maximal inputted energy. Diagnostic discharge chamber is applied for in-depth study of parameters of pulsed arcs using optical techniques. Electrodischarge installation of preliminary adiabatic gas compression is used for studying the arcs at ultra-high pressure. "Big" test bench with the programmable capacitive power supply CPS-25 voltage of 25 kV, energy 9 MJ, discharge current of up to 10 MA, and the world's highest current integral action is used for power supplying of powerful pulsed plasma generator PPG-6 for research of arcs, burning at a current up to 2 MA, and at ultra-high initial pressure.

2.1 Construction and Equipment of the Test Benches

To perform the investigations of mega-ampere pulsed electric arcs and low-temperature plasma, the test benches besides plasma generators and power supplies must include:

- Gas preparation and supply system
- Control and measuring system for the initial parameters of the experiment
- Equipment for measuring and registering

V. Kolikov et al., *Powerful Pulsed Plasma Generators*,
Springer Series on Atomic, Optical, and Plasma Physics 101,
https://doi.org/10.1007/978-3-319-95249-9_2

- Devices for registering, processing, and storage of experimental data
- Supporting equipment.

Gas preparation and supply system includes containers for transportation and storage of gas, high pressure compressor, and pipelines with valves.

Control and measuring system includes control room devices for measuring voltage of the capacitor bank, initial pressure of gas, thyristor regulator of capacitor bank charging current.

Vacuum pumps are used, if necessary, for vacuuming the discharge chamber before filling it by gas.

Measuring and registering equipment includes pulses current sensors, voltage dividers, pulsed pressure sensors, sensors of X-rays, high-speed shooting camera, shadow and interferometry devices, and spectrographs.

Supporting equipment are analytical balances for measuring weight of electrodes before and after experiment to determine the erosion, and some other devices.

2.2 Power Supplies for Pulsed Plasma Generators

Most expensive and oversized component of the test bench is the power supply. At relative diversity of power supplies systems for pulsed plasma generators and other electrophysical devices, such as electric machines [1–3], cryogenic and superconducting inductive energy storages [4, 5], compulsators [6, 7], MHD-generators [8–10], ME-generators [11–13], and even nuclear reactors [14]. MHD-generators there are some very complex devices with low energy conversion efficiency. ME-generators are single-acting sources of supply, in fact, is a powerful bomb. Thus, the most prevalent and easy-to-use are capacitive power supplies. There are several reasons for those:

- Capacitive power supplies are consisted mainly of standard components (capacitors, cable, vacuum switches, et al.)
- Energy of capacitive power supply is determined only by the number of capacitors
- Voltage of capacitors is varied from few to hundreds of kV
- Capacitive power supplies can be fitted with option of programmable input of energy in arc
- Capacitive power supplies as opposed to the electric machines do not mechanical impact on test bench foundation.

At the same time, the capacitive power supplies have disadvantages, the main of which is the big weight and size indicators. However, this drawback as improvement of capacitors is eliminated. Already capacitors with energy density 2–3 J/cm^3 available are available [15–18], and now capacitors are developed with energy density 10–30 J/cm^3 and more [19, 20]. These capacitors allow developing the compact power supplies for pulsed generators and other electrophysical devices.

With that unlimited increase of stored energy of capacitive power supplies increases, the probability of the capacitor explosion, which at energy ~100 MJ, is very high.

2.2.1 Capacitive Power Supply CPS-25

The power supply CPS-25 of voltage 25 kV and stored energy of 9 MJ (design energy is 17 MJ) is used to supply powerful pulsed plasma generators and to study phase transitions in carbon at very high pulsed pressure.

CPS-25 is modular capacitive energy storage (Fig. 2.1). The half-cycle of the discharge current in short-circuit mode is 65 μs, and maximal discharge current is 10 MA. Design of energy storage allows simultaneous supply of several loads. Energy storage control system provides the ability to operate both in the single-pulse mode, i.e., simultaneous switching-on of all modules, and in the programmable mode, when module is switched on in required time. Under programmable mode, the required pulsed current form and increase of the discharge duration up to 3 ms are achieved.

CPS-25 characteristics:

- Stored energy—9 MJ
- Voltage—25 kV

Fig. 2.1 Power supply CPS-25 [21]

- Maximum discharge current—10 MA
- Frequency of discharge current at short circuit—7.7 kHz
- Number of modules—12
- Stored energy of module—750 kJ
- Number of cells in the module—8
- Capacity of storage—53 mF
- Electric charge of storage—1325 C
- Dimensions—17 × 12 × 9 m.

At the IEE RAS has developed the capacitive energy storages differing in stored energy, voltage, discharge current, and its rise rate [22, 23]. Stored energy is from 0.1 to 9 MJ, voltage from 5 to 50 kV, maximal discharge current from 0.1 to 10 MA, and discharge duration from few μs to 3 ms.

The design of the power supply CPS-25 is modular one (Fig. 2.2). Currently in operation 12 modules placed in three stories. The capacity of all modules is 0.0288 F, at charging voltage of 25 kV the stored energy is 9 MJ. In short circuit, the discharge current is 10 MA, and duration of the half-cycle is 65 μs.

Each of twelve modules includes eight capacitor cells, ignition unit (1), charging (5), and discharging (6) contactors. Control pulses from synchronization unit (2) are

Fig. 2.2 CPS-25 power supply modules [21]

delivered to the ignition unit (1). Charging and discharging of capacitive storage are carried out by the remote control unit (3) and computer (4) located in the shielded chamber (Fig. 2.3).

Capacitor cell (Fig. 2.4) is the basic unit of power supply; its circuit diagram is shown in Fig. 2.5.

The cell concludes the parallel connected capacitors C_1–C_{10} each of 30 μF, vacuum spark-gap (1) with ignition system, and charging and discharging resistors R_1, R_2. The capacity of the cell is 300 μF. Spark-gap (1) switches on discharge current up to 150 kA.

The elements C_{11} and R_3 are used to improve the stability of the spark-gap operation. The cell is connected to the collector by means of four coaxial cables (5). The frequency of short-circuit current of the cell is 25 kHz. The control pulse (Ignition) is delivered to the ignition transformer (2) from the ignition unit (in diagram it is not shown).

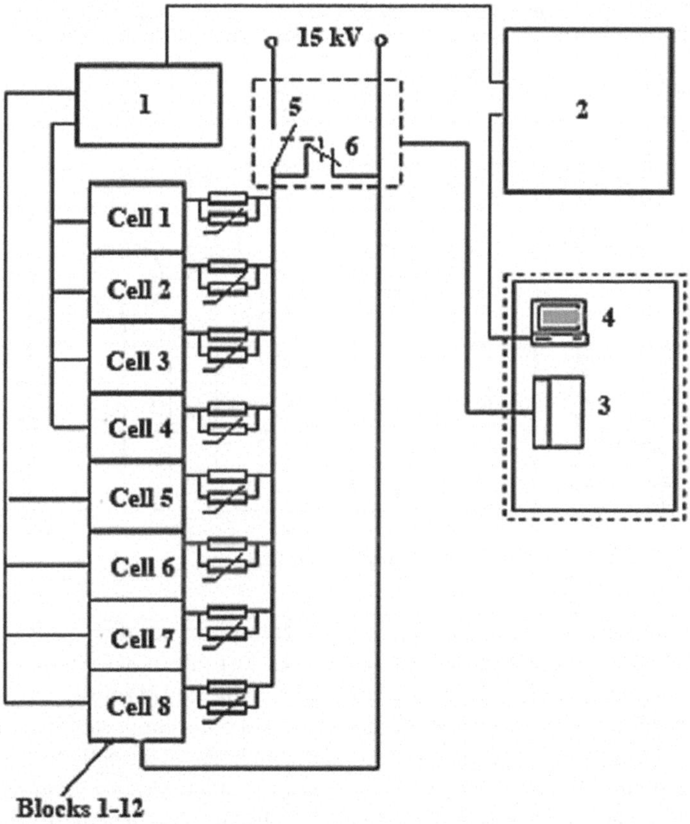

Fig. 2.3 Circuit diagram of the capacitive storage CPS-25 [21]

Fig. 2.4 Capacitor cell of the CPS-25 [21]

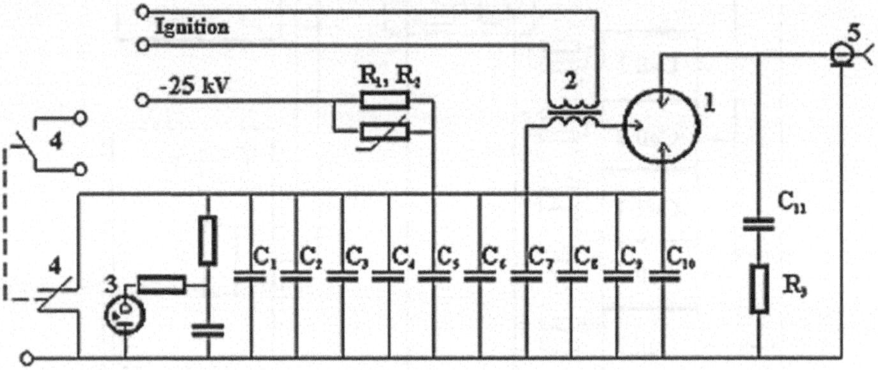

Fig. 2.5 Circuit diagram of the CPS-25 capacitor cell [21]

The resistors R_1, R_2 limit the charging current and in accident limit the current between the cells, increasing the reliability and safety of the power supply operation. The cell has a protective fence, when which removed the key (4) closes the capacitors. The discharge current is measured by the Rogowski coils connected in series.

Module. Module of the capacitive storage concludes eight cells by capacitance 2400 μF and stored energy 750 kJ. The control system allows charging independently any module to the required voltage. Cells are connected to high-voltage feeder separately by two parallel resistors: wire resistor R_1 of 1 kΩ and power of

500 W; and assembly of silicon oxide nonlinear resistances R_2 (Fig. 2.5). These resistors prevent both discharges of the cells to trouble one and recharge between the cells at programmable mode. The module is equipped with the high-resistant voltage divider for measurement the initial and residual voltages.

Module includes two high-voltage keys: normally open to connect the module to the feeder; normally closed to neutralize the electric charge. At neutralizing of the capacitors charge, most of the energy is released in the assembly of nonlinear resistors R_2. At charging, voltage of 25 kV temperature of resistor R_2 at discharge is increased by 40 °C, and the heating of the wire resistor R_1 is negligible.

Control system of the capacitive storage concludes the synchronization block and twelve blocks of spark-gaps initiation. Each initiation block has two pulsed generators, producing four synchronous pulses of 1.2–1.6 kA, voltage of 8 kV, and duration of 5 μs.

Pulsed generator is controlled by the synchronization block, produces pulses of 12 A, 600 V, and 5 μs.

Synchronization block generates two modes:

- Synchronous, when all 24 pulse generators forming single pulse;
- Asynchronous, when pulses are generated with delays controlled by computer [24, 25].

At programmable mode, the value and shape of the current are determined not only by the discharge circuit parameters and initial voltage of capacitors, but also by the time between switching-on of the modules in the discharge circuit. Varying these time intervals one can form the required shape and amplitude of the current, and realize optimum energy input in the load.

Current collector. The collector connects the load to the power supply at the discharge current up to 10 MA and voltages up to 25 kV (Fig. 2.6). The collector consists of six blocks (1), each block connects 80 coaxial cables (3). Insulation of the block tested at twice operating voltage, i.e., 50 kV. Now to each block, 64 cables are connected from 16 cells, i.e., from two modules.

Thus, the module of the capacitive storage can be divided into two parts, each part can supply the own load, that enables to supply 6 independent loads. At operating on a single load all blocks are connected to the collector output buses (2). Inductance of collector is 1.6 nH. Blocks and output buses are placed on the supporting frame. Output buses are isolated from each other by multi-layer polyethylene insulation of 1 mm thickness (5). To compensate the electrodynamic forces occurring during the discharge current, the six loads (6) by total weight of 6000 kg are used.

Elements mentioned above together with the output cables (4) form the discharge circuit with the following parameters:

- Inductance 13.3 nH
- Wave resistance 0.68 mΩ
- Frequency of the discharge current at short circuit 7.7 kHz.

Fig. 2.6 Collector of the CPS-25 [21]

To the output buses, the different loads can be connected or directly to the collector buses, either by coaxial intermediate current collector (Fig. 2.7).

Charge unit. The device is used to charge the capacitor bank by rectified and stabilized current to 25 kV. The main element of the device is the three-phase oil transformer of power 160 kW, voltage on the secondary winding of 25 kV, and voltage on the primary winding 380 V. Rated current of the primary winding is 557 A, and secondary current is 23 A. The charging time of the capacitor bank to voltage of 25 kV is ~5 min.

Connecting of the modules to the charging circuit is carried out by means of high-voltage contactors (4) (Fig. 2.5).

Control system. The control system is used for registration of the voltage of the capacitor bank and gas initial pressure in the discharge chamber of the plasma generator. For measuring voltage of the capacitor bank, the high-voltage dividers are used. The number of measuring circuits is 24. The voltage from the low-voltage shoulder of the voltage dividers is delivered to the control unit. Visual control of the charging process is carried out by the digital voltmeters. After the charging, voltage at the module is registered by computer. For each of the 24 channels, several countdowns are performed, as long as the difference between countdowns not more

Fig. 2.7 Intermediate current collector (left) connected to the generator PPG-5

than 0.1%. Voltage of the modules is recorded in the experiment file. The residual capacitor bank voltage is measured after the experiment in a similar manner. Voltage measurement error does not exceed 0.5%. Registration of the initial gas pressure in the discharge chamber of the generator is performed by potentiometric pressure sensor mounted on the feed gas pipeline.

Data registration and processing system. Registration of the experiment parameters by the two-channel digital oscilloscopes is performed. Oscilloscopes run by clock pulse from the capacitor bank synchronization system. Recording, processing, and storage of data are performed by the computer.

During the experiment are recorded as follows:

- Gas initial pressure in the discharge chamber
- Charge and remaining voltage on the modules
- Electric charge of the modules
- Time delay of the modules switch-on
- Signals of the measuring devices at the time of start-up
- Discharge current of the modules and total discharge current
- Voltage drop across the arc
- Flowed electric charge
- Pulsed pressure in the discharge chamber of the generator
- Signals of the magnetic probes and other sensors
- Clock pulses of the photorecording devices.

After the experiment, the results of direct measurements are processed and are calculated: power of the arc; energy inputted in the arc; arc resistance; current integral, etc.

2.2.2 Capacitive Power Supply CPS-10

Power supply CPS-10 of 6 MJ energy, and voltage of 10 kV as well as the CPS-25 is used to carry out the researches of the pulsed plasma generators (Fig. 2.8).

Capacitance, inductance, internal resistance and charging voltage of supply, resistance and inductance of the discharge circuit and arc define uniquely the amplitude-frequency characteristics of the current pulse. However, it is valid when the load is linear (in our case it is not so), and power supply and discharge circuit parameters are constant.

The arc is nonlinear load, which has variable both resistance and inductance. Parameters of the discharge circuit elements such as current collector, current input, connecting cables, and spark-gaps are also vary in dependence on the shape of the discharge current and its rise rate.

To determine inductance, wave and active resistance of the discharge circuit, and the capacitor bank of the power supply CPS-10, a series of experiments at short circuit was carried out. The current I in oscillatory circuit with capacitance C, inductance L, and active resistance R_{dc} is determined as

$$I = U_{\mathrm{i}}\sqrt{\frac{C}{L}}\mathrm{e}^{-\delta t}\sin\omega t. \tag{2.1}$$

Fig. 2.8 Power supply CPS-10 [26]

where U_o is charge voltage, δ is damping constant, ω is circular frequency of the current, t is time.

Knowing the period of oscillation T, we find circular frequency $\omega = 2\pi/T$. Decrement damping δ is the ratio of the adjacent current amplitudes I_1/I_2

$$\delta = \frac{2\ln(I_1/I_2)}{T}. \tag{2.2}$$

Based on ω, δ, and bank capacity c, we determine the resistance R_{dc}, inductance L, and wave resistance Z_{dc}

$$R_{dc} = 2\delta L, \tag{2.3}$$

$$L = \frac{1}{C(\omega^2 + \delta^2)} \tag{2.4}$$

$$Z_{dc} = \sqrt{L/C}. \tag{2.5}$$

During studies, some modernizations of the power supply were carried out; together with those its characteristics were changed.

At first embodiment of the battery, all capacitors were connected parallel that allowed working at voltage up to 5 kV. At this configuration, the power supply was used in the first test, and then, based on results of obtained data, it was decided to connect the capacitors at series–parallel manner.

After that, two significant modernizations of the power supply were performed, resulted in doubling of capacity of the battery in comparison with the initial value.

Now the power supply CPS-10 consists of 4320 capacitors of 100 μF, 50 nH, and voltage of 5 kV. The stored energy in one capacitor at rated voltage is 1.25 kJ. Power supply circuit is shown in (Fig. 2.9). Main characteristics of the power supply CPS-10 are as follows:

- Capacity 0.113 F
- Energy stored at voltage of 10 kV—5.6 MJ
- Resistance of the discharge circuit 1.46 ± 0.08 mΩ
- Inductance of the discharge circuit 0.248 ± 0.015 μH
- Impedance 1.5 mΩ
- Current rise rate 3.2×10^{10} A/s.

The capacitor bank is divided into six independent modules C_1–C_6, each may independently be charged to required voltage by means thyristor voltage regulator TC, high-voltage transformer Tr, rectifier R, charging resistors $R_1{}^3$–$R_6{}^3$, and vacuum switches k_1–k_6.

By means of air trigatron dischargers P_1–P_6, the modules C_1–C_6 are connected to the load R_l that allows controlling the duration, shape, and amplitude of the discharge current. The neutralization of capacitor bank charge is carried out by

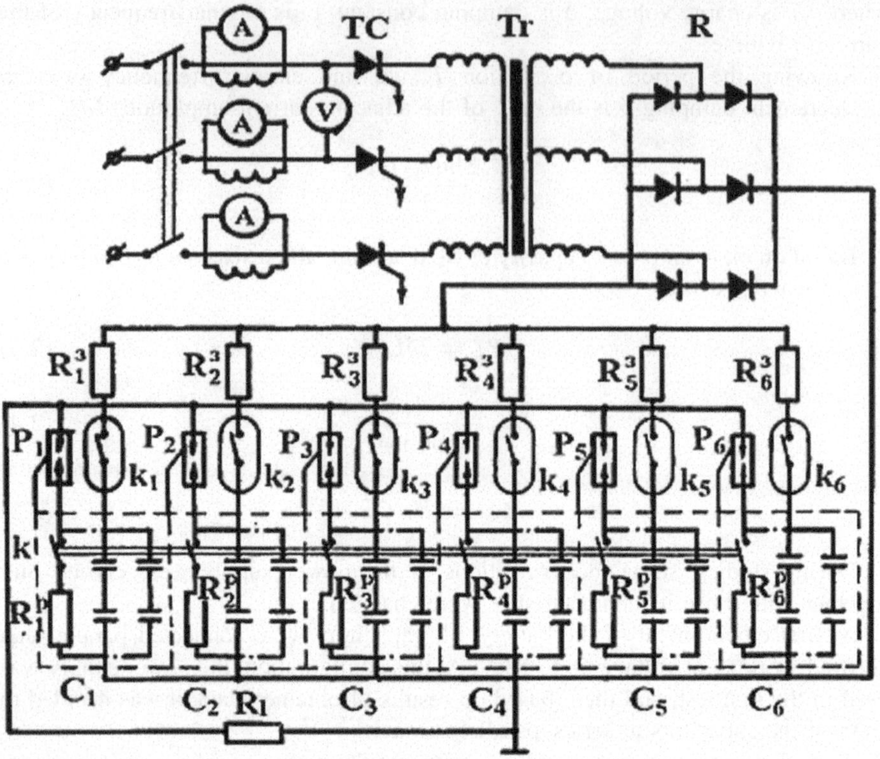

Fig. 2.9 Power supply circuit of the CPS-10 [26]

closing of the high-voltage six-pole contactor *k* through a discharge resistors $R_1{}^p$–$R_6{}^p$ to the ground.

The capacitor bank concludes the 144 cells, each with the 30 capacitors. The cell by size of 70 × 80 × 120 cm is closed by three sides of the steel sheets of 1.5 mm thickness, which prevent scattering of capacitors' debris at the accident. To the front side of the cell, two copper buses are attached, which by means of copper wires are connected to capacitors terminals. Copper wires at the accident of capacitor act as the fuses, preventing the discharge of the all capacitors on the emergency one. Shells of the capacitors and cells are connected together and grounded.

Initially, the capacitor bank has been developed for operating at voltage of 5 kV. At this, its capacity was 0.24 F, energy 3 MJ, and electric charge 1200 C. The first experiment showed poor agreement with the arc. It was decided to rewire capacitor bank to voltage of 10 kV. For this collector, buses of two cells were connected in series, and pair of cells by means of coaxial cable is connected to the collector-switch (Fig. 2.10). The collector-switch in its turn by means of coaxial cables is connected to the current collector.

Fig. 2.10 Connection of the capacitor banks' cells [26]

Increasing of the initial capacitor bank voltage from 5 to 10 kV was intended to:

- Achieve better match of the power supply and arc, as typical voltage drop across the arc is 6–8 kV
- Improve the efficiency of energy transfer from the arc into the gas
- Halve the electric charge of the capacitor bank, which is reduced proportionally the electrodes erosion.

Experience on power supply CPS-10 operating shows that the reliability of its operation is possible at voltage ≤ 8 kV.

Figure 2.11 shows the graph of the ratio of the accidents number *Ne* to the number of all experiments *Nt* versus charge voltage.

Some rise of curve at 6 kV due to that the most number of experiments was performed at this voltage, and therefore the probability is more accurate compared to the point corresponding to the voltage of 7 kV.

Thus, the level of the stored capacitor bank energy at relatively safe operation is $\sim$4 MJ.

2.3 Components of Power Supply

Since the design of the capacitive power supplies is fundamentally similar, but have some structural differences, therefore, let's give a brief description of the CPS-10 as more simple.

Power supply CPS-10 is consisted of the next elements:

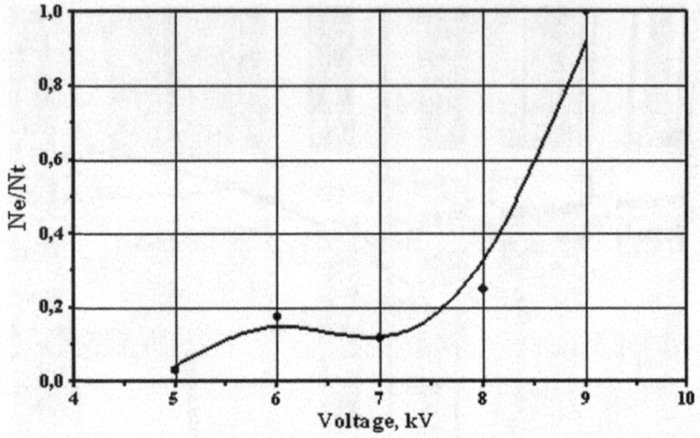

Fig. 2.11 Ratio of the accidents number *Ne* to the number of all experiments *Nt* versus the charge voltage [26]

- Charge unit
- High-current switching system (cables, collector-switch, current collector)
- Discharge unit
- Control system.

Charge unit (Fig. 2.12). The basic element of the charge unit of the power supply CPS-10 is a three-phase transformer power of 400 kW, voltage on the primary winding is 380 V, and at the secondary one 10 kV. Rated primary current is 557 A, and secondary is 23 A. In the primary coil of the transformer is switched the thyristor voltage regulator. Regulator is applied to maintain in the primary winding constant by value current, which is not exceeded the maximum permissible current in the secondary winding of the transformer.

As mentioned, if necessary, it can be use the separated ("module by module") charging of the capacitor bank, which is performed by connecting the required module to the charging circuit by means of high-voltage switches at the current-free mode. Charging time of the capacitor bank up to 8 kV is ~60 s.

High-current switching system. The system is consisted of the collector-switch with six air trigatron dischargers, current collector, and coaxial cables.

Collector-switch (Fig. 2.13) is the intermediate element between the capacitor bank and the load. The collector-switch performs switching of the pulsed current by amplitude up to 2 MA at the voltage up to 10 kV.

The capacitor bank is connected with the collector-switch by means of seventy-two coaxial cables and the collector-switch with the collector current by twelve coaxial cables. Switching of the current is carried out by six air trigatron spark-gaps. Ignition of the spark-gaps is performed by pulses of 16 kV voltage, and duration of 10 μs from the multi-channel generator. It is possible to control spark-gaps to ensure their optimal performance at different voltages on the capacitor

Fig. 2.12 Charge unit of the power supply CPS-10 [26]

Fig. 2.13 Collector-switch with the air spark-gaps [26]

bank. To measure pulsed current, six current coils on the collector-switch are installed.

Current collector (Fig. 2.14) is the end element of high-current switching system of the power supply CPS-10.

Current collector of outer diameter 250 mm connects the braids and cores of twelve coaxial cables with its collets connecting them to the coaxial current inlet of the plasma generator. Current collector operates at pulsed current of up to 2 MA, and electric charge up to 800 C.

Discharge unit (Fig. 2.15). Function of the discharge unit is neutralization of the capacitor bank electric charge at voltage of up to 8 kV. The apparatus includes the six-phase high-voltage normally closed contactor with electromagnetic drive and six load resistor blocks. Each load block is consisted of ten series-connected resistors and has resistance ~ 1 Ω. Time constant of the capacitor bank discharge at voltage is 8 kV ~0.1 s.

In addition to the discharge unit, the capacitor bank is equipped by two earthing rods: one is the high-resistant and other is short-circuiting, stationary contactors, which constantly closed and opened only at experiment.

Control system (Fig. 2.16). The control system of the power supply CPS-10 drives the discharge unit, contactors connecting the modules to the charge unit, ignition generators of the spark-gaps, etc.

Start of two ignition generators and oscilloscopes are performed by six-channel pulsed generator with adjustable delay up to 6 ms.

Control devices of the power supply parameters, oscilloscopes, and computer are located in the shielded control room.

Fig. 2.14 Coaxial current collector [26]

Fig. 2.15 Discharge unit [26]

Fig. 2.16 Control system [26]

2.4 Diagnostic Methods and Measurement Equipment

The measurement equipment includes the registering devices of initial, current, and final parameters of the experiment.

Initial parameters are:

- Capacity and charge voltage of the capacity bank
- Parameters of the discharge circuit
- Pressure and kind of gas in the discharge chamber
- Volume of the discharge chamber and its configuration
- Interelectrode gap
- Weight of the electrodes.

Current parameters and processes are:

- Discharge current
- Voltage drop across the arc and some parts of the discharge circuit
- Pressure in the discharge chamber
- Evolution of the arc
- Electrode jets and shock waves
- Emission spectra
- Intensity of the arc luminescence
- Intensity of X-ray irradiation of the arc, and others.

Final parameters of the experiment are:

- Residual voltage of the capacitor bank
- Weight of the electrodes
- Traces of arc binding to the electrodes.

2.4.1 Initial Parameters of Experiment

One of the main initial parameters of the experiment is the energy stored in the capacitive power supply. For its evaluation, in addition to charging voltage, you must know power supply capacitance.

Capacitance of the capacitor bank module was measured by digital device. Average capacitance of one module of the CPS-10 is 18 mF. Measurement error of the digital device at the three-digit counting is $\pm 0.5\%$. Measured "module by module," the total capacitance of the capacitor bank is 113 ± 0.6 mF. Thus, storage energy of the capacitor bank is 5.65 MJ at charging voltage of 10 kV. To calculate the outputted energy, the system that measures the charging and residual voltage on each module has been developed.

Charging voltage on the modules of the CPS-10 is measured by the high-voltage resistive dividers consisting of high-voltage shoulder of 40 MΩ, and low-voltage shoulder of 4 kΩ, thereby dividing ratio is 1/10,000. Voltage from dividers is delivered to the switching unit, then to digital voltmeter, and finally to PC. After capacitor bank charging, the switching unit sequentially connects the measuring circuit to the voltmeter input. On each measuring circuit, several countings were performed until the difference between them is reduced to 0.1%. Voltages on the modules are registered by the PC.

Residual voltage on the modules is measured after the experiment in similar manner. Measurement error does not exceed 0.5%. In addition, to the voltage dividers are connected pointer voltmeters for visual control of the capacitor bank charging.

Gas initial pressure in the discharge chamber is measured by means of high accuracy static pressure sensor with resistance output. Measurement error of the pressure sensor in the range 20–40 MPa is no more than ±0.4%. To display the sensor readings, there is the digital voltmeter with the measurement error ±0.5%.

Volume of the discharge chamber is calculated by dimensions of its elements. Changing of the discharge chamber volume was carried out by set of anode inserts.

Weight of the gas is calculated by the volume of the discharge chamber and initial gas pressure. Due to that even at pressures of 20–40 MPa, there is the gas nonideal, the calculation of the gas weight is carried out taking into account covolume [27, 28].

Erosion of the electrodes is determined by weighing the cathode and anode before and after the experiment on an analytical balance with error ±10 mg. Knowing the capacitance c, difference between the charging and residual voltages ΔU, it was determined electric charge $Q = c\Delta U$. Ratio of electrodes weight loss to electric charge Q gives the specific erosion of the electrode metal.

2.4.2 Pulsed Current

When capacitor bank with energy of several mega-joules at voltage of 10–25 kV is discharged on the arc or other loads with impedance 5–10 mΩ, discharge current is 1–5 MA and rise rate 10^9 to 10^{11} A/s. Measurement of such current is not easy task. The most common in practice are two methods: the Rogowski coils and low-inductance shunts.

Measurement of discharge current at the CPS-10 is carried out by six series-connected Rogowski coils. Each of them is encircled the central core of cables connecting the collector-switch and current collector.

Configuration of the current measuring system was carried out by means of the reference oscillating circuit, whose parameters were measured with error $\leq 1\%$. Capacitance of the circuit is 30 μF, fundamental frequency 25 kHz, and charging voltage 15–20 kV. By comparison of the measured signal with the calculated curve, the measurement error was determined. Based on obtained data, the signal processing program for PC was developed, which provides the discharge current measurement error $\leq 3\%$.

As alternative configuration of the current measurement system, the low-inductance reference shunt was used. The coaxial shunt has the constantan measuring element of 0.2 mm thickness.

Load capacity of the shunt, i.e., its resistance change due to the heating by the current, was evaluated according the formula

$$R_{\mathrm{f}} = R_{\mathrm{i}}\left(1 + \frac{I_{\mathrm{a}}^2 R_{\mathrm{i}} \alpha t_{\mathrm{cur}}}{C_{m_{\mathrm{s}}}}\right). \tag{2.6}$$

where R_{f} is final shunt resistance, R_{o} is initial shunt resistance, I_{a} is rated current, t_{cur} is time current flow, α is temperature coefficient, C is heat capacity, m_{s} is weight of the measuring element.

Evaluations have shown that heating of the shunt by discharge current results in change of its resistance $\leq 0.2\%$, which is significantly less than the measurement error of the shunt resistance.

To ensure high accuracy of measurements in the whole range of frequencies of the recorded signal is necessary that the thickness of the measuring element was much less than the thickness of the skin-layer at the maximal signal frequency. The ratio of active shunt resistance at alternating current R_{a} to its resistance at direct current R_{c} can be estimated by the formula

$$\frac{R_{\mathrm{a}}}{R_{\mathrm{c}}} = 1 + \frac{1}{3}\left(\frac{\delta}{b}\right)^4. \tag{2.7}$$

where δ is thickness of the conductor, b is thickness of the skin layer.

Value of b is calculated by formula

$$b = \sqrt{\frac{2}{\pi \gamma \mu \omega}}. \tag{2.8}$$

where μ is magnetic permeability of the shunt metal, γ is specific conductivity of the shunt metal, ω is circular frequency of the current.

Considering that the circular frequency of the current is less than 10^{-4} s^{-1}, for constantan b = 5 mm. At the thickness of the shunt δ = 0.2 mm from the formula for the resistances ratio must be that a change of the shunt resistance $\leq 0.1\%$.

2.4.3 *Voltage Drop Across Arc*

To determine the energy inputted in the arc, arc resistance, and power of discharge, it is necessary to perform accurate measurement of the voltage drop across the discharge gap. For such reliable registration, you must have the measuring devices and line with a bandwidth ≥ 1 MHz.

As the voltage sensors for the capacitor bank of the CPS-10, the ohmic dividers were used. Dividers are connected to the measured parts of the discharge circuit through the coupling capacitors. The division factor of the dividers is 1/1000. Measurement error of the dividing factor at direct current does not exceed ±1%. From the low-voltage shoulder of the dividers, signals are delivered to the digital oscilloscopes and then to the PC.

Bandwidth of the divider f_d and reaction time T_r are determined by the distributed capacitance of divider elements c_g against its grounded housing, by the resistance R_{hv} of the high-voltage shoulder and can be calculated by formulas [29]

$$f_d = \frac{1.46}{R_{hv}c_g}, \quad T_r = \frac{R_{hv}c_g}{6}. \tag{2.9}$$

Capacitance of the divider against ground

$$c_g = \frac{2\pi\pi_0\varepsilon\, l_d}{\ln(r_2/r_1)}. \tag{2.10}$$

where l_d is divider length, r_1 is outer radius of the divider' resistor, r_2 is inner radius of the divider housing.

Estimated value of c_g is 10 pF, for bandwidth f_d of 1 MHz, resistance R_{hv} of the divider' high-voltage shoulder must not exceed 140 kΩ. At resistance R_{hv} of 100 kΩ, reaction time T_r is 0.16 μs.

In our case at measurement of voltage drop across the arc, it is necessary to take into account those on current lead, its holder, current collector housing, outer sheath of cables connecting discharge chamber to collector-switch, and potential difference between grounds of collector-switch and oscilloscopes.

Additional measurements at short circuit have shown that main voltage drop under the discharge current of 1 MA is across the current lead ~470 V, and the rest voltage drops no more than few tens of volts. These voltage drops were accounted by experimental data processing.

2.4.4 Pulsed Pressure

Important data of the plasma generator operating such as internal energy of the gas, motion of the arc, and shock waves can be obtained by pulsed pressure sensors. Pulsed pressure is one of the main parameters that characterize the properties of the generated plasma. For measuring the pulsed pressure in gases up to 1000 MPa, the most commonly used are piezoelectric sensors.

At our experiments, it is necessary to measure gas pulsed pressure that higher than strength limit of the piezoelectric elements. Produced commercially sensors to measure such pressure under required time resolution are film dielectric and strain ones. However, they are unsuitable for operating under high thermal, electromagnetic, and acoustic interferences.

To measure pulsed pressure under these conditions at the initial stage of our researches, we have developed the piezoelectric sensor with time resolution 0.6 μs (Fig. 2.17). The sensor measures the pulsed pressure up to 500 MPa at current rise rate $\sim 10^{12}$ A/s.

For measuring the pressure at lower current rise rate, the piezoelectric sensors "Kessler" T-500 and T-6000 were used. These sensors allow to measure pulsed pressure up to 600 MPa and have cutoff frequency of 250 kHz, sensitivity of the sensors is 0.2 ± 0.02 pC/MPa, and the nonlinearity is $\pm 0.5\%$. The signals from sensors are delivered to the charge amplifier, then to digital oscilloscopes, and further to the PC. Bandwidth of the charge amplifier is 1 MHz at nonlinearity of $\pm 1\%$. To improve measurement accuracy, the calibration curve of each sensor is registered in the computer. Thus, total pressure measurement error is $\pm 1.5\%$.

During discharge current, the potential difference up to several tens of volts between the plasma generator housing and ground is occurred. This potential, if the pressure sensor is not insulated from the generator, is delivered to the charge amplifier and distorts the measured signal. To prevent these distortions, we use insulating adapters with reliable sealing and shortest length of channel.

Protection from electromagnetic interference of measuring circuits of pulsed pressure transducers by the screening of both the sensors and their amplifiers was performed.

The peculiarity of measurement of pulsed pressure in the discharge chamber of the plasma generator is the channel connecting the discharge chamber with a pressure sensor. Channels' length of some generators is from 5 to 15 cm and diameter is 4 mm.

To determine the pressure waveform amplitude-frequency distortion due to those channels, several experiments were carried out under lower pulsed pressure. At the experiments, two sensors were used: main sensor with the long (15 cm) channel, and control one with minimal (5 mm) length of the channel. Comparison of signals

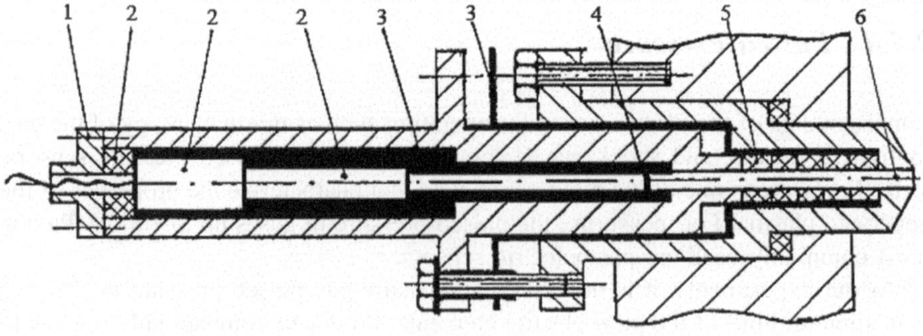

Fig. 2.17 Rod pulsed pressure sensor [33] 1 is cable connector, 2 is sound absorber, 3 is insulation, 4 is tourmaline piezoelement, 5 is sealing, 6 is ceramic rod

from the main and control pressure sensors showed no significant amplitude distortion; however, the signal from the main sensor comes with time delay Δt

$$\Delta t = a_o \Delta l. \tag{2.11}$$

where a_o is speed of sound in cold gas, Δl is difference between the channels lengths of the main and control sensors.

Further those delays and amplitude distortions are taken into account at signal processing.

Rod piezoelectric sensor. It was established experimentally that the effective protection of the sensor from electromagnetic fields is the sensor housing itself, however, in some works the double shielding of the sensor is applied [30, 31].

Acoustic interference immunity of rod sensors is performed by absorption of load wave by the sensor rod, acoustic matching of the rods and other designs addressed to adequate transfer of the initial perturbation [31, 32]. Attempts to eliminate the parasitic acoustic waves by electrical filtering, compensating sensors connected by the differential circuit, are low effective.

Time resolution of the rod sensor is determined by increasing of pressure pulse front at its passing through the face rod and by the time of sound wave passing the distance between the plates of the piezoelement. Time resolution of the sensor with a beryllium face rod registered by voltage pulse generated by piezoelement is 0.54 μs [34].

Pressure pulse rise time t, due to the pressure front dispersion in the face rod, is determined by [34]

$$t = 2.3\mu^{\frac{2}{3}}\left(\frac{x}{d}\right)^{\frac{1}{3}}\frac{d}{c}. \tag{2.12}$$

where μ is Poisson's ratio of the rod material, x is distance from the face end of the rod to the piezoelement, d is diameter of the rod, c is speed of sound in the rod material.

Pressure pulse rise time due to the dispersion of pressure front in the face rod is $\leq$0.1–0.2 μs. Passing time of the sound wave through the piezoelement is determined by its length and the speed of sound in it.

At best, when the piezoelement is tourmaline 0.2 mm in length, time of sound wave passing is 0.03 μs. If consider that at registration of the voltage on the piezoelement's plates it must be repeated passage of a sound wave, the rise time of the pulse pressure due to that is 0.1–0.2 μs.

However, the rise time of the pulse pressure can be substantially reduced, when registered not voltage ($U = Q/c$, where Q is charge of the piezoelement under pressure impact, c is capacity of the piezoelement and registration circuit), but the current in the short-circuit piezoelement [35, 36]

$$I = S\frac{\mathrm{d}I}{\mathrm{d}t} = KS\frac{a_s}{l}(P_0 - P_1). \tag{2.13}$$

where S is cross-sectional area of the piezoelement, $\mathrm{d}I/\mathrm{d}t$ is bias current, a_s is speed of sound in the piezoelement, l is piezoelement length, P_0 and P_1 are pressures on the face and back plates of the piezoelement, K is characteristic of the piezoelement material.

Piezoelement switched on under this circuit, according to data of work [36] at piezoelement length of 0.5 mm registers the pressure rise time of 0.07 μs. At this registering circuit, when piezoelement length is 0.2 mm, its resolution time is 0.03 μs. Thus, rise time of the pulse pressure for the rod of piezoelectric sensors is determined mainly by the pressure front dispersion in the face rod and is of 0.2 μs.

The data presented above are more informative and indicate the problems are faced us in the initial stage of the research. Superfluous to say that there are now commercially available pulse pressure sensors with time resolution of a few nanoseconds, allowing investigating the ultrafast phenomena such as the shock waves generated by pulsed electrical discharges in dense gases and liquids.

2.4.5 *Optical Registration*

Shadow photography of the processes in the discharge chamber was carried out on the diagnostic discharge camera (Fig. 3.16) as shown in Fig. 2.18. It was used the method of the visualizing diaphragm in focus [37]. The method involves the location of the circular visualizing diaphragm in focus of the receiving lens of the shadow device. The diaphragm of a given diameter is used for qualitative research, and for quantitative measurements is used the spatial filter with a known transmission function.

The choice of this method is based on high decreasing of own discharge irradiation, and at the same time on allocation of required interval of the rays deflection in the light beam of the transmission source. The light source is the argon laser with spectral intensity of radiation exceeding the spectral intensity of any other available light source that allows selecting the probe radiation on the background of the arc radiation. Power of the laser radiation of 4 W is distributed in five wavelengths: 514.5, 496.5, 488.0, 476.5, and 457.9 nm. The highest intensity is in 514.5 and 488.0 nm wavelengths.

High-speed photography was performed in single frame mode and in photographic scan mode. High-speed camera photographs the arc through the neutral filters with optical density from 0.5 to 2. Such increase in optical density corresponds to decrease in intensity of the radiation source up to 100 times. The high-speed camera makes 300 images at shooting frequency up to 10^6 frames/s, and resolution 25 lines/mm.

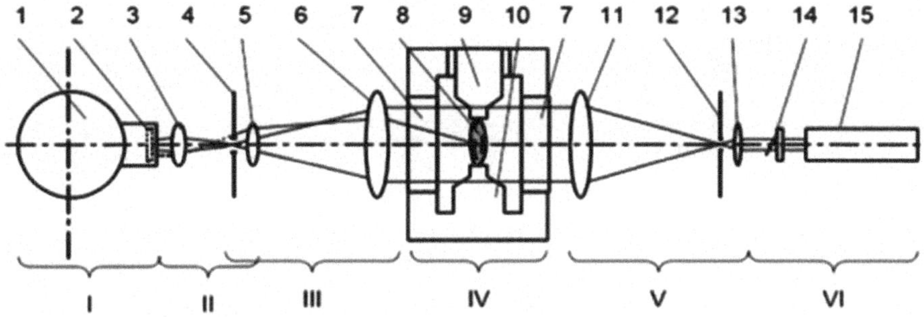

Fig. 2.18 Scheme of the shadow device [33]. I is high-speed camera, II is matching lens, III is receiving part of the shadow device, IV is discharge chamber, V is collimator of the shadow device, VI is light source, 1 is speed camera, 2 are filters, 3 is second component of the matching lens, 4 is visualizing diaphragm, 5 is first component of the matching lens, 6 is receiving lens, 7 are diagnostic windows, 8 is arc, 9 is cathode, 10 is anode, 11 is collimator lens, 12 is beam cleaning device, 13 is condenser, 14 is electrodynamic gate, 15 is argon laser

At photographing, the frequencies of 125,000, 250,000, and 500,000 frames/s were used. For photography the arc, spectrometric, and pyrometric measurements at pulsed pressure up to 600 MPa and current rise rate 10^{10} to 10^{11} A/s, the polycarbonate optical window was used.

Determination of the shock waves and arc light radiation impacts on the optical properties of windows was carried by means the extensions allowing varying the distance from the arc to the window from 40 to 150 mm. At distances ≥ 60 mm, the windows deformation is small and a noticeable impact on the images quality was not detected. At location of the windows ≤ 50 mm and at higher energy input in the arc, the considerable residual deformation of the polycarbonate windows was observed. Therefore, windows at the distance ≥ 65 mm from the arc were located.

The photographic of the arc glowing by the camera with optical filters provided registration in the spectral range near the wavelength of ~ 500 nm was made. The photographic slit is located perpendicular to the axis of the arc in the middle between the anode and cathode. When slit width 1 mm: time resolution is ~ 1 μs.

Measuring of the brightness temperature on the diagnostic discharge chamber was performed.

Figure 2.19 shows the scheme of the measurements.

The brightness temperature of the various sections of the arc by the film's blackening density was estimated. Photographing by using of various density filters was performed. As the standard, the film's blackening density of Podmoshensky capillary source with brightness temperature of 4×10^4 K was accepted [38].

By the plasma generator PPG-6, the arc brightness temperature registered by two identical monochromatic pyrometers with the calibrated semiconducting radiation detectors with maximal sensitivity at wavelengths 550 and 694 nm, and time resolution ~ 1 μs. Pyrometers by the standard Podmoshensky capillary source were calibrated.

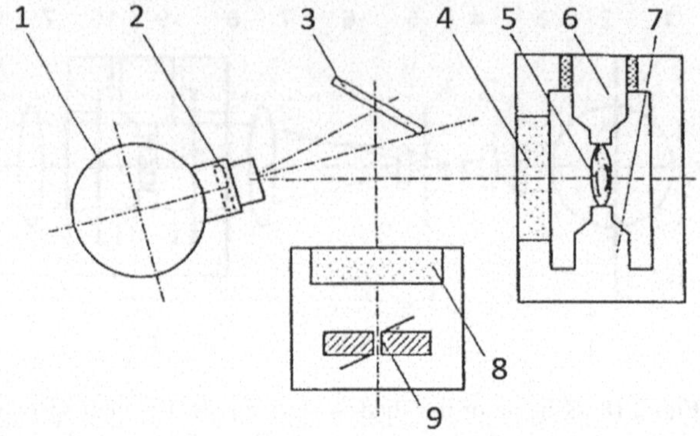

Fig. 2.19 Scheme of the arc brightness temperature measuring [33]. 1 is camera, 2 are neutral filters, 3 is mirror, 4 and 8 are windows, 5 is arc, 6 is cathode, 7 is anode, 9 is standard light source

2.4.6 X-ray Registration

Because of the arc's plasma opacity, are particularly important the X-ray diagnostic techniques. It should be noted that the study of X-ray radiation from the arc is complicated by high density of the gas surrounding the arc, and large thickness of the optical windows. At transmitting of the arc by soft X-ray radiation to those difficulties, the problem choosing of the suitable X-ray source is joined.

The results of the arc burning in argon at pressure of 40–60 MPa and current of several tens of kA by the X-ray sensing are presented in papers [39, 40]. As radiation source is standard, X-ray tube for structural analysis was used. At higher currents, to explore the structure of the arc in argon was not possible due to the metal vapor of electrodes.

In papers [41–43], sensing of the low pressure pinch discharges plasma, by the X-pinch as source of soft X-ray, was carried out.

In order to study the density of metal-halogen arc sections in work [44] was used the fluorescence induced by X-ray radiation.

For our study, the special X-ray tube with detector protected from the high gas pressure has been developed.

Registration of the arc soft X-ray radiation. The challenge for the registration of the soft X-ray radiation from the arc was in need to protect the radiation detector from impact of high (up to 300 MPa) pulsed pressure, temperature up to 3000 K, and shock waves. For this, the special camera with the X-ray detector has been developed [45]. As a radiation detector, the X-ray diode SPD-8UVHS was used. Figure 2.20 shows the scheme of the soft X-ray measuring.

X-ray detector (1) placed in a special buffer chamber (2) connected to the housing of the plasma generator's discharge chamber (4). The buffer chamber in volume of ~ 1.5 dm^3 is protected the X-ray detector from high pressure, temperature, and shock waves in the discharge chamber (6). The buffer and discharge chambers are connected by the measuring channel (3) of diameter 3 mm. The distance from the axis of the arc initiated by the explosion of the wire (10) and burning between cathode (7) insulated from the housing (4) by insulator (5) and the

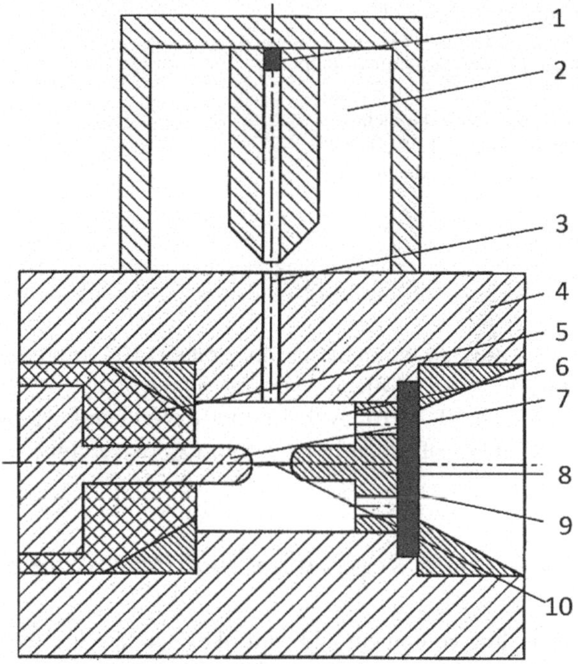

Fig. 2.20 Scheme of the soft X-ray radiation from the arc measuring [33]

anode (8) to the detector is ~25 cm. The heated gas after the rupture of the diaphragm (9) is flowed from the discharge chamber. As the filter the aluminum foils of 20–25 μm thickness were used. For estimation of the temperature by foils method, the two identical diagnostic units with filters of various thicknesses were simultaneously used.

Diameter of the detector core is 3.2 mm. Spectral range of the SPD-photodiodes is continuous and is 0.02–1100 nm, and sensitivity in the range of irradiation from 20 to 20,000 eV is ~0.25 A/W.

X-ray sensing. To determine the distribution of a concentration of the metal vapor eroded from electrodes (3, 4), one of which (4) insulated (8) from the housing of the discharge chamber, and metal vapor of the ignition wire (6) in the high-current arc (7), the X-raying was applied (Fig. 2.21) [46].

The scheme is consisted of the pulsed X-ray source with energy of 20–50 keV (1) and X-ray CCD-camera (5). To power the pulsed X-ray source, two high-voltage nanosecond pulsed generators with pulse duration of 10–20, and 50 ns were used.

According to our estimates, the metal vapor at concentration of 10^{19} to 10^{20} cm^{-3} is located in area of diameter 1–2 cm (7) near the arc axis [47]. The absorption of the X-ray radiation to measure the concentration of metal vapors of 10^{19} to 10^{21} cm^{-3} against background absorption in polycarbonate diagnostic windows (2) and in area filled by hydrogen at pressure of about 10 MPa is optimal at the quanta energy range of 30–40 keV.

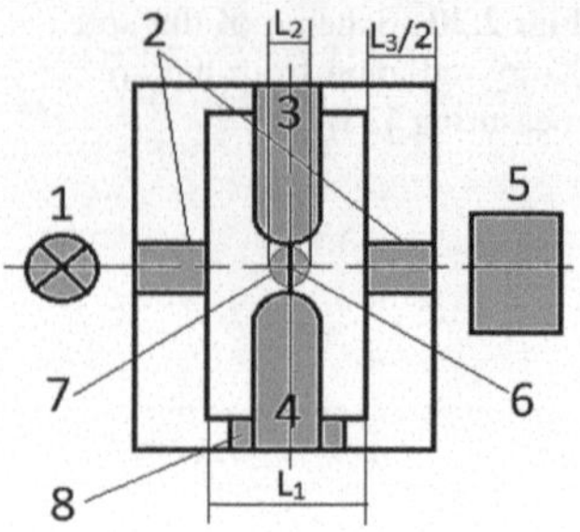

Fig. 2.21 Scheme of X-raying the arc [33]. L_1 = 13 cm is area filled with hydrogen under high pressure, L_2 is area with the metal vapors, $L_3/2$ = 1.25 cm is thickness of the diagnostic windows (2)

To provide the radiation dose rate $\sim 10^8$ R/s required for the registration of the concentration of metal vapor at tube voltage 40–50 kV, electron beam current should be 10^2 to 10^3 A that in our experiments was accomplished by means of the X-ray tubes of properly type with explosive emission cathode.

2.4.7 *Data Registration and Processing System*

Recording equipment and control system are located in the control rooms shielded from electromagnetic fields (Fig. 2.16).

Registration of the experimental parameters by two-channel digital oscilloscopes is carried out. Start of the oscilloscopes and spark-gaps is performed by pulses from six-channel generator synchronously or with delays required by the experiment conditions.

At the experiments are recorded:

- Initial and residual voltage of the capacitor bank
- Discharge current
- The voltage drop across the arc
- Primary and pulsed gas pressure in the discharge chamber of the generator.

For personnel safety, measuring elements and sensors are isolated from the plasma generator and power supply or connected to the oscilloscopes via the coupling capacitors and transformers.

Recording, processing, and storage of experimental data are carried out by the PC connected to the oscilloscopes, control system, and voltmeters used for measuring voltage of the capacitor bank modules. The program for the PC has been developed that allows changing quickly the settings of the oscilloscopes and processing the experimental data. Based on the experimental data, it was developed the database containing information of all experiments.

References

1. J.H. Gully, E.G. Cotes, W.A. Walls, W.F. Weldon, IEEE Trans. Magn. **20**, 203 (1984)
2. R.D. Ford, D. Jenkins, W.H. Lupton, L.M. Vitkovitskiy, Rev. Sci. Instrum. **52**, 694 (1981)
3. R.L. Laughlin, J.H. Gully, K.E. Natly, R.C. Zowarka, IEEE Trans. Magn. **22**, 1578 (1986)
4. M.L. Spann, S.B. Pratap, J.H. Gully, W.F. Weldon, H.H. Woodson, IEEE Trans. Magn. **20**, 215 (1984)
5. E.R. Johnson, W.G. Chen, IEEE Trans. Magn. **22**, 1558 (1986)
6. M.D. Driga, M.W. Ingram, W.F. Weldon, IEEE Trans. Magn. **25**, 147 (1989)
7. M.J. Spann, S.B. Pratap, W.G. Srinkman, D. Perkins, R.F. Thelen, IEEE Trans. Magn. **22**, 1538 (1986)
8. M. Cowan, IEEE Trans. Magn. **18**, 125 (1982)
9. G.A. Shvetsov, V.M. Titov, J.L. Bashkatov, I.A. Stadnichenko, A.V. Orlov, Russ. J. Comb. Explos. Shock Waves **20**, 111 (1984) (in Russian)
10. G.A. Shvetsov, V.M. Titov, J.L. Bashkatov, I.A. Stadnichenko, A.V. Orlov, *Superstrong Magnetic Fields. Physics, Application* (Moscow, 1984) (in Russian)
11. C.M. Fowler, E.L. Zimmerman, C.E. Cummings, R.F. Davidson, E. Foley, R.S. Hawke, J.F. Kerrisk, J.V. Parker, W.M. Parsons, D.K. Peterson, N.M. Schnurr, P.M. Stenly, IEEE Trans. Magn. **22**, 1475 (1986)
12. C.M. Fowler, E.L. Zimmerman, C.E. Cummings, R.F. Davidson, E. Foley, R.S. Hawke, J.F. Kerrisk, J.V. Parker, W.M. Parsons, D.K. Peterson, N.M. Schnurr, P.M. Stenly, in *IV International Conference on Megagauss Magnetic Fields Generation and Related Topics* (Santa Fe, 1986)
13. R.S. Caird, C.M. Fowler, in *IV International Conference on Megagauss Magnetic Fields Generation and Related Topics* (Santa Fe, 1986)
14. J. Powell, T. Bottes, C.M. Stickley, S. Metch, IEEE Trans. Magn. **18**, 135 (1982)
15. F.W. McDougall, P.E. Hardy, V.P. Winsor, X.H. Yang, in *4th European Symposium on Electromagnetic Launch Technology* (Celle, 1993)
16. K.M. Slenes, P. Winsor, T. Scholz, M. Hudis, IEEE Trans. Magn. **37**, 324 (2001)
17. H.G. Wisken, F. Podeyn, T.H.G.G. Weise, IEEE Trans. Magn. **37**, 332 (2001)
18. K.M. Slenes, L.E. Bragg, IEEE Trans. Magn. **41**, 326 (2005)
19. G.L. Bullard, H.B. Sierra-Alcazar, M.I. Lee, J.L. Morris, IEEE Trans. Magn. **25**, 102 (1989)
20. K.M. Burke, W.J. Sarjeant, J.L. Zirnheld, in *15th International Pulsed Power Conference* (Monterey, 2005)
21. A.F. Savvateev, *Doctoral Thesis* (St.-Petersburg, 2003) (in Russian)
22. D.A. Andreev, M.O. Znesin, V.A. Kolikov, B.P. Levchenko, A.V. Radjushin, P.G. Rutberg, *Power Supplies of Short-Term and Pulse High-Power Loadings* (Leningrad, 1981) (in Russian)
23. B.E. Fridman, P.G. Rutberg, Izv. RAS Energy **2**, 46 (1998) (in Russian)
24. B.E. Fridman, in *9th Symposium on Electromagnetic Launch Technology* (Edinburgh, 1998)
25. B.E. Fridman, P.G. Rutberg, in *5th European Symposium on Electromagnetic Launch Technology* (Toulouse, 1995)
26. V.A. Kolikov, *Doctoral Thesis,* (St.-Petersburg, 2005) (in Russian)
27. V.P. Kopushev, V.V. Hrustalev, Russ. J. Appl. Mech. Tech. Phys. **1**, 122 (1980). (in Russian)
28. J.R. Baker, H.F. Swift, J. Appl. Phys. **43**, 950 (1972)
29. M. Bejer, V. Beck, K. Meller, V. Tsangel, *Technics of High Voltage: Theoretical and Practical Bases of Application* (Moscow, 1989) (in Russian)
30. V.A. Godonjuk, B.V. Zuravlev, I.P. Shed'ko, Russ. J. Instrum. Exp. Tech. **5**, 214 (1984). (in Russian)
31. E.M. Golubev, N.N. Ogurtsova, Sov. J. Tech. Phys. **48**, 732 (1978). (in Russian)
32. K.V. Regland, R.E. Kullen, Sov. J. Review Sci. Instr. **38**, 18 (1967). (in Russian)
33. A.A. Bogomaz, *Doctoral Thesis* (St.-Petersburg, 2013) (in Russian)
34. I.P. Jons, Sov. J. Rev. Sci. Instrum. **37**, 74 (1966). (in Russian)

35. R. Kiler, E. Rois, *Physics of High Density of Energy* (Moskow, 1974) (in Russian)
36. N.P. Hohlov, V.N. Mineev, A.G. Ivanov, *I All-Union Symposium on Pulse Pressure* (Moskow, 1973) (in Russian)
37. L.A. Vasil'ev, *Shadow Methods* (Moskow, 1968) (in Russian)
38. M.I. Demidova, N.N. Ogurtsova, I.V. Podmyshenskiy, V.M. Shelemina, J. Appl. Spectrosc. **23**, 1545 (1975)
39. R.V. Mitin, *Properties of Low-Temperature Plasmas and Diagnostics Methods* (Novosibirsk, 1977) (in Russian)
40. V.I. Petrenko, R.V. Mitin, Sov. J. Tech. Phys. **XLV**, 1225 (1975) (in Russian)
41. T.A. Shelkovenko, S.A. Pikuz, S.A. Mishin, A.R. Mingaleev, I.N. Tilikin, P.F. Knapp, A.D. Cahill, C.L. Hoyt, D.A. Hammer, Plasma Phys. Rep. **38**, 359 (2012)
42. G.S. Volkov, E.V. Grabovskii, K.N. Mitrofanov, G.M. Oleinik, Plasma Phys. Rep. **30**, 99 (2004)
43. E.V. Grabovskii, K.N. Mitrofanov, G.M. Oleinik, I.Y. Porofeev, Plasma Phys. Rep. **30**, 121 (2004)
44. J.J. Curry, H.G. Adler, S.D.W.-K. Shastri, J. Appl. Phys. **93**, 2359 (2003)
45. A.A. Bogomaz, A.V. Budin, V.V. Zabrodskiy, I.V. Kuznetsova, S.Y. Losev, M.V. Petrenko, M.E. Pinchuk, P.G. Rutberg, Instrum. Exp. Tech. **51**, 744 (2008)
46. M.E. Pinchuk, A.A. Bogomaz, A.V. Budin, L.A. Shirochin, M.A. Polyakov, A.G. Leks, S.Y. Losev, P.G. Rutberg, Instrum. Exp. Tech. **53**, 723 (2010)
47. P.G. Rutberg, A.V. Budin, M.E. Pinchuk, A.A. Bogomaz, L.A. Shirochin, M.A. Polyakov, A.G. Leks, S.Y. Losev, IEEE Trans. Plasma Sci. **39**, 394 (2010)

Chapter 3
Pulsed Plasma Generators

Abstract Applying of the pulsed plasma generators as powerful X-ray sources or sources of forplasma for fusion involves usage of plasma-forming gases with low molecular/atomic weight, such as hydrogen or helium at pressure up to hundreds of megapascals. Pulsed plasma generators are relatively simple devices, but their developments require novel technologies and materials. This is due to the parameters of the generated plasma and the characteristics of the electrical energy used to power them. Plasma parameters and characteristics of modern plasma generators are next:

- Mean weight temperature of the gas up to 10^4 K
- Gas pressure up to 1000 MPa
- Discharge current up to 2 MA
- Voltage up to 25 kV
- Energy input in the arc up to 10 MJ.

Combination of these parameters shows the complexity of the problems those must be solved at the developing and study of these devices. Basic component of the pulsed plasma generators is electrodischarge chamber.

3.1 Types of Electrodischarge Chambers

Active works on pulsed plasma generators were begun in the mid-twentieth century and were originally addressed to developing hypersonic wind tunnels simulating flight of the spacecrafts. Even at that time, the electrodischarge chambers of the pulsed plasma generators were based on almost all design ideas, which according to the novel technologies are used in today's designs.

To navigate the various designs of the electric discharge chambers of the pulsed plasma generators, we will present their classification on the basis of shape and arrangement of electrodes formulated in works [1, 2].

At this, almost all names of the electrodischarge chambers' types presented in these works were corrected in the book. This is made to bring the type's names of

V. Kolikov et al., *Powerful Pulsed Plasma Generators*,
Springer Series on Atomic, Optical, and Plasma Physics 101,
https://doi.org/10.1007/978-3-319-95249-9_3

the electrodischarge chambers to exact match in English with classification criteria and their design features.

And so, almost all, at rare exceptions, designs of the electrodischarge chambers are divided into four types as follows:

- Discharge chambers with coaxial electrodes (Fig. 3.1a)
- Coaxial discharge chambers (Fig. 3.1b)
- Discharge chambers with coaxial opposing rod electrodes (Fig. 3.1c)
- Discharge chambers with coaxial rod and ring electrodes (Fig. 3.1d).

All electrodischarge chambers have common elements: two electrodes (1, 5), insulators (3, 4), locking diaphragm (7), nozzle (6), and fuse (2).

Let us describe briefly the designs of the discharge chambers types. Here we note that now and further a comparison of the characteristics and conclusions on the advantages and disadvantages of various types of discharge chambers we will make on the basis of their applicability to the powerful plasma generators and working gases such as helium or hydrogen and in specified cases other gases.

Discharge chambers with coaxial electrodes. In these discharge chambers (Fig. 3.1a) central–axial (1) and external–coaxial (5) electrodes form an electrode system. The electrodes are insulated from each other and the discharge chamber housing by insulators (3, 4). Normal polarity of the electrodes: axial is cathode and coaxial is anode. Initiating of the arc, as in majority of the plasma generators' electrodischarge chambers is performed by fuse (metal wire) (2).

Characteristic feature of the arcing in such electrodischarge chamber is arc's mobility, or as they call this phenomenon in some works—"magnetic blowing" [4]. Nature of the phenomenon is in moving of the arc, which is ignited in point of the

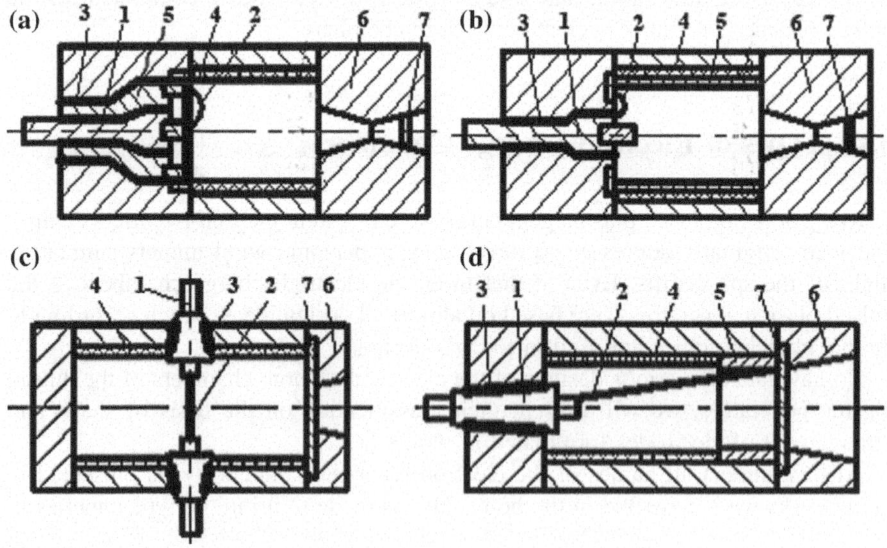

Fig. 3.1 Types of electrodischarge chambers [3]

fuse location and moves then along the discharge chamber toward the nozzle while increasing own length. The reason of this phenomenon is the action of the magnetic field of the arc current on carriers of current, so and gas pressure gradient in the discharge chamber caused by the uneven gas heating. The velocity of the movement of the arc, in dependence on the value of current, its rise rate, type of gas, and its initial density is several hundred meters per second and higher.

This phenomenon has positive so negative aspects:

- The positive is the intensive flowing of the gas in the chamber caused by the movement of the arc and by the electrode jets, which perform efficient heat transfer from the arc to the gas, and accelerate the thermodynamic equilibrium in the discharge chamber
- The negative is the fast extinction of the arc, i.e., short duration of a single current pulse and a relatively low level of inputted energy in the arc.

Based on the results of a number of the experiments carried out at the IEE RAS, such discharge chamber design in relation to powerful pulsed plasma generators can be characterized by negatively rather than positively. This conclusion is not definitive, since subsequent experience with similar—coaxial discharge chambers leads us to conclusion that this type of camera was not studied enough, that not allowed revealing their potential.

In the early sixties, a number of electric heaters for pulsed wind tunnels were developed, including the super power pulsed generator with coaxial electrodes of the Mac Donnell Douglas (Fig. 3.2) [5].

The discharge chamber is powered by a capacitor bank of energy is 7 MJ, voltage 12 kV, and the inductance of the discharge circuit $\sim 10^{-6}$ H. Operating gas is the air at a pressure up to 70 MPa.

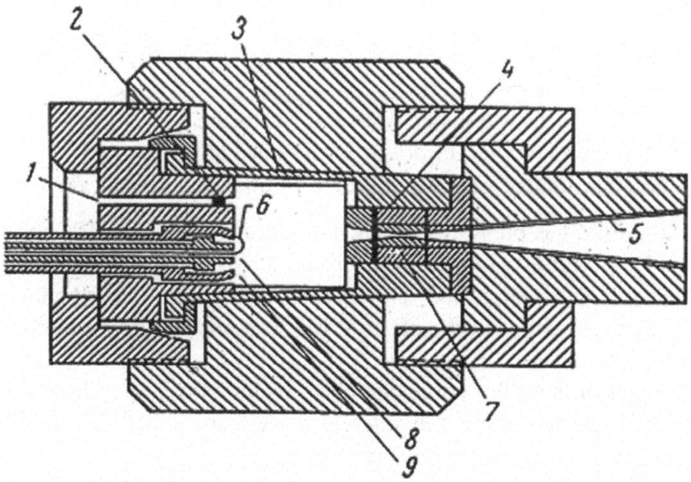

Fig. 3.2 Electrodischarge chamber with coaxial electrodes of the McDonnell Douglas [5]

The housing of the electrodischarge chamber of weighing ~3500 kg is made of stainless steel and had been calculated for operating at an internal pressure up to 700 MPa. At the development of the discharge chamber the successful design was used with the easily removable electrode assembly with conical split surface (3) concluding the electrode system (8, 9), nozzle block with the replaceable nozzle insert and the diaphragm (4, 5, 7), and the drainage device for gas outlet (1, 2). The weight of the electrode assembly is 1050 kg, volume of the discharge chamber ~17 dm^3, and limit internal pressure of the electrode assembly is 42 MPa.

The electrode system consists of two insulated from each other coaxial electrodes (8, 9) made of beryllium bronze and designed to operate at a current up to 4 MA. Initiating the arc by explosion of the wire was carried out.

The electrode system is located in the locking nut eccentrically in order to an area of the arcing, which is determined by the place of installation of the fuse, was cogruent with the axis of the discharge chamber.

As the result of the experiments, carried out on this discharge chamber, it was discovered the phenomenon of arc stretching under the action of the own current magnetic field and the pressure gradient, it was revealed the growing dependence of the voltage drop across the arc on the initial density of the air, it was estimated the dependence of the efficiency of the system operation on the initial gas pressure and capacitor bank voltage, in the first case is growing one, in the second—falling [4].

To this type of chambers is owned the discharge chamber of the wind tunnel SKB-8A of the AEDC, USA [6].

Coaxial discharge chambers (Fig. 3.1b) have as prototype the discharge chambers with the coaxial electrodes and are differed in that they have the length of the coaxial electrode is equal to the length of the discharge chamber, and this electrode is under the potential of the discharge chamber housing. Thus, in this design is necessary and sufficient the one insulator between the axial electrode and the housing that makes it the simplest and most reliable in comparison with any other types of discharge chambers.

In addition, the coaxial discharge chambers have next advantages:

- High voltage drop across the arc
- High coefficient of heat transfer from the arc to the gas
- Fast the thermodynamic equilibrium in the discharge chamber
- Stable the arcing under virtually unlimited initial and pulsed gas pressures.

Initial and pulsed gas pressures for such devices are 40–120 MPa and 600 MPa, respectively. Limit pulsed pressure is determined by the mechanical strength of the discharge chamber housing and electrode's insulation. On coaxial discharge chambers, the energy input in the arc to 2 MJ was reached at the possibility of its multiple increasing.

Disadvantages of these discharge chambers include the strong dependence of the arcing and gas parameters on the interelectrode gap, and the instability of the discharge current and voltage drop across the arc.

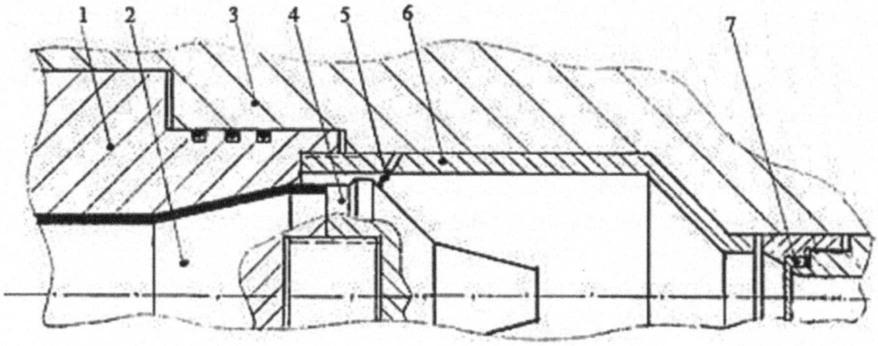

Fig. 3.3 Discharge chamber of the plasma generator PPG-4 [3]

To these discharge chambers are owned the plasma generator PPG-4, IEE RAS (Fig. 3.3). The generator consists of the housing (3), inside which is placed the electrode system consisting of the current lead (2) and cathode (4) placed in the holder (1), anode (6), and fuse (5). The discharge chamber is sealed by the diaphragm (7) and the rubber seals.

Volume of the discharge chamber is 0.7–1.2 dm^3, pulsed gas pressure up to 400 MPa. The inner surface of the steel anode is coated by the plasma sprayed tungsten of thickness $\sim$0.3 mm. This coating makes it possible the operation of the discharge chamber at high reproducibility of the plasma final parameters and moderate contamination of the gas by eroded metal. Elements of the discharge chamber were operated under pulsed currents up to 1.5 MA and gas pressure up to 320 MPa.

This type of discharge chambers has chambers 50-0 and 50-0M of the wind tunnel "Hotshot-2" [7], discharge chamber of the wind tunnel "Tunnel-F" [8], and discharge chamber of the plasma generators PPG-1, PPG-2, and PPG-3 (IEE RAS).

Thus, the coaxial discharge chambers have possibilities to control the parameters of the discharge current and gas. By the sum of the advantages and disadvantages, this type of discharge chambers is most preferred for high-power pulsed plasma generators operated with hydrogen of ultra-high initial density.

Discharge chamber with coaxial opposing rod electrodes. The chambers of this type (Fig. 3.1c) have electrodes (1) placed on the one axis (their polarity is irrelevant), each electrode has the insulator (3), in addition, in some designs, the inner surface of the discharge chamber is covered by the heat-resistant insulator (4), which preventing the breakdown electrode $\rightarrow$ housing $\rightarrow$ electrode and protects the inner surface of the discharge chamber from heat flow arc radiation. Ignition of the arc is carried out usually by the metal fuse (2).

Some cameras use a third additional igniting electrode (trigger), which is powered from a separate supply, usually by two–three orders of magnitude lower than energy of the main power supply. A fuse is connected to igniting and main electrodes. Explosion of a fuse ionizes the interelectrode gap and initiates the arc between main electrodes. Application of this scheme eliminates the switch of total discharge current, but complicates the design of the discharge chamber.

Discharge chambers with rod electrodes are characterized by the predominance of disadvantages to advantages. Main disadvantages are next:

- Big numbers of insulators and, therefore, higher probability of the breakdown in comparison with other types of chambers
- Small arc length, at increasing of the length of the arc, there are instabilities
- Inefficient heat transfer in the discharge chamber
- Low voltage drop across the arc
- Low energy input into the arc
- High current density and fixed spot of the arc connection to the electrodes.

All mentioned above shows that the discharge chambers of this type are not used now in high-power generators of the megajoule level, but used at the studies of the pulsed arc where is needed the high stability of the arcing under fixed initial conditions of the experiments.

Let us present the discharge chamber of this type developed at the L'Institut Franco-Allemand de Recherches de St. Louis, France [9].

The discharge chamber of the plasma generator (Fig. 3.4) is a thick-walled steel vessel (1) with two opposing electrodes (2), which are fixed in the housing by the nuts (3).

Fig. 3.4 Discharge chamber of the L'Institut Franco-Allemand de Recherches [9]

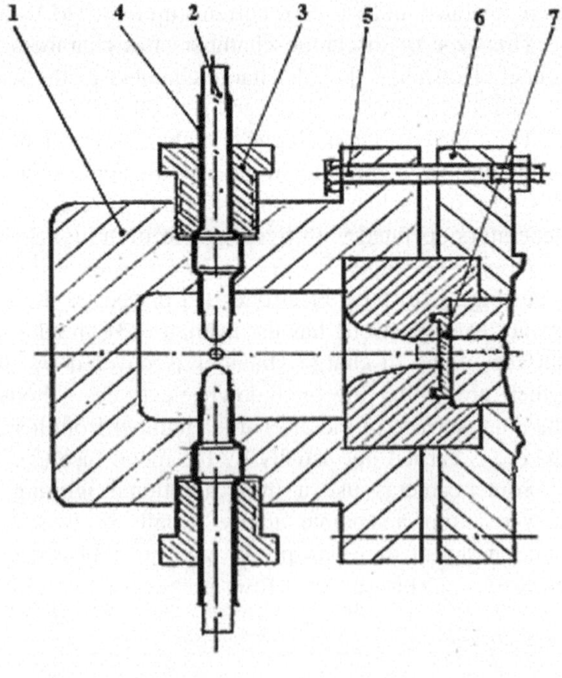

Noteworthy is the absence of the high-temperature insulator that, in our opinion, is a reasonable design, the absence of insulator is compensated by the increased length of the isolators (4). Volume of the discharge chamber is 80 cm^3, diameter is 35 mm, and length is 85 mm. Sealing of the hydrogen under initial pressure up to 14 MPa is performed, by diaphragm (7), which is fixed to the housing (1) by the flange (6) and bolts (5). Pulsed pressure is up to 600 MPa, and gas temperature up to 8000 K.

Generator is powered by the capacitor bank of energy 800 kJ, voltage 16 kV, and pulse duration 40 μs. Efficiency of the generator is very low that is agreed with the above disadvantages of this type of discharge chambers. It should be noted, however, that a significant increase in efficiency and reduce heat load to the elements of the discharge chamber can be achieved by increasing the initial pressure of hydrogen and a reasonable choice of power supply parameters, in particular, voltage.

Content of papers published on this subject at the time leads us to the conclusion that these opportunities had been not used by researchers, and as a result, the developments of high-power pulsed plasma generators with this type of discharge chambers have been stopped because of seeming hopelessness.

Among the devices of this type, it can be mentioned the discharge chamber of the wind tunnels of the AEDC [10], and plasma generator PG-3 of the IEE RAS [11].

There are designs of the discharge chambers that did not satisfied to the classification. For example, discharge chamber of the wind tunnel of the AEDC (Fig. 3.5) [12] has the parallel arrangement of the electrodes (1). Here as well as in other discharge chambers, the electrodes are insulated from the housing (3), the arc is initiated by fuse (2), inner surface of the discharge chamber is covered by high-temperature insulator (4). However, this arrangement of the electrodes substantially changes the arcing process and brings this chamber to the type of discharge chambers with coaxial electrodes.

Discharge chambers with coaxial rod and ring electrodes (Fig. 3.1d) were widespread in the past. The chamber consists of a rod axial cathode isolated from the housing (1) and spaced from cathode the ring coaxial anode (5). Anode is at a potential of the discharge chamber housing. The difference between these discharge chambers and discharge chambers with coaxial electrodes is in insulating bushing.

This bushing defines the characteristics of the arcing; the most important is next:

- High and almost constant the arc length
- High and stable voltage drop across the arc
- High degree of filling by the arc of the discharge chamber
- Active gas mixing and intensive heat transfer from the arc to gas
- Unlimited tangential mobility of the arc.

The arc in all similar discharge chambers is initiated by metal wires, the shape of which in some cases may even be spiral [13].

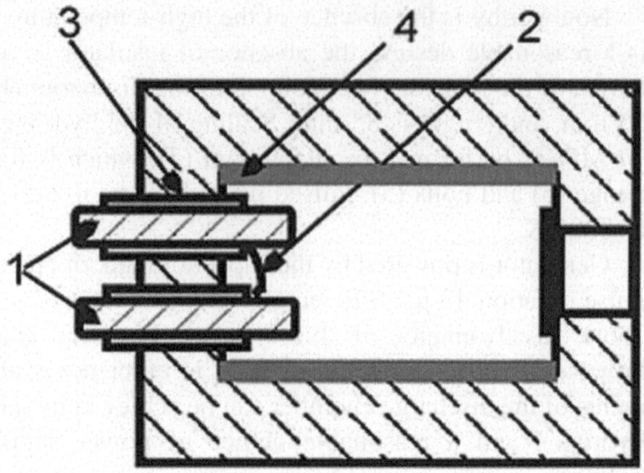

Fig. 3.5 Discharge chamber of the wind tunnel of the AEDC [12]

However, having a number of advantages these discharge chambers have some disadvantages; the main is a heat-resistant insulator which because of its low strength and fragility make it impossible repeated operating of the discharge chamber under high rise rate of the pressure. Another disadvantage under high initial pressures of the gas is the instability of the arcing at the low power supply voltage 5–10 kV.

These discharge chambers operate rather well at hydrogen, helium, argon, and nitrogen at initial pressure of 0.1–6.0 MPa, currents up to 350 kA, pulsed pressure up to 120 MPa, and high energy input. In some cases, these chambers operate at liquids, particularly at methanol [14]. At PPG-2 generator when powered by the shock generator was reached the highest energy input of 10 MJ [1].

Devices of this type are the discharge chamber of the high-pressure shock tube [15], and discharge chamber of the plasma generator of the TZN, Germany [16].

3.2 IEE RAS' Pulsed Plasma Generators

At the IEE RAS over the last fifty years is conducted study of the high-power pulsed arcs in hydrogen, helium, nitrogen, argon, and other media under the initial pressure up to 250 MPa, current of up to 2 MA and arc temperature up to 50 eV. For these purposes, the reliable and high-performance pulsed plasma generators had been developed. Among them, the diagnostic discharge chamber developed for producing the plasma of the supercritical parameters for thermonuclear fusion, and to determine the possibility of the powerful X-ray source's development.

3.2.1 Design of Plasma Generators

Plasma generator PG-1 was the first of the pulsed plasma generator developed at the IEE RAS. The electrode system of the generator consists of the axis rod and coaxial ring electrodes. By means of this generator, the basic elements of design: current leads, electrode system, diaphragm unit, energy and gas supply systems, control and measuring devices were worked out. Power supply of the generator was the capacitor bank of 100 kJ and voltage of 5 kV. The initial pressure of the hydrogen is up to 14 MPa. The discharge current was measured by the low-inductance shunt and Rogowski coil, voltage drop across the arc by the resistive divider.

Plasma generator PG-2 was developed for the same operating parameters, but is differed from the first generator by the type of electrode system. Two rod electrodes were arranged coaxially along the axis of the discharge chamber in front of each other. The generator had two optical windows for sensing plasma by the helium–neon laser.

Plasma generator PG-3 was used to continue the research the pulsed arc's characteristics based on the discharge current, voltage drop across the arc, and gas pulsed pressure data obtained earlier. In addition, the optical spectra of arc plasma radiation were recorded, and photographs of the arc were made [11]. Generator PG-3 (Fig. 3.6) as well as PG-1 and PG-2, of low-carbon steel were made, which is characterized by the low hydrogen fragility.

By the type, the generator PG-3 is owned to the discharge chambers with coaxial opposing rod electrodes. Generator PG-3 is the thick-walled vessel (1), designed to pulsed pressures up to 200 MPa. Discharge chamber is of 5 cm diameter and volume of 130 cm^3. Two copper or tungsten rod electrodes (3) of diameter 7–10 mm are insulated from the housing (1) and arranged coaxially in the current leads (4). Current leads are attached to the housing of the discharge chamber by the nuts (5) and sealings (6).

Inner surface of the discharge chamber is protected by thermo-insulating bushing (2) made of aluminum oxide. With bushing and current leads, the volume of the discharge chamber of the generator was $\sim$75 cm^3. Locking diaphragm (8) in some experiments was made of plexiglass to perform optical measurements.

Plasma generator PPG-1 (Fig. 3.7) is designed to further enhance the plasma parameters by increasing the power and energy input into the arc.

Generator's housing consists of the sections fastened together by flanges and bolts. Housing was calculated to the pulsed pressure of 200 MPa. By type of electric discharge chamber is matched to the chambers with coaxial rod (1) and ring (4) electrodes. Volume of the discharge chamber is 0.8–5.6 dm^3 and interelectrode gap is 2.5–50 cm. Inside the discharge chamber was inserted high-temperature ceramic insulator (liner) (3) with inner diameter of 9 cm for fixing the length of the arc and protect the inner surface of the chamber from the radiation of the arc and the heat flow from the high enthalpy gas. The arc was initiated by the explosion of tungsten or constantan wires (2) of 0.15–0.7 mm diameter.

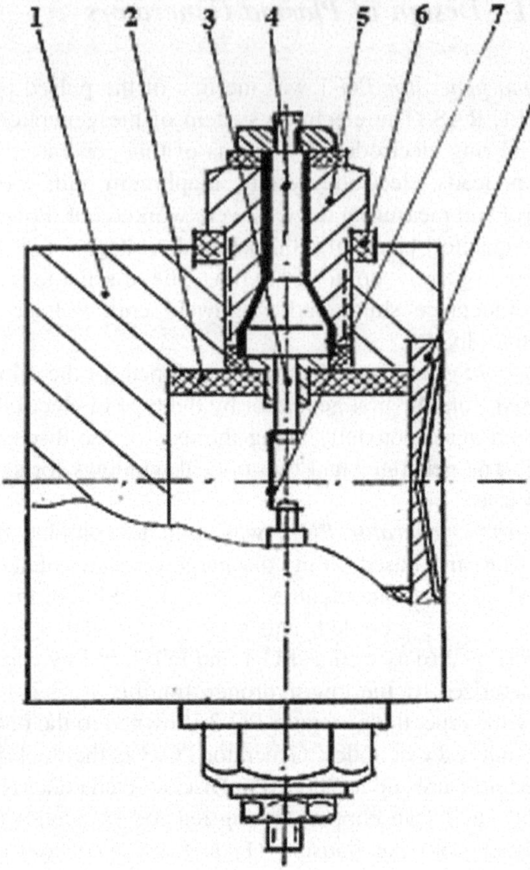

Fig. 3.6 Pulsed plasma generator PG-3 [11]

When the gas pressure is increased, the rupture of the diaphragm (5) and outflow through the nozzle take place. The measurement section (not shown in the diagram) with two windows of quartz or plexiglass provide the diagnostics of the optical parameters of the arcing. Generator was used for researches at the current up to 200 kA, discharge duration ~1 ms, current rise rate 10^8 A/s, and initial pressure 0.1–4.0 MPa [17]. Maximal energy input in the arc is 10 MJ, pulsed pressure up to 150 MPa. Power supplies of the generator are the capacitor bank, shock generator with flywheel, and inductive storage [1, 18, 19].

Plasma Generator PPG-2 (Fig. 3.8) is the modernization of the generator PPG-1 [20]. Generator had the discharge chamber with the coaxial rod (6) and ring (9) electrodes. The aim of the modernization was to facilitate operation of the generator. The problem was solved by the use of a hydraulic jack (2), which provided the axial force of 106 N, needed for sealing the between sectional copper rings.

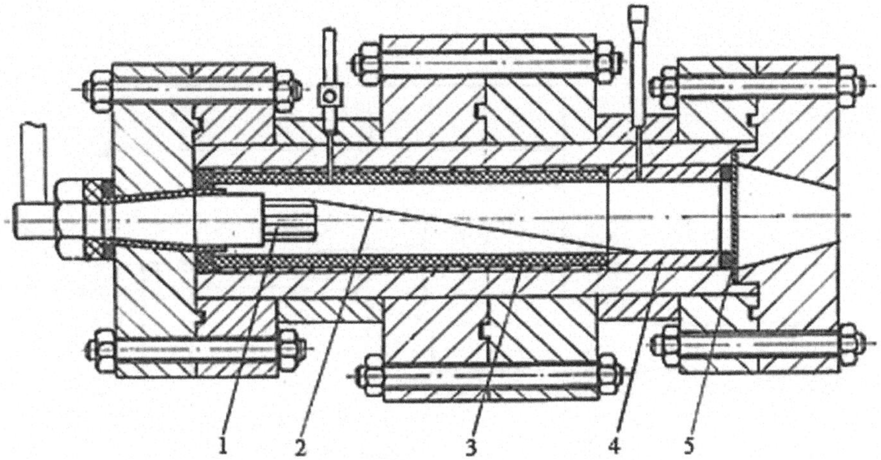

Fig. 3.7 Powerful pulsed plasma generator PPG-1 [17]

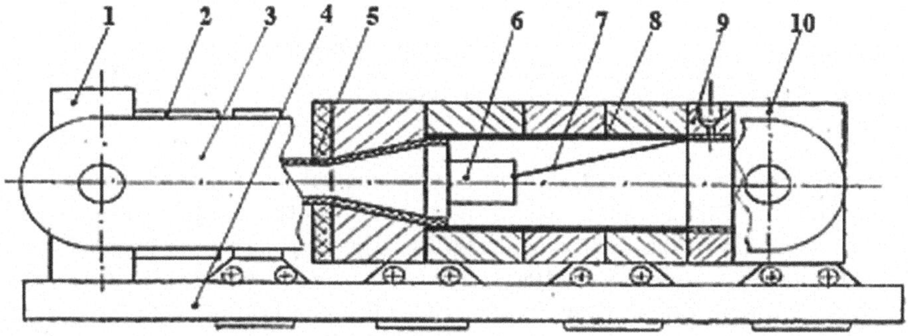

Fig. 3.8 Powerful pulsed plasma generator PPG-2 [20]

Generator is mounted on the berth (4), where the units of the discharge chamber and the movable stop (10) are moved on the lunettes. Fixed (1) and movable (10) stops are connected by two steel rods (3), which provided the force of the hydraulic jack to the between units sealings. The arc is initiated between the cathode (6) insulated from the housing (5) and anode (9) by the explosion of the fuse (7). The arc length is determined by the number of the units and length of the insulating liner (8).

Designs of both the generators are typical for these devices where the distinguishing feature is multisection, which allowed changing the volume of the discharge chamber and interelectrode gap in dependence on the experiments conditions. At these generators, a lot of researches were performed, including at the highest energy input of 10 MJ at powered by the shock generator with flywheel.

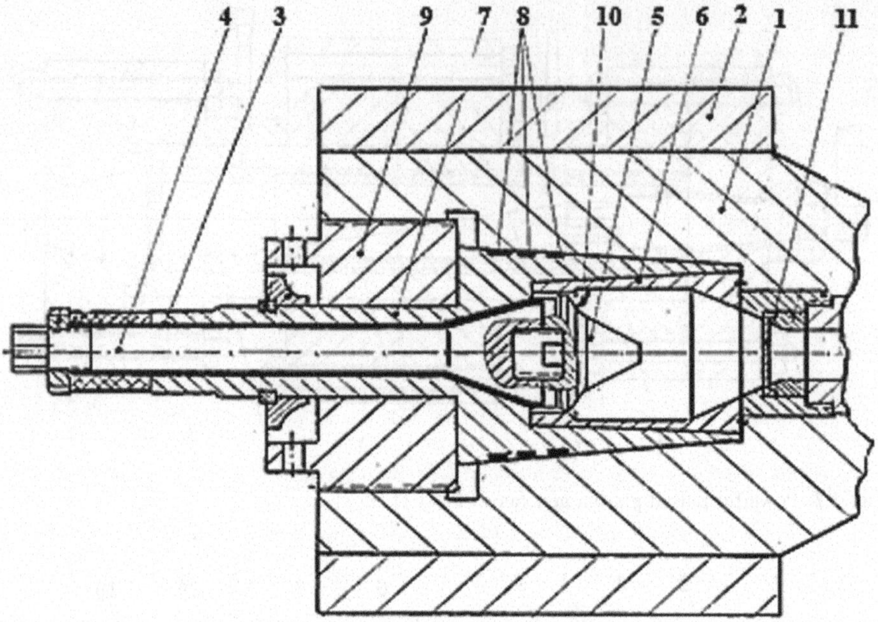

Fig. 3.9 Powerful pulsed plasma generator PPG-3 [3]

Plasma generator PPG-3 (Fig. 3.9) has the coaxial discharge chamber of volume from 0.5 to 2.0 dm^3. Generator consists of the steel housing (1) fastened by the steel bandage (2) of the total weight 900 kg. The housing of the generator was tested by static pressure of 800 MPa. Inside the housing is placed easily removable electrode unit with insulated by fiberglass (3) electrode lead (4) with the cathode (5), anode insert (6), holder (7) with three rubber sealing rings (8), and the back nut (9).

Initiation of the arc is carried out with copper plates (10) by thick $\sim$0.5 mm and width 3–4 mm. After the plate's explosion, the arc itself and binding spots to the electrodes are moved toward the diaphragm (11). Single pulse duration is 500–700 μs, and energy input during this time is $\geq$1 MJ.

Plasma generator PPG-4 as well as generator PPG-3 has a coaxial discharge chamber (Fig. 3.3). The generator consists of the housing (3), inside which is placed the electrode system consisting of the current lead (2) and cathode (4) placed in the holder (1), anode (6), and fuse (5). The discharge chamber is sealed by the diaphragm (7) and the rubber seals.

Volume of the discharge chamber is 0.7–1.2 dm^3, pulsed gas pressure up to 400 MPa. The inner surface of the steel anode is coated by the plasma-sprayed tungsten of thickness $\sim$0.3 mm. This coating makes it possible the operation of the discharge chamber at high reproducibility of the plasma final parameters and moderate contamination of the gas by eroded metal. Elements of the discharge chamber were operated under pulsed currents up to 1.5 MA and gas pressure up to 320 MPa.

This type of discharge chambers has chambers 50-0 and 50-0M of the wind tunnel "Hotshot-2" [7], discharge chamber of the wind tunnel "Tunnel-F" [8], and discharge chamber of the plasma generators PPG-1, PPG-2, and PPG-3 (IEE RAS).

Plasma Generator PPG-5 (Fig. 3.10) was designed to study high-current arcing in hydrogen and helium, between the steel electrode of diameter 6–20 mm, at interelectrode gap 0.5–9.0 mm, initial pressure 10–40 MPa, pulsed pressure up to 1.5 GPa, and weight average temperature up to 4000 K. During the researches, the voltage of the power supply CPS-25 ranged from 8 to 16 kV, energy from 0.5 to 2.0 MJ, discharge current 2 MA, current rise rate $\sim 2 \times 10^{10}$ A/s, duration of the current pulse ~ 100 μs, voltage drop across the arc 3–5 kV, energy input 150–500 kJ, and energy transfer efficiency from the arc to gas $\leq 90\%$.

Generator PPG-5 is the thick-walled vessel (1) made of high-strength steel and is fastened together by means of the steel bandage (2). Electrodischarge chamber with the coaxial opposing rod electrodes includes current lead (5) with the cathode (6) insulated (4) from the housing (1), and anode (7) at the housing potential. Arc is initiated by explosion of copper wire diameter of 0.5 mm (11). Current collector (3) connects the power supply and the electrode system of the generator. Diaphragm (8) seals the gas, which after the diaphragm rupture is expired through the nozzle (9), pressure of the diaphragm opening is ranged from 50 to 300 MPa. Gas inlet into the discharge chamber through the valve (10) is performed.

The generator is designed to operate at the discharge current up to 2 MA. Volume of the discharge chamber is 250 cm^3. In the discharge chamber, there are two coaxial windows (not shown in the diagram) for recording the optical parameters of the arc and gas.

Plasma Generator PPG-6 (Fig. 3.11). The task of raising the gas parameters for researches at hypersonic speeds of the gas expiration required the development of the discharge chamber operating under high gas pressures. One of these devices is

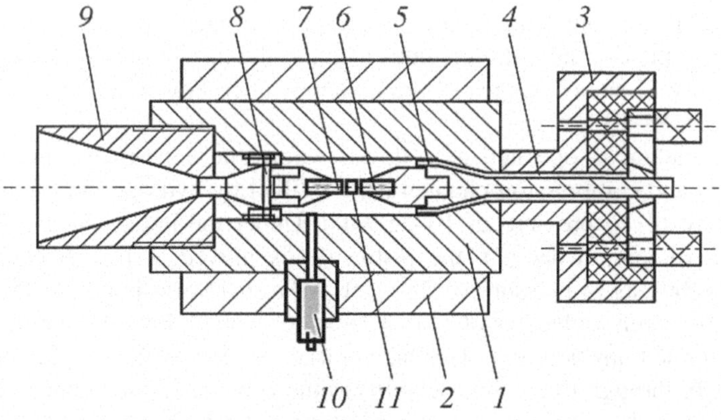

Fig. 3.10 Powerful plasma generator PPG-5 [21]

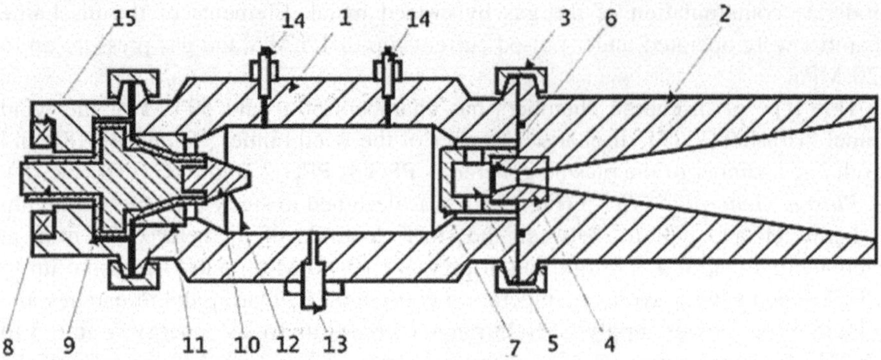

Fig. 3.11 Powerful plasma generator PPG-6 [21]

the plasma generator PPG-6. From the other ones, the plasma generator is differed by the current lead design.

The generator housing (1) of volume 500 cm^3 is made of high-strength steel. The cathode unit consists of the electrode holder (8) with cathode (10), and insulator (9). Sealing of the cathode unit is performed by movable bushing (11). Arc initiation by the explosion of the fuse (12) is carried out. Inlet of the gas into the discharge chamber through the valve (13) is performed. For measuring the pulsed gas pressure and recording the shock waves, two piezoelectric transducers (14) mounted at the distance of 120 mm from each other are used.

Supersonic nozzle (2) is connected to the housing (1) by the clamps (3). In the critical section of the nozzle (2), the tungsten insert (4) is placed. Upstream of the nozzle the diaphragm unit (5) with the diaphragm (6) and the protective screen (7) are placed. Protective screen is designed to reduce pollution of the gas by the metal particles and vapor of electrodes and the discharge chamber walls. To measure, the discharge current Rogowski coil (15) is used. Generator is powered from the power supply CPS-25.

Two-stage electrodischarge installation with d*I*/d*t of* 10^{10} A/s (Fig. 3.12) consists of the piston adiabatic compression of hydrogen channel (1), discharge chamber with the coaxial rod electrodes (5), and the nozzle unit (10) connected to channel (1) by means of flanges (9), and clamps (12). Power supply of the installation is the CPS-10.

The installation is operated as follows: before the experiment, the discharge chamber (5) and compression channel (1) are filled by hydrogen through the inlet (7) and valve (8) to pressure of 10–25 MPa. Under the impact of gas formed at combustion of gunpowder (2), the piston (3) is moved in the channel (1) and compresses the gas to pressure of 250–350 MPa that corresponds to the concentration of hydrogen molecules of (2.0–3.3) × 10^{22} cm^{-3}. Braking and stopping of the piston in the transition unit (4) is taken place. When voltage is applied from the power supply through the current lead (6) to the cathode (7) and anode (13), there are explosion of the copper wire and arc initiation. Hydrogen heated to pressure of

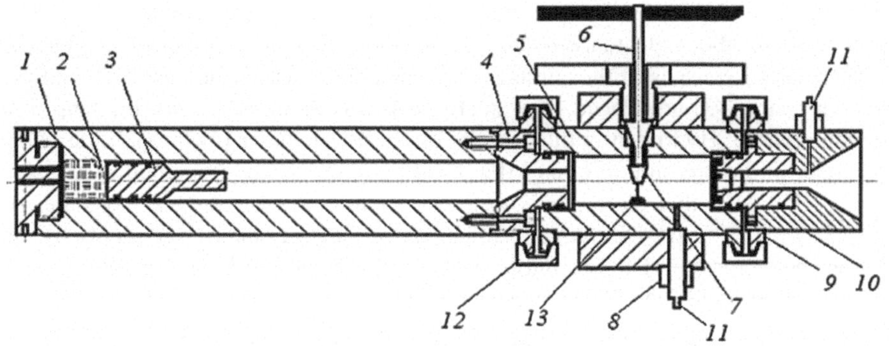

Fig. 3.12 Two-stage electrodischarge installation with d*I*/d*t* of 10^{10} A/s [21]

200–600 MPa and temperatures of 3000 K breaks the diaphragm (9) and is expired through the nozzle (10). Diameter of the discharge chamber and compression channel is 6 cm. Volume of the discharge chamber is 70 cm^3, and compression channel is 640 cm^3. Such ratio of the volumes gives the maximal ratio of gas compression is ~ 10.

Pressure sensors (11) are installed in the compression channel (1) (are not shown in the diagram) in the discharge chamber (5) and in the nozzle (10). Sensor installed in the nozzle generates a signal that controls the switching of the power supply to the discharge circuit. The interelectrode gap is varied for sustainable arcing and efficient energy transfer from the arc to the gas.

Two-stage electrodischarge installation with d*I*/d*t of* 10^{11} A/s (Fig. 3.13). Design of the installation is similar to the previous one (Fig. 3.12). The difference is that the discharge chamber has lower inductance, as a combination of the coaxial

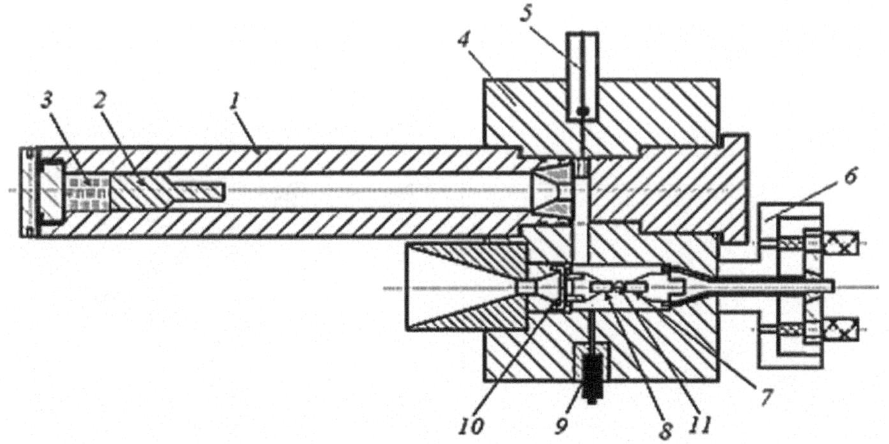

Fig. 3.13 Two-stage electrodischarge installation with d*I*/d*t* of 10^{11} A/s [21]

discharge chamber and chamber with coaxial opposing rod electrodes (7, 8). The discharge chamber inductance reduction increases d*I*/d*t* on an order of magnitude.

Installation consists of the compression channel (1) where the piston (2) and the powder charge (3) are placed. The steel piston was accelerated with the help of a powder charge. Hydrogen is filled in the discharge chamber (4) through the valve (5) under initial pressure of 6–20 MPa. Piston compresses hydrogen up to pressure of 150–350 MPa that corresponds to the concentration of hydrogen molecules of $(1.5–3.3) \times 10^{22}$ cm^3. Discharge chamber is connected to the power supply by the current collector (6). The arc burns between the cathode (7) and anode (8) at a current of 80–500 kA. Pulse pressure is measured by the sensor (9). Optical measurements are performed through the window (11). Diaphragm (10) at pressure build-up is ruptured and hydrogen flows from the discharge chamber.

Low-inductance electrodischarge installation (Fig. 3.14). Installation consists of eight low-inductive capacitors of 3.7 μF, and voltage 50 kV. The total stored energy is 37 kJ.

Capacitors are combined in pairs in the four units. The units are connected to the discharge chamber (1) by two flat copper bars (2) of 3 mm in thickness. Copper bars are isolated from each other by the sixteen layers of polyethylene film (4). Current collector (5) is the connecting element of the discharge chamber and copper bars.

Figure 3.15a shows the discharge chamber design. The inductance of the installation discharge circuit is determined by the inductance of the discharge chamber and is 80 nH. The current rise rate is 6×10^{11} A/s. The discharge current measured by the Rogowski coil (6) and at the maximal number of the capacitors and voltage of 50 kV is 600 kA.

Housing of the discharge chamber connected to the top plate of the current collector (6) is the reverse current lead. To the bottom plate of the current collector

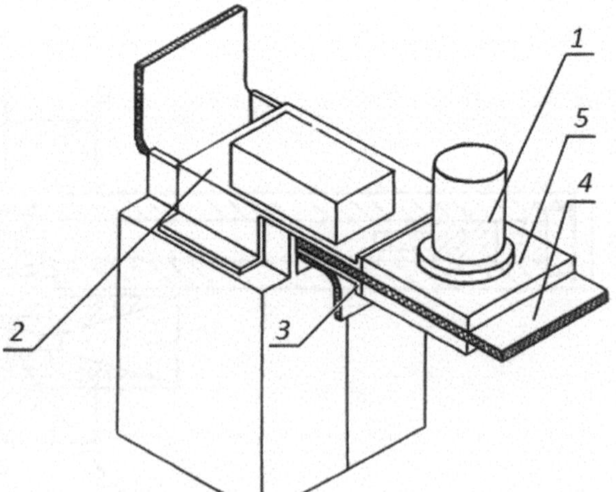

Fig. 3.14 Unit of the low-inductance electrodischarge installation [22]

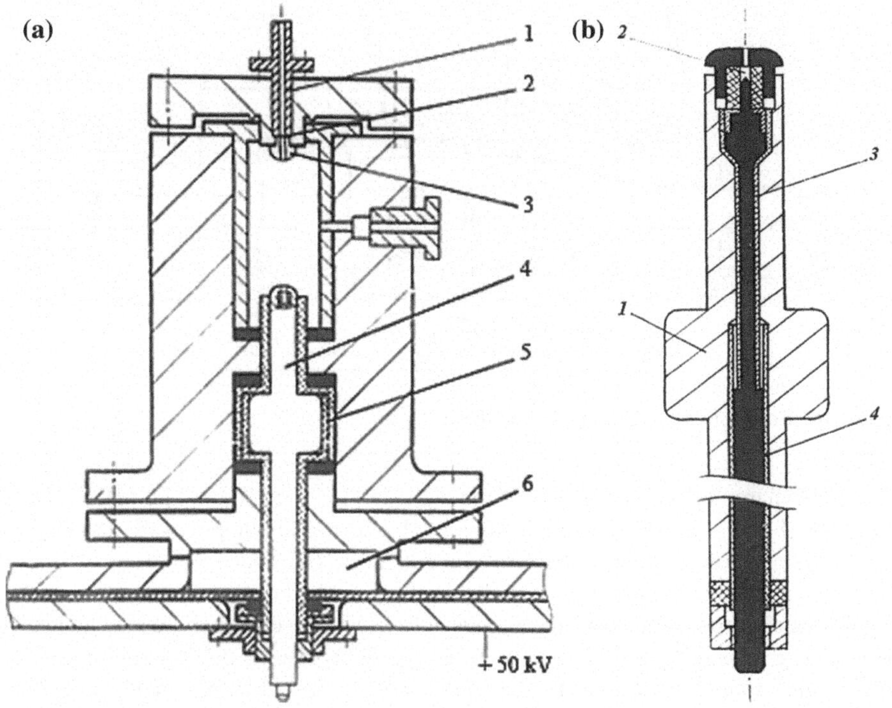

Fig. 3.15 Discharge chamber (**a**) and anode-injector (**b**) [22]

(+50 kV) is connected the anode (4) and insulated from the housing of the discharge chamber by the insulator (5). The cathode holder (1), cathode (3), and the locking diaphragm (2) are the exhaust valve. The inner diameter of the discharge chamber is 4 cm, and with the insulator is 2–3 cm. The interelectrode gap is 1–5 cm.

In order to initiate an arc in helium at initial concentration of normal atoms of (1.4–4.0) × 10^{21} cm^{-3} the plasma injector located in the anode housing (Fig. 3.15b) was developed. The injector consists of the housing (1) (anode), discharge chamber of the injector (2), cathode (3), and insulator (4).

At study of the arcing in the air under atmospheric pressure, the initiation of the arc by means of the additional electrode was carried. To register the pulsed pressure distribution in the cross section of the discharge chamber, the piezoelectric sensor has been developed. One of these sensors is placed in the anode housing (1), i.e., on the axis of the arc, the second one on the wall of the discharge chamber.

Diagnostic electrodischarge chamber (Fig. 3.16) has been developed for the study of the processes in the discharge chambers of the power pulsed plasma generators with the coaxial discharge chamber.

Housing (5) of the diagnostic chamber is made of heat-treated high-strength steel and has the removable element (2) for contacting with the current collector. Coaxial

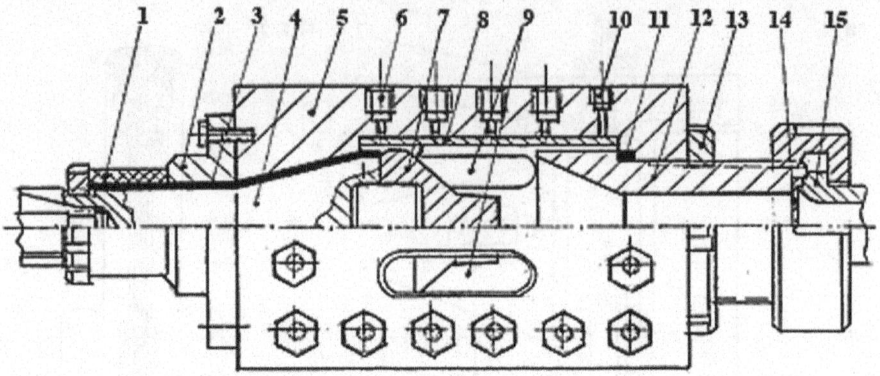

Fig. 3.16 Diagnostic electrodischarge chamber [3]

electrodischarge chamber consists of the cathode (7), electrode holder (4) insulated from the chamber housing by the insulator (1), and two anode plates (8) attached to the upper and lower inner surfaces of the discharge chamber.

You can change the initial type of the discharge chamber to the chamber with the axial opposing rod electrodes. To the side surfaces of the chamber are attached steel lids having depending on the conditions of the experiments, two longitudinal slots (9) or the circular windows. In the upper wall of the chamber in the openings (6) can be placed up to four piezoelectric sensors. The gas fills the discharge chamber through the opening (10). The nozzle (12) has the sealing (11), sealing nut (13), diaphragm (14), and outlet channel (15). Volume of the discharge chamber is 1.37 dm^3. The chamber is designed for pulsed pressure up to 40 MPa. Diagnostic discharge chamber is powered by the one module of the CPS-10 of capacity 0.018 F at voltage of 3.5–5.5 kV.

Maximal parameters of the experiments under initial hydrogen pressure of 2–5 MPa are next:

- Discharge current up to 400 kA
- Energy inputted in the arc up to 200 kJ
- Pulsed pressure up to 25 MPa.

The arc was initiated by the copper wires of diameter 0.15–2.0 mm. Interelectrode gap, depending on the type of the discharge chamber was ranged from 1.5 to 30 mm. The discharge current was measured by Rogowski coil, voltage drop across the arc by the resistive voltage divider or by the inductive sensor.

Additionally, the arcing through the optical windows by the high-speed camera at frequency up to 10^6 frames/s was performed. The recording of the processes by the laser, and recording of the integral spectrum of the arc radiation by the diffraction spectrograph were performed. At the optical measurements, the reference light source of Podmoshensky with a brightness temperature of 4×10^4 K was used.

Parameters of some IEE RAS generators are presented in Table 3.1.

Table 3.1 Parameters of the IEE RAS pulsed generators

Pulsed plasma generator	PG-3	PPG-1	PPG-2	PPG-3
Discharge chamber volume (cm^3)	75	870–2600	2000–3300	600–1600
Power (MW)	200	300	1,000	6600
Maximal current (MA)	0.36	0.18	0.35	1.9
Maximal energy input (MJ)	0.07	4	10	2.1
Gas	H_2	H_2, N_2	H_2, N_2, He, Ar	H_2, N_2
Initial pressure (MPa/gas)	2.8/H_2	4/H_2	11/H_2	42/H_2
Pulse pressure (MPa)	160	70	80	520
Average temperature ($\times 10^{-3}$ K)	20	10	20	4.5

3.3 Components of Plasma Generators' Electrodischarge Chambers

The first variant of the power pulsed plasma generator PPG-3 had the discharge chamber with coaxial rod and ring electrodes (Fig. 3.1d) and was powered by the power supply CPS-10. Electrodischarge chambers of this type were well-proved at operation on the nitrogen and hydrogen under initial pressure up to 12 MPa, pulsed pressure 80–100 MPa, and current up to 350 kA.

However, the first experiments under much higher initial pressure of hydrogen showed the main disadvantage of these chambers—the unstable arcing, and as a consequence, the impossibility to achieve the required ≥ 1 MJ input of energy into the arc. The reason for this was the large interelectrode gap that prevents the repeated breakdown of the interelectrode gap and initiation of the new arc under high initial pressure of hydrogen. It will be shown later that the repeated breakdowns provide a full battery discharge and maximal energy input into the arc.

Figure 3.17 shows the first discharge chamber of the generator PPG-3. Its main features were the bipolar current lead (2, 7), two heat-resistant ceramic (Al_2O_3) insulators (10, 12), and unsupported coaxial anode (9) (due to the ring gap between the inner surface of the housing (13) and its (anode) outer surface. Cathode (8) consists of the tungsten rods of diameter 10 mm soldered in the copper cup, which is screwed into the electrode (2) isolated from the outer current lead (7) by the fiberglass insulator (6). Sealings (1, 3, and 5) were made of polyurethane. Initiation of the arc by explosion of metal wires (11) was performed.

Bipolar current lead (2, 7) and the outer ceramic insulator (12) were intended to prevent the contact of the discharge chamber housing (13) with the arc. However, after the first experiment, the ceramic insulator (12) was destructed completely and the inner surface of the discharge chamber was melted off. This all indicate the futility measures adopted are because of the lack of negative effects. Cathode (8) was destroyed completely under the impact of shock waves and electrodynamic forces.

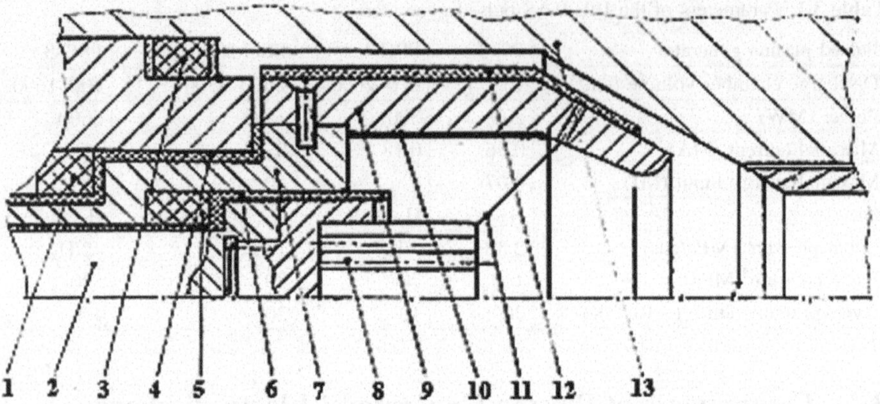

Fig. 3.17 First discharge chamber of the generator PPG-3 [3]

In addition, there was a partial destruction of the ceramic bushing (10) resulting in the shunt arcs appearance and disruption of the camera operation, so this element was removed, changing thereby the type of the discharge chamber.

However, changes in the discharge chamber (Fig. 3.18) did not have satisfied our requirements; i.e., at too big (26 mm) interelectrode gap, the repeated breakdowns of the interelectrode gap and ignition of the arc were absent.

Modified new discharge chamber (Fig. 3.18), the type of the previous discharge chamber was saved. Bipolar electrode system consisting of the central electrode (3, 6) and insulated from the outer electrode (7) by the insulator (4) was saved. The outer electrode with the sealing (1), against the housing of the discharge chamber by the insulator (2) has been insulated. In addition, the protective ring (5) was added to protect the surface of the insulator (4) from the arc's radiation and shock waves. As usually for ignition of the arc, the metal wires (8) were used.

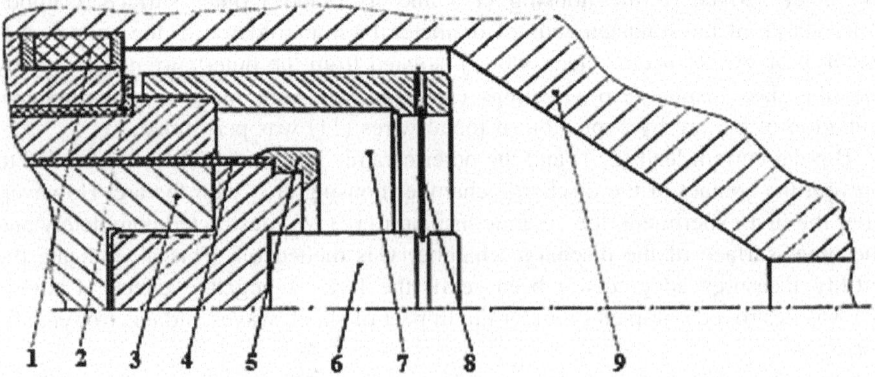

Fig. 3.18 Discharge chamber with coaxial electrodes of the generator PPG-3 [3]

The modernization of the discharge chamber in order to maximize the transferring of the power supply energy into the arc, ended in failure. In fact, the high-current switch has been developed, where arc extinction occurs up to 100–150 μs and the energy transferred into the gas did not exceed 300 kJ.

The main task, in these circumstances, was an increase of the duration of the discharge pulse t_p, when the electric arc power of 2.5 GW, was close to its maximal value. Thus, if we exclude the discharge current I and the voltage drop across the arc U_a, the energy input W can be increased only by increasing the discharge pulse duration

$$W = \int_0^{t_p} I U_a \, dt. \tag{3.1}$$

The problem was solved by applying the "coaxial" type of discharge chamber (Fig. 3.1b). By the way, this type of discharge chamber has appeared as a result of modernizations performed. One of the discharge chambers of this type is shown in Fig. 3.19.

Figure 3.19 shows that the design of the discharge chamber was simplified greatly. Firstly, the electrode system (2, 3, 6, and 7) became the unipolar one. Second, the cathode (3, 6) is placed in the electrode holder (2), which is at the potential of the discharge chamber (8). Thirdly, the number of sealings was reduced to one (1). The inner surface of the brass anode (7) was coated with the molybdenum insert (4).

The main difference of these discharge chambers against previously used is more stretching of the arc. At these conditions, the arc and its attachment spots to the anode (7) and cathode (6) are moved axially to the nozzle for more time. All this affected the increase in the duration of the discharge pulse. Pulse duration in its turn

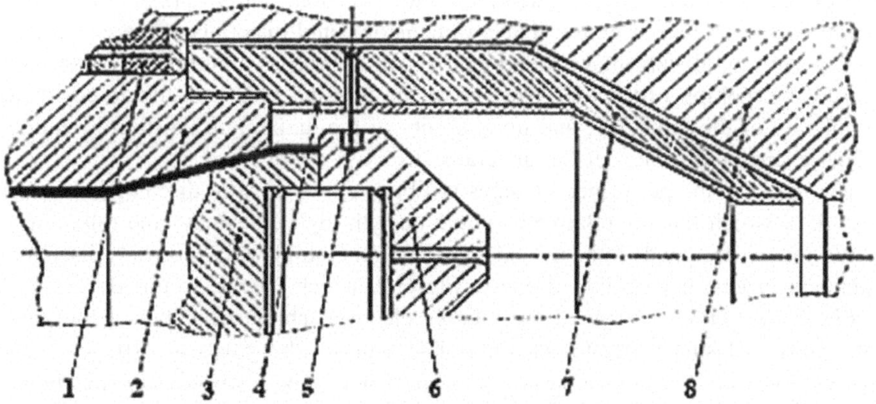

Fig. 3.19 Coaxial discharge chamber of the generator PPG-3 [3]

determines the characteristics of the arcing and heat transfer from the arc in gas in the discharge chamber.

In particular, at the transition to the new discharge chamber, the duration of the current pulse was increased from 100–150 to 500–700 μs, i.e., in average by ~500 μs, and the energy of the arc from 200–300 to 1.5–2.1 MJ. Thus, the task of ensuring the required energy inputted into the arc was solved.

However, a number of new equally complex problems due to the high energy inputted into the arc are appeared:

- Reliability of the most loaded elements of the discharge chamber, such as insulator of the current lead, electrodes, diaphragm unit, and sealings
- Searching of the optimal arcing modes
- Reproducibility of gas parameters under the fixed initial parameters of the experiment
- Minimization of gas pollution by the electrodes erosion
- Increasing the efficiency of energy transfer at all stages of its transformation.

As mentioned above, at the transition to the new type of discharge chamber, the bipolar current lead has been changed to the unipolar one. The reason for this was that the potential on the discharge chamber housing of the generator during operation does not exceed a few tens of volts. The main share of the voltage drop is across the arc, and other small voltage drops are on the switching system, current lead, cables, etc.

As a result of the changes, the discharge took place between the insulated cathode (6) and the anode (7) having ground potential of the housing of the discharge chamber. This simplified greatly the design of the generator, increased the reliability, and safety of its operation.

The weakest element of the new discharge chamber was the unsupported anode (7), which has been designed such way to prevent its jamming in the housing of the discharge chamber (8) under the impact high gradient of the pulsed gas pressure.

Without radical changes of the discharge chamber design, more than one hundred of the experiments were performed with over than thirty variants of the discharge chamber differed by the shape, dimensions, and materials of the electrodes.

However, the disadvantage mentioned above as well as a number of other ones such as the melting of the inner surface of the housing of the discharge chamber, complexity of the replacing, and unreliability of the diaphragm unit have required radical changes in design of the generator.

The result of the performed changes has become the coaxial discharge chamber (Fig. 3.9), which has the relatively simple design, high efficiency, and reliability.

We have designed four pulsed plasma generators with coaxial discharge chambers structurally similar to that in Fig. 3.9, which operate to this day.

Electrodes. Development of the reliable electrodischarge chamber of the powerful pulsed plasma generator is impossible without the practical experience for their exploitation. Significant part of this experience is the knowledge of specific for such devices aspects of the materials science, engineering, and operating conditions of the elements of the electrodischarge chambers. In our opinion, it will be useful, at least briefly, to dwell on these aspects and to present the main results of the work on improvement of the elements of the electrodischarge chambers.

Electrode system of the plasma generator operates in the dense high-temperature gases under the impact of the strong shock waves, and bending moments by the electrodynamic forces of $\geq 10^6$ N $\times$ m. Besides, the cathode and anode are exposed to arcing at current up to 2 MA and electric charge up to 700 °C. These factors lead to melting, erosion, deformation, and destruction of the elements of the electrode system that requires the necessary measures to ensure their reliability operation.

Under the typical conditions of the experiments, the active arc resistance is ≥ 6 mΩ, which causes the unipolar discharge of the capacitor bank and uniquely determines the polarity of the electrode system. Wherein the central electrode is the axial cathode connected to the negatively charged plates of the capacitor bank, and coaxial is anode.

The first electrode system of the PPG-3 plasma generator consists of the cathode recruited from 19 tungsten rods of 10-mm diameter and of 50 mm length (Fig. 3.20a), and anode made of high-strength steel (Fig. 3.20b).

At first experiment, the generator PPG-3 fed from the CPS-10 before reswitching to 10 kV. At the experiment, the capacitance of the power supply CPS-10 was 0.24 F, stored energy is 3 MJ, and electric charge 1200 °C. During the experiment, the cathode was completely destroyed (in Fig. 3.20a, the cathode shown before the experiment) under impact of shock waves and bending moments, and anode was melted significantly (Fig. 3.20b).

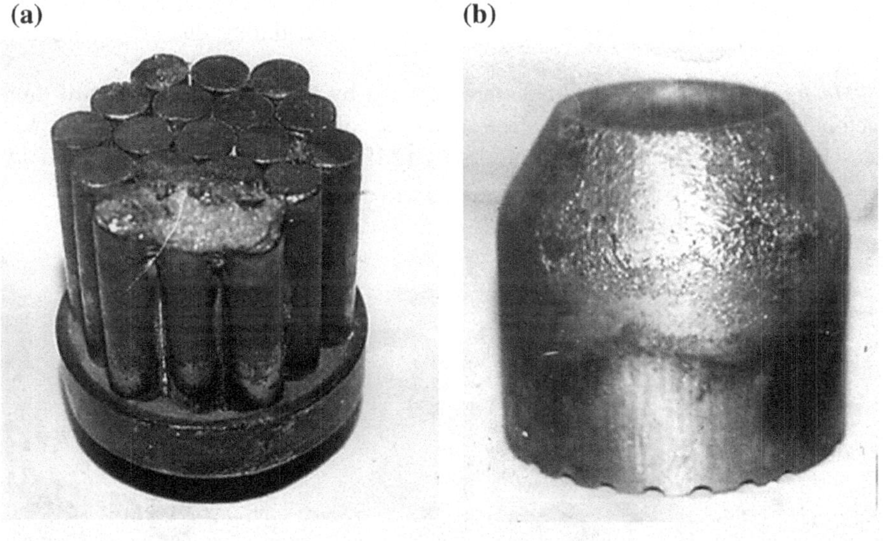

Fig. 3.20 Electrodes of the first electrode system generator PPG-3 [3]

Later, the cathodes were made of various materials: tungsten, molybdenum, steel, and pseudoalloys based on tungsten and molybdenum.

For the producing of the anodes steel, copper, brass, molybdenum, reinforced by the steel pseudoalloys WNF (W + Ni + Fe), WNC (W + Ni + Cu), MWC (Mo + W + Cu), and a number of other materials coated by copper, tungsten, and chromium with ultradispersed diamond powder were used.

To determine the optimal design of the electrodischarge chamber and electrode material, over thirty variants of the discharge chambers, large number of cathodes (Fig. 3.21a), and anodes (Fig. 3.21b) have been developed and tested.

At that, some anodes and cathodes, for example molybdenum, are destroyed at the first experiment, but others ones, such as copper anodes, withstood up to ten or more cycles and were replaced because of the discharge chamber modernization.

As mentioned above, the processes in the discharge chamber depend essentially on the sizes and shape of the electrodes. One should also emphasize that the properties of the electrode material play a significant role in the arcing. The main ones are: electric and thermal conductivity, atomic weight, density, melting and boiling temperatures, and work function of the electron.

Let us represent some characteristics of the electrode materials used at the experiments. There are several criteria for selection of electrode materials, the main of which are:

- Mechanical strength
- Erosion durability
- Manufacturability
- Availability and cost.

In this list, the sequence of a listing does not reflect the priority of one over the other criteria, so any of them could become the determining ones in a particular case.

Mechanical strength. Electrodes are exposed by the combined impact of three factors: bending moment of the electrodynamic force, which at a current of 1 MA is $\sim 10^6$ N $\times$ m, gas pulsed pressure ≤ 620 MPa at rise rate of 10^6 to 10^7 MPa/s,

(a) (b)

Fig. 3.21 Electrodes of the generator PPG-3 [3]

and shock waves generated in forming of the arc. Thus, the electrode material must have not only of high strength but also of high ductility.

Erosion durability. Application area of the pulsed plasma generators depends essentially on the pollution of the gas by the products of electrode erosion, which increases the molecular weight and changes other physicochemical properties of the plasma. Since the electrode erosion gives main share of the impurities, but in most cases, it is desirable that the concentration of impurities in the gas was minimal, it requires the usage of electrode materials with the highest erosion durability, such as alloys and pseudoalloys of tungsten.

Manufacturability. There are a number of materials with unique properties: very high erosion durability and melting temperature, such as tungsten–rhenium alloy and pseudoalloy MWC (Mo + W + Cu), but their hardness is so high that tooling by means of conventional cutting tools is almost impossible.

Availability and cost. Some materials suitable for the electrode systems are not produced by the required size, and other materials are very expensive for regular use. Therefore, at the choice of electrode material, we are forced to take into account these circumstances. Using the criteria listed above, we present a brief description of some materials used for the manufacture the electrodes.

Tungsten and its alloys are some of the most expensive materials. Materials of this group have a high melting point, maximum erosion durability, and high mechanical strength. During the researches, six different tungsten alloys, including pure tungsten have been tried.

Pure tungsten is used as the coatings on substrates by means of vapor deposition, and plasma spraying. Despite the attractiveness of these methods and their indisputable advantages, such as low consumption of expensive tungsten and the possibility of its application to the electrodes of any shape and size, these coatings have common disadvantage—low adhesion to the substrate material. This makes it impossible multiple uses of these electrodes, and became the reason that these electrodes were not use at our researches any more. However, when we need the one-off electrode, then such coating may be quite acceptable.

Alloy of the tungsten and rhenium with obvious advantages—highest erosion durability, melting point, and strength—has a very high hardness, which makes its tooling is almost impossible. This fact makes its use in the high-power plasma generators is limited.

Significantly greater opportunities for development of the discharge chambers have pseudoalloys based on tungsten and molybdenum manufactured by powder metallurgy. Among them, we used:

- WNF (80% W, 10% Ni, 10% Fe)
- WNC (90% W, 1.5% Ni, 5–10% Cu)
- MWC (50% Mo, 45% W, 5–10% Cu).

These pseudoalloys are differed from the tungsten–rhenium alloy by the lower hardness. However, their common feature—the fragility—requires put a band of the high-strength steel. Since these pseudoalloys are rather expensive and have highest erosion durability, then their use is expedient only when the discharge chamber should generate the plasma with minimal concentration of metal vapor.

Molybdenum has a relatively high melting point, high erosion durability, is very fragile (not withstand shock loads), and rather expensive—so in sum not be recommended as an electrode material for high-power pulsed plasma generators.

Copper has two main advantages: the availability and relative cheapness. For this reason, and despite the relatively low erosion durability, low melting point and strength, copper anodes have been used in large number of experiments, especially at the initial stage of research.

Steel is characterized by high strength, high shock toughness, manufacturability, low cost, and availability. However, it has low erosion durability. If the average temperature of the gas does not exceed the boiling temperature of steel $\sim$3000 K, when the effervescence of the surface layer of the electrode takes place, the use of steel is quite acceptable.

Aluminum alloys and brass are characterized by extremely low erosion durability, low melting temperature, and low strength, which make their use impossible.

Current leads. These elements of the discharge chamber had required the most efforts at working out and operation, in particular, their isolation. Considering the combination of the operating conditions, the current leads are the most loaded elements of the discharge chamber. The next factors impact to them:

- Voltage up to 25 kV
- Mechanical loads of the gas pressure exceeding the strength of any dielectrics except diamond and ceramics
- Bending moments $\geq 10^6$ N $\times$ m of the electrodynamic forces
- Heat flow from the gas at 4500 K
- Radiation from the arc
- Shock waves
- Heating from the surface layer of a current lead due to the skin effect.

During the operation with the discharge chambers, three types of current leads were used:

- Rectangular current lead (Fig. 3.22a)
- Conical current lead (Fig. 3.22b)
- Combined current lead (Fig. 3.22c).

Omitting the details of the long searches for the insulation material of the electrodes, it is necessary to say that, ultimately, the choice was made in favor of the composite materials, in particular, the fiberglass—all the other materials for various reasons were rejected. From that time where it was possible, the insulation was performed of the high-modulus fiberglass impregnated with epoxy compound. Current leads were made of heat-treated high-strength steels. The surface of the current leads was coated by the layer of electrolytic copper thickness of $\sim$0.3 mm in order to reduce its active and contact resistance.

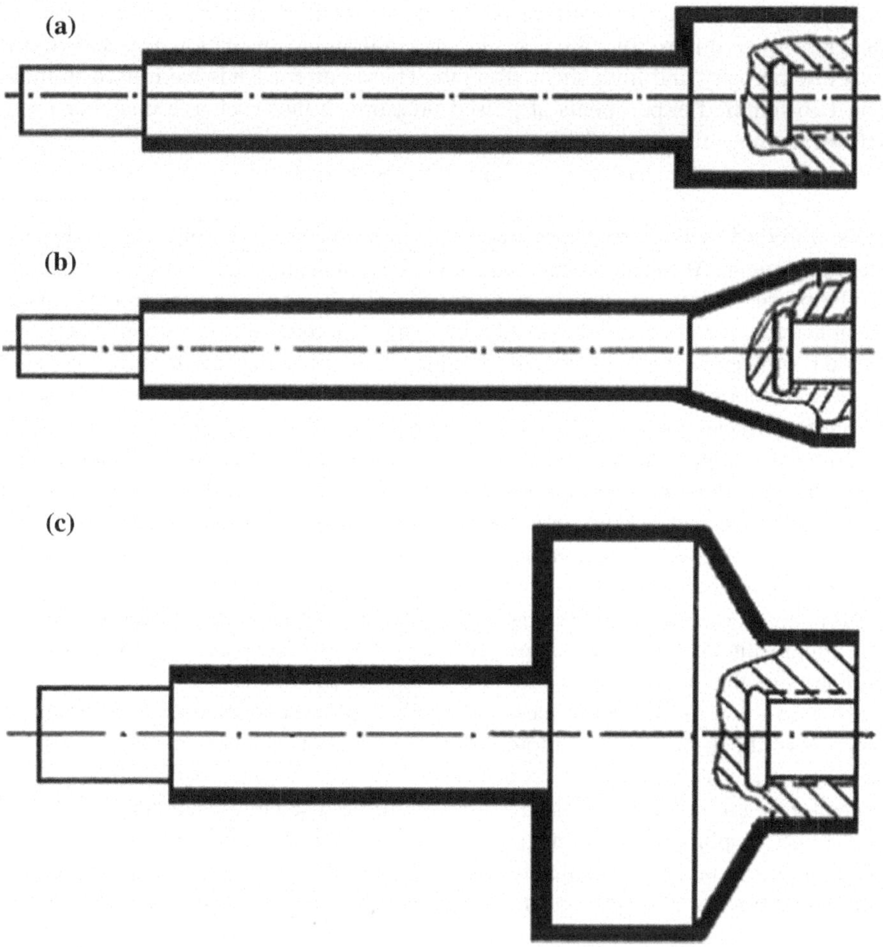

Fig. 3.22 Types of the current leads [3]

Rectangular current lead (Fig. 3.22a) is characterized by the supporting surface perpendicular to its axis. Such shape of the electrode was unsuitable because of:

- Low ratio of supporting area to the cross-sectional area of the electrode, which is exposed the gas presses (factor determining the stress in the insulation material)
- Impossibility to perform the optimal winding, when the fiberglass is imposed at the angle of 45° to the current lead's axis, thus achieving its spatial fixation.

Glass fiber tangentially imposed on these current leads had low mechanical strength. As the result, the insulation material in the place of the sealing rings was broken under high load because of gas pulsed pressure.

Conical current lead (Fig. 3.22b) is differed from the rectangular one by the shape of the support surface, which is a cone. Changing the shape of the current lead increased the support surface and made it possible to apply cross-scheme of the glass fiber imposition. As a result, the monolithic insulation was developed with low porosity and high shear strength. These current leads were used in more than two hundred experiments at pulsed pressure in the discharge chamber up to 510 MPa, and voltage of 12 kV.

Combined current lead (Fig. 3.22c). Further increase of the pulsed pressure in the discharge chamber of the plasma generators requires the development of current leads, differing by the larger supporting surface. This design requires the additional sealing element. At the same time to provide the necessary mechanical strength, the insulation of the current lead was reshaped. The latter circumstance made inapplicable cross-winding of the insulation, and required the constituent elements bonded with each other by means of glue. This all complicated the process of insulation imposition and somewhat reduced reliability due to the joints between the insulating elements.

Three discharge chambers with the current leads of this type had been developed. The schematic diagram two of those are shown in Figs. 3.11 and 3.12. Some current leads of this type operated at pressure up to 620 MPa and voltages up to 15 kV.

Recently, in pulsed technique and technology of high pressures, a large spreading has polycarbonate—material with the dynamic reinforcement. Dynamic reinforcement is unique property of certain substances in several times to increase its strength under the impact of pulsed load. Polycarbonate is used by us for the insulators of the compact high-pressure discharge chambers.

Diaphragms. An important part of the pulsed plasma generators performing the function of the high-velocity output valve is the locking diaphragm (Fig. 3.23). On the functioning of this element depends such operating characteristics of the plasma generator as the initial and pulsed pressures in the discharge chamber.

To the diaphragm, a number of often conflicting requirements are made. Diaphragm should be mechanically strong, therefore thick in order to withstand the initial pressure of the gas under several tens of megapascals and high pulsed

Before opening After opening

Fig. 3.23 Locking diaphragms [3]

pressure during the arcing. At the same time, the diaphragm should be low-inertial for fast opening and ductile to prevent the tearing of its blades. Thus, the searching for the optimal sizes and material of the diaphragms is the searching for compromise between what is desirable and possible.

Properties of the diaphragm significantly depend on the shape and size of the grooves—stress concentrators on the surface of the diaphragm, which are the rupture lines. The grooves depth is about one-third of the diaphragm thickness and placed radially from the center. In the center, the hole of depth slightly greater than the depth of the groove hubs, but less than the thickness of the diaphragm is drilled. These grooves provide the full opening of the diaphragm of 5 mm thickness under the pulsed pressure of 300–500 MPa for a time of 60–70 μs.

Material of the diaphragms must have a high mechanical strength and dynamic viscosity. The low-carbon stainless steels satisfy to these requirements more than other ones. The elements of the diaphragm unit are made of high-strength steel heat treated to ≥ 1200 MPa.

For the effective operation of the plasma generator, the particular importance has a "right"—with no separation of blades and without significant narrowing of the output cross section of the diaphragm unit, as the separated blades make the plasma unusable, and nozzle narrowing reduces the speed characteristics of the plasma flow.

To the correct opening of the diaphragm, great importance has the nozzle form, i.e., the place where after the opening, the diaphragm blades are laid. Exactly the form of the nozzle, sizes of the diaphragm, properties of its material, value and rise rate of the pulsed gas pressure, and gas temperature determine the integrity of the diaphragm after its opening.

Large number of factors impacting on the opening of the diaphragm difficult the searching the unambiguous recommendations on the optimal form of the nozzle, so in each case the question must be solved by trial and error, which requires additional time and efforts.

There is the different diaphragm unit, where the number of inner faces is equal to the number of blades [23]. Thus, in these constructions, the flat flexure of the diaphragm blades takes place that is reduced the stress concentrations, which is their major advantage. The disadvantages of this design are the relative complexity of manufacturing these bushings, and that they have stress concentrators.

For some applications, it is essential knowing the opening time of the diaphragm. The method of the opening time determination is based on the registration of the arc light radiation. The light from the arc passes through the opening diaphragm, is reflected from the mirror, and directed onto the photomultiplier. The same other photomultiplier at the same time registers the background illumination. The signals from the photomultipliers are delivered to the oscilloscope, where background illumination of the additional photomultiplier partially compensates the background illumination of the main photomultiplier. Obtained by this method, the radiation waveform contains the information of the start and end of the diaphragm opening.

The characteristic time of full diaphragm opening thickness of 5 mm at the pulsed pressure of 300–500 MPa is 60–70 μs [24].

References

1. I.A. Glebov, F.G. Rutberg, *High-Power Plasma Generators* (Moscow, 1985) (in Russian)
2. Ph. Rutberg, *Physics and Technology of High-Current Discharges in Dense Gas Media and Flows* (New York, 2009)
3. V.A. Kolikov, Doctoral Thesis, St.-Petersburg (2005) (in Russian)
4. J.F. Blanshetta, K.P. Sivjer, *Modern Technics of Aerodynamic Rresearches at Hypersonic Speeds* (Moskow, 1965) (in Russian)
5. R. Rotert, K.P. Sivjer, *Technics of Hypersonic Researches* (Moskow, 1964) (in Russian)
6. D. Kaizer, *Technics of Hypersonic Researches* (Moskow, 1964) (in Russian)
7. D. Lukashevich, U. Garris, R. Jekson, D. van-der-Blick, R. Miller, *Technics of Hypersonic Researches* (Moskow, 1964) (in Russian)
8. J.A. Blick, *Modern Technics of Aerodynamic Researches At Hypersonic Speeds* (Moskow, 1965) (in Russian)
9. K. Lekont, *Physics of Hypervelocity Processes*, **II** (Moskow, 1971) (in Russian)
10. Д. Лукашевич, У. Гаррис, Р. Джексон, Д. Ван-дер-Блик, Р. Миллер, *Техника гиперзвуковых исследований* (Moskow, 1964) (in Russian)
11. G.G. Antonov, V.S. Borodin, A.I. Zaitsev, P.G. Rutberg, Sov. J. Tech. Phys. **XLII**, 2121 (1972) (in Russian)
12. J.A. van der Blick, *in 2nd Symposium on Hypervelocity Techniques, New York* (1962)
13. R. Dannenberg, A. Silva, Sov. J. Aerosp. Sci. **10**, 13 (1972). (in Russian)
14. T.H.G.G. Weise, in *4th Symposium on Electromagnetic Launch Technology, Celle* (1993)
15. J.G. Kelley, M.A. Levine, A.L. Besse, A. Tatarian, Rev. Sci. Instr. **38**, 641 (1967)
16. T.H.G.G. Weise, R. Dormeval, P. Noiret, in *5th European Symposium on Electromagnetic Launch Technology, Toulouse* (1995)
17. PhG Rutberg, B.P. Levchenko, Sov. J. High Temp. **9**, 215 (1971). (in Russian)
18. D.A. Andreev, M.O. Znesin, V.A. Kolikov, B.P. Levchenko, A.V. Radjushin, P.G. Rutberg, *Power Supplies of Short-Term and Pulse High-Power Loadings* (Leningrad, 1981) (in Russian)
19. E.A. Azizov, A.A. Bogomaz, B.P. Levchenko, P.G. Rutberg, Sov. J. Tech. Phys. **XLIX,** 441 (1979) (in Russian)
20. B.P. Levchenko, P.G. Rutberg, *Generators of Plasma Jets and High-Current Arches* (Leningrad, 1973) (in Russian)
21. A.F. Savvateev, Doctoral Thesis, St.-Petersburg (2003) (in Russian)
22. A.A. Bogomaz, Doctoral Thesis, St.-Petersburg (2013) (in Russian)
23. N.A. Zlatin, G.I. Mishin, *Ballistic Installations and Their Application in Experimental Researches* (Moskow, 1971) (in Russian)
24. I.P. Makarevich, PhG Rutberg, V.A. Kolikov, A.V. Budin, V.V. Leontiev, B.P. Levchenko, N.A. Shirokov, Tech. Phys. **41**, 227 (1996)

Chapter 4
Parameters of the Arc

Abstract The main characteristics of the pulsed arc, operating gas, and discharge process, which are available for immediate registration during the full-scale experiment on the powerful pulsed plasma generators, are:

- Discharge current
- Voltage drop across the arc
- Initial and pulsed gas pressure in the discharge chamber of the generator
- Erosion of the electrodes
- Discharge duration.

On the basis of these data, the following parameters are calculated:

- Peak and average power of the discharge
- Arc resistance
- Energy inputted into the arc
- Electric charge
- Internal energy of the gas
- Average gas temperature.

These parameters are used to determine such characteristics of the generator:

- Coefficient of the power supply energy conversion into the arc
- Coefficient of the arc electrical energy conversion into the internal energy of the gas
- Specific erosion of the electrodes
- Apparent molecular weight of the gas
- Speed of sound in the gas, etc.

The nature and the parameters of the arc for the particular type of the discharge chamber are determined by many factors which are from the power supply parameters to the chemical composition of the electrode material. During researches, some factors that mainly determine the nature of the arcing and arc's parameters were identified. These factors are next:

- Capacitance and inductance of the power supply
- Inductance and resistance of the discharge circuit

V. Kolikov et al., *Powerful Pulsed Plasma Generators*,
Springer Series on Atomic, Optical, and Plasma Physics 101,
https://doi.org/10.1007/978-3-319-95249-9_4

- Charging voltage
- Gas and its initial pressure
- Interelectrode gap
- Material and shape of the electrodes
- Configuration and volume of the discharge chamber.

At this, we must remember of the significant impact of nonlinear arc parameters (resistance and inductance) on the amplitude-frequency characteristics of the discharge current. Thus, there is a multifactorial interaction of the initial conditions of the experiment and the load, and their joint impact on the characteristics of the discharge process.

4.1 Resistance

Oscillograms of the discharge current and voltage drop across the arc allow us to draw a graph of the impedance of the arc on time and determine the average arc resistance and active resistance at the current maximum. The resulting data can be used to analyze the arcing and for searching the ways to improve the efficiency of the plasma generator, since the resistance of the arc is a key parameter that determines the efficiency of energy transfer from the power supply into the arc.

Figure 4.1 shows the oscillograms of the discharge current and voltage drop across the arc, the graph of the arc resistance, and dynamic volt–ampere characteristics of the discharge in the experiment, which was carried out at the electrodischarge chamber of 1.12 dm^3 of the generator PPG-3. Discharge chamber had steel cathode and tungsten anode at interelectrode gap of 8 mm, copper fuse of diameter 1.2 mm, charging voltage of 8.5 kV, and initial hydrogen pressure 29 MPa.

Oscillogram of the voltage (Fig. 4.1b) shows that in the 10th μs, the explosion of the wire and peak of voltage of 5 kV took place. By comparison of the oscillograms of the experiments carried out at various interelectrode gaps but at the same electrode's material, it can be concluded that the peak of the voltage is proportional to the interelectrode gap.

Further, there is a steady increase from 2 to 8 kV, while at 100th μs the characteristic recession and the peak of the voltage are observed, which will be analyzed below.

At 145th μs at the arc voltage of 10.5 kV, there is a sharp decline in the voltage up to the voltage of the power source, which may be due to self-breakdown of the arc. During this time, the current at the 80th μs reaches its maximum of 920 kA and then is decreased at 150th μs to 560 kA. At this, the rise rate of the current at period from 0 to 50 μs is $\sim 1.5 \times 10^{10}$ A/s.

This time interval (0–150 μs) is characterized by almost constant (with a slight rise in the 140th μs) resistance of 5–8 mΩ. Later on from the 160th μs, there is an exponential increase in resistance until the extinction of the arc.

(a)

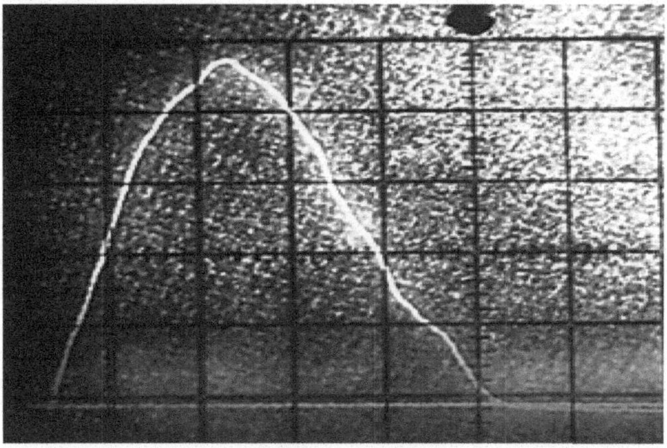

200 kA/div; 50 μs/div

(b)

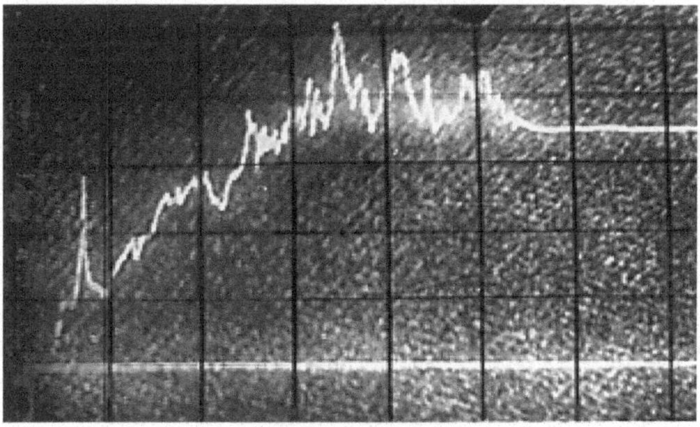

2 kV/div; 50 μs/div

(c)

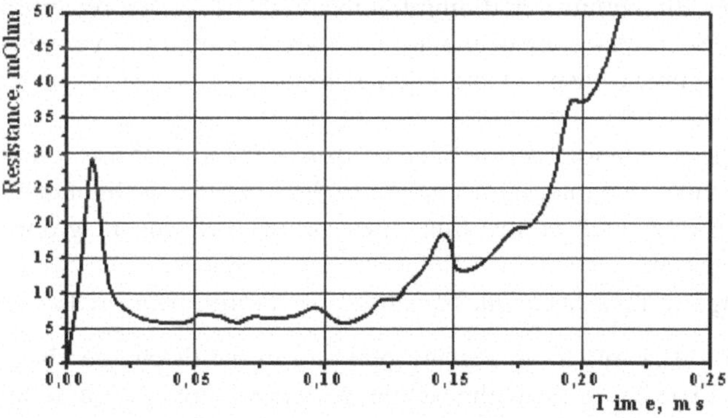

Fig. 4.1 Oscillograms of the current (**a**), voltage drop across the arc (**b**), graphs of the arc resistance (**c**), and dynamic volt–ampere characteristics (**d**) [1]

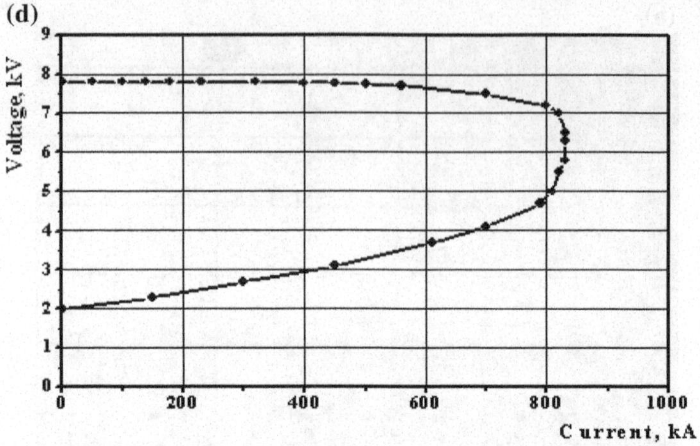

Fig. 4.1 (continued)

Based on the analysis of the oscillograms of this and other experiments performed under similar conditions, it can be concluded that such character of the arc resistance variation is typical for situations, when the arc is initiated in a wide interelectrode gap, and when one or both electrodes are made of highly erosion-resistant material; in this case, the anode material is tungsten.

When the electrodes are made of material with low erosion resistance, such as stainless steels, the character of the arcing and changing of resistance are significantly different from the case presented above.

Figure 4.2 shows the oscillograms of the discharge current and voltage drop across the arc, and graph of the arc's resistance in the experiment under similar initial conditions: voltage of 7.8 kV, initial pressure of 19.5 MPa, interelectrode gap of 12.5 mm, and volume of the discharge chamber of 1.3 dm^3.

Graph of the arc resistance has a sawtooth shape, i.e., characterized by the large number of the maximums and minimums at the average repetition frequency of $\sim$10 kHz. Such character of the arc resistance change (curve–initial) is caused by the multiple breakdowns of the interelectrode gap, which occur approximately every 100 μs.

The average resistance (curve–filtered) from 0.05 to 0.45 ms is increased from 7 to 18 mΩ, in the time interval from 0.45 to 0.75 ms after a slight decline is almost constant $\sim$16 mΩ, and in the time interval of 0.75 ms is growing until the extinction of the arc.

Such change in resistance can be caused by the following reasons:

- from 0.05 to 0.4 ms, by increasing pressure in the discharge chamber;
- from 0.4 to 0.75 ms, by the stable parameters of gas at a relatively low power $\sim$700 MW;

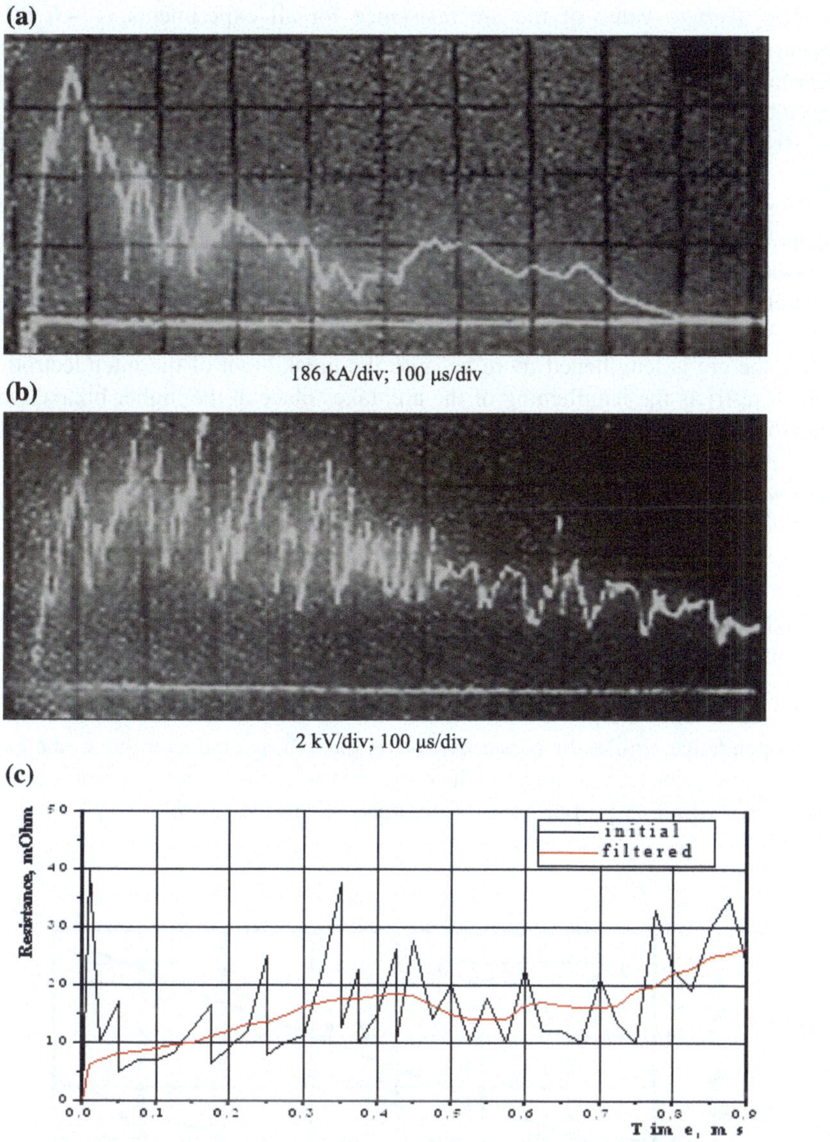

Fig. 4.2 Oscillograms of the current (**a**), voltage drop across the arc (**b**), and graph of the resistance (**c**) [1]

- from 0.75 ms till the arc extinction, by decreasing supply voltage, by decreasing discharge power, by intense outflow of gas, and by decreasing the arc and gas temperature.

The average value of the arc resistance for all experiments is ~6 mΩ that confirms earlier conclusion of that the arc parameters, in particular its active resistance, are critical in the battery discharge, since the resistance of the discharge circuit is four times less ~1.5 mΩ. At the current maximum when length of the arc is 10 cm and its radius is 2 cm, the average specific resistance of the arc is 6.2×10^{-5} Ω m, and the specific conductivity of 1.6×10^4 $(\Omega\ \mathrm{m})^{-1}$.

Figure 4.3 shows the dependences of the arc resistance versus initial gas pressure at fixed interelectrode gaps.

Both dependences are growing; at this, the curve corresponding to large (6–16 mm) interelectrode gaps is substantially higher than curve corresponding to the small (0.5–3.0 mm) gaps. This difference can be explained by that at larger interelectrode gaps, the arc is lengthened more because the breakdown of the interelectrode gap, which restricts the lengthening of the arc, takes place at the higher breakdown and post-breakdown voltages.

Figure 4.4 shows the dependence of the arc resistance versus interelectrode gap at low and high initial pressures of hydrogen.

The dependency of arc growing can be explained by the greater voltage drop across the arc in the moments of repeated breakdowns of the gap at higher interelectrode distances. In addition, the curve corresponding to the high initial pressure is higher than curve corresponding to the low pressure, which is explained by the growing nature of the dependence of the resistance versus pressure (Fig. 4.3).

Figure 4.5 shows the dependence of the arc resistance versus mass of the eroded electrode metal.

Dependence, unlike the presented above, has falling character due to the increase in the arc conduction at the growth of the charge carrier's concentration (electrons and ions), which is proportional to the mass of eroded metal.

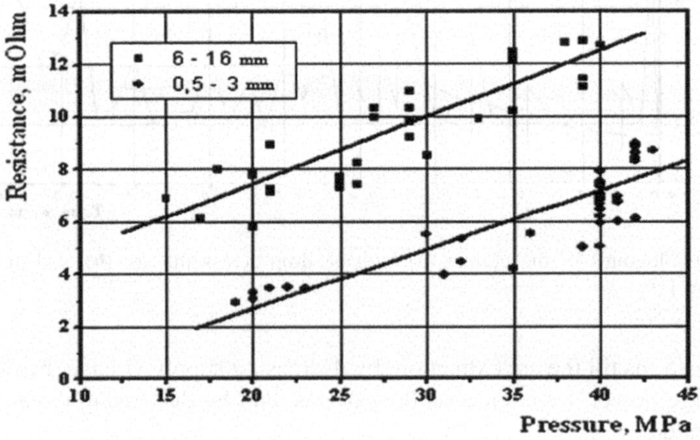

Fig. 4.3 Resistance of the arc versus initial gas pressure [1]

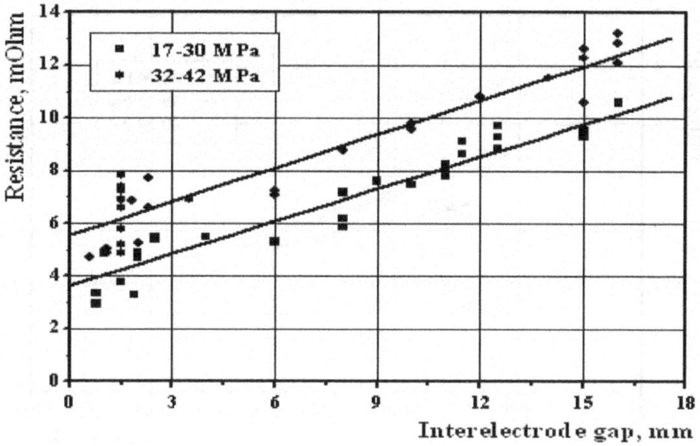

Fig. 4.4 Resistance of the arc versus interelectrode gap [1]

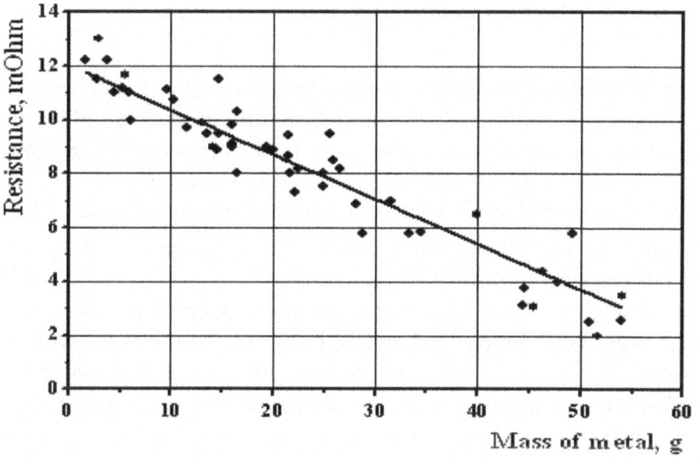

Fig. 4.5 Arc resistance versus mass of the eroded metal [1]

Figure 4.6 shows the statistical volt–ampere characteristics of the arc at initial hydrogen pressure of 20, 30, and 40 MPa.

Graph shows that all characteristics are falling, at this the voltage drop across the arc at the same discharge current the above, than is higher the initial pressure of hydrogen.

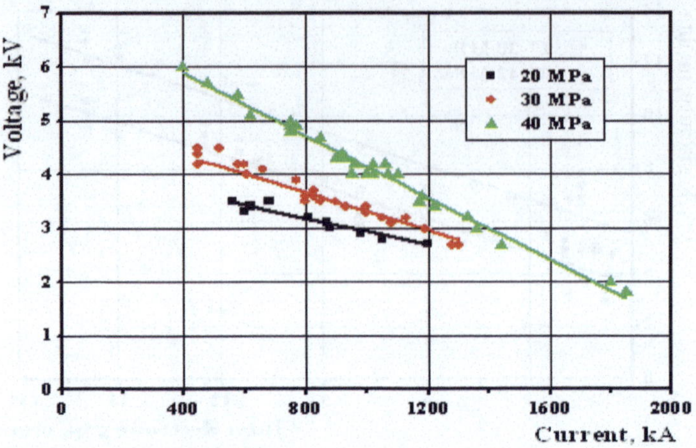

Fig. 4.6 Volt–ampere characteristics of the arc at various initial pressures of hydrogen [1]

4.2 Electric Field Intensity in Generator PPG-3

To determine the electric field intensity in the arc, it is necessary to know the arc length L_a, the voltage drop across the arc U_a, and the near-electrode voltage drops U_{ne}. The arc length, with some error, can be determined by the speed of its movement v_a, which in the time from zero to maximum current is ~1300 m/s at average dI/dt ~ 9 × 10^9 A/s. We must estimate the arc length at the current maximum, when the inductive component of the arc resistance is zero. The average time of the current maximum is 100 μs.

Since the increment of the arc length depends on the dimensions of the discharge chamber and angle between the surfaces of the anode and cathode, you need to set this parameter. As most of the experiments are performed on the discharge chambers where the angle between the surfaces is 60–90°, taking it of 75°, we make the error of ±20% that is quite acceptable for such assessments. Additionally, by "autograph" method it was found that the path length of the arc attachment to the surface of the anode is average of 10 cm, and that to the cathode is 5–7 cm. Thus, the arc length La, obtained at these assumptions, is ~10 cm and in most cases cannot be larger because it is limited by the sizes of the discharge chamber.

The average sum of the near-electrode voltage drops can be estimated from the plot of post-breakdown voltage on the interelectrode gap L_{ig} (Fig. 4.7) when extrapolated to zero.

The dependence is obtained based on the voltage across the interelectrode gap at time close to the maximum of the first current pulse. If we assume that for the time ~10 μs, when the current after the breakdown at rise rate of ~1.5 × 10^{10} A/s reaches 150 kA, and the arc voltage during this time at dU/dt ~3.7 × 10^7 V/s is remained practically unchanged, then at velocity v_a ~1300 m/s the arc moves from

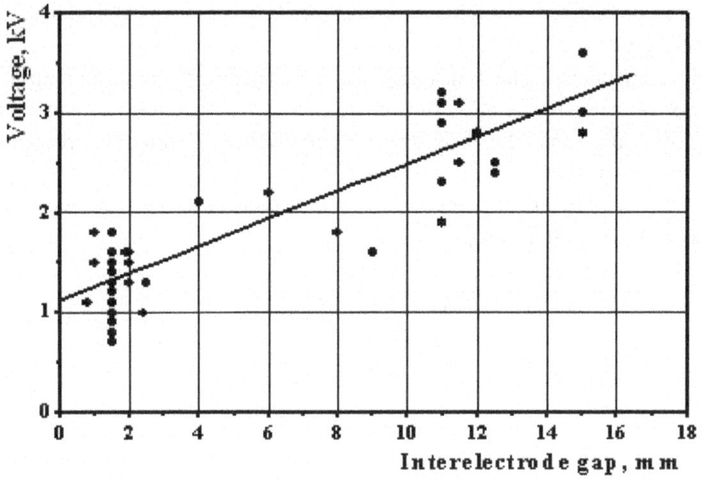

Fig. 4.7 Post-breakdown voltage versus interelectrode gap [1]

the start point no more than 1.0–1.5 cm, and thus, the increment of its length depending on electrode system configuration will not exceed 1.5 cm.

Considering that the electric field intensity E in the arc under similar conditions is ~300 V/cm [2], the near-electrode drops U_{ne} are somewhat lower than those obtained from the graph, and taking into account the above considerations, it is ~1 kV at ~150 kA. Therefore, at these near-electrode drops $U_{ne} \approx 1$ kV, the electric field intensity E in the arc is ~230 V/cm

$$E = (U_a - U_{ne})/L_a. \tag{4.1}$$

Electric field intensity E in the arc can also be estimated from the voltage rise rate dU/dt and growth rate of the arc length v_a

$$E = \frac{dU}{dl} = \frac{dU}{v_a dt} = \frac{1}{v_a}\frac{dU}{dt}. \tag{4.2}$$

Typically for these experiments dU/dt = 3.7×10^7 V/s and $v_a \approx 1.3 \times 10^3$ m/s, the electric field intensity E calculated by the formula (4.2) is 280 V/cm, i.e., 20% higher than the formula (4.1).

Parameters of the arc are depended strongly on the type of the discharge chamber. This is clearly seen in the comparison between the above data and data of the arcing process in the discharge chamber with coaxial rod electrodes, obtained in experiments at the diagnostic discharge chamber (Fig. 3.16) [12].

Assuming that the plasma in this case is formed by electrode jet of the second type, the concentration of tungsten ions n and the average ion charge $\bar{z}$ were determined at the time of 70 μs on brightness plasma temperature T near the electrode, and pressure P in the front of the shock wave, formed under the action of the electrode jet.

According to data of the work [3], the concentration of metal vapor near the cathode spot is 10^{19}–10^{20} cm^{-3}, the plasma must be in a state of local thermodynamic equilibrium, and the brightness temperature T must correspond to the equilibrium temperature.

Concentration of tungsten ions n and their average charge $\bar{z}$ were determined from the system of equations [4]

$$
\begin{aligned}
I\left(\bar{z}+\frac{1}{2}\right) &= kT\ln\frac{AT^{3/2}}{\bar{z}n} \\
P &= n(1+\bar{z})kT \\
A &= 6\times 10^{21}\,\mathrm{cm}^{-3}\mathrm{eV}^{-3/2}
\end{aligned}
\tag{4.3}
$$

where I is ionization potential of the tungsten in the point ($\bar{z}$ + 1/2), P is pressure in the front of the shock wave, A is constant, and k is Boltzmann constant.

Pressure in the front of the shock wave P, formed by the action of the electrode jets from the cathode (1), was assessed by its coordinate (2) (Fig. 4.8), using the data of the work [5]

$$
P = P_{\mathrm{ud}} \times \left(\frac{l}{0.64d}\right)^2 \tag{4.4}
$$

where P_{ud} is the gas pressure in the unperturbed medium, l is distance from the cathode to the front of the shock wave, and d is diameter of the cathode.

At l/d = 3.6, the pressure in the front of the shock wave is 126 MPa. For more correct assessment of P, we assume the equality of the magnetic and gas-kinetic pressures in the point of the expiration of the jet, based on the observed contraction of the jet [7, 8].

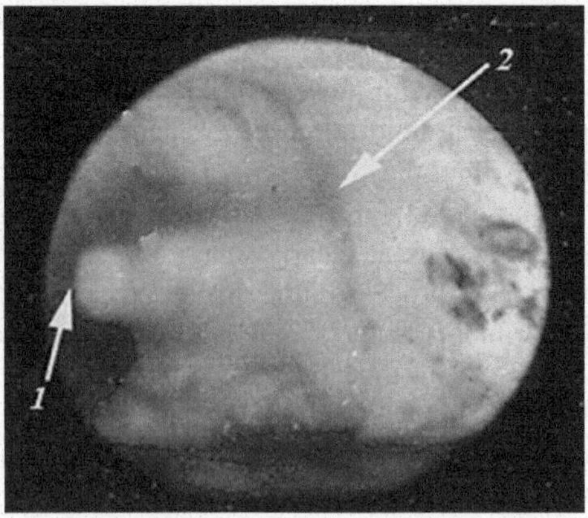

Fig. 4.8 Shock wave, formed by the jet from the cathode [6]

The magnetic pressure

$$P = 1.6 \times 10^{-4} \frac{I^2}{r^2}. \qquad (4.5)$$

where I is discharge current and r is arc radius.

At $J = 316$ kA and $r = 3$ mm, the pressure P in the arc and thus in the front of the shock wave is 177 MPa.

Solution of the system (4.3) at $T = 5.9 \times 10^4$ K, $P = 177$ MPa and tungsten electrodes gives $\bar{z} = 3.1$ and $n = 5.3 \times 10^{19}$ cm^{-3}.

At the 90 μs, the brightness temperature of the arc near the anode is 5.2×10^4 K, pressure 190 MPa, average ion charge 2.6, and concentration of ions 7.4×10^{19} cm^{-3}.

Higher ion concentration n near the anode in comparison with cathode is due to its higher specific erosion [9] under equality of the speeds of the anode and cathode jets [10].

Values of $\bar{z}$ and n determined by the system (4.3) according to experimental data T and P for tungsten cathode and anode are close to the values of $\bar{z}$ and the n, obtained by calculations in [11] for plasma aluminum vapor, titanium, and gland. Therefore, the averaged results of these calculations are used to approximate estimates of the plasma conductivity and the length of the photons run Rosseland.

Near-electrode voltage drops. At the values of T, $\bar{z}$, and n obtained above, the specific conductivity σ of the plasma in the arc is ~400 (Ω cm)$^{-1}$ and that at the current density $j \sim 10^6$ A/cm^2 corresponds to the electric field intensity $E \sim 2.5 \times 10^3$ V/cm. The current density $j \sim 10^6$ A/cm^2 at 70–90 μs had determined that the calculated diameter of the arc is equal to the diameter of the luminous zones near the electrodes and is close to the diameter of the electrodes. The total voltage drop U near the electrodes is ~1 kV [12–14]. The arc length l, on which this value of the voltage drop is determined from the U/E ratio, is 4 mm at the interelectrode gap of 1.7 cm.

Thus, the sections of the arc, in which the maximal energy is outputted, are located near the cathode and anode and have a length of ~2 mm, and the photon run length there by the Rosseland of the same order of magnitude. Therefore, we can conclude that the plasma for radiation with these length waves is translucent and that most energy is transferred by the radiation.

In the case of completely transparent plasma under the same conditions, the recombination radiation power would exceed electrical power IU by several orders of magnitude. At the translucent plasma, the recombination radiation power is comparable to the electric power.

Power of the recombination radiation from a unit volume of completely transparent plasma with charge Z according to data of the work [15]

$$Q = 5 \times 10^{-31} \cdot Z^4 \cdot n_e n_i T^{-1/2}, \quad (4.6)$$

where Z is ion charge (3.1), n_i is ion concentration (5.3×10^{19} cm^{-3}), n_e is electron concentration (Zn_i), and T is temperature in the arc (5.1 eV).

The radiation power Q from the near-electrode areas with a radius of 3 mm and a length of 4 mm at these data is 2×10^9 W.

In the case of the fully opaque plasma emitting as a blackbody, the radiation from the ends and the side surface of the electrodes is $(2\pi r^2 + 2\pi rl)/\sigma T^4 = 9.1 \times 10^7$ W. Electrical power of those areas is $\sim 3.2 \times 10^8$ W and is between these values.

Existence of the intensively radiating electrode sections of the arc of ~2 mm length at 70 μs is supported by the next facts:

- Photograph of these areas near the ends of the cathode (1) and anode (3) (Fig. 4.9)
- Voltage drop of ~1 kV across the interelectrode gap of 2–4 mm [16, 17]
- Voltage peak of 400 V at the moving of the arc attachment spot from the side surface of the anode to its end (Fig. 4.10).

Figure 4.10 shows that the peak of the voltage drop $\Delta U = 400$ V near the anode at the electric field intensity in the arc $E = 2500$ V/cm is corresponded to the part of the arc of 1.6 mm length.

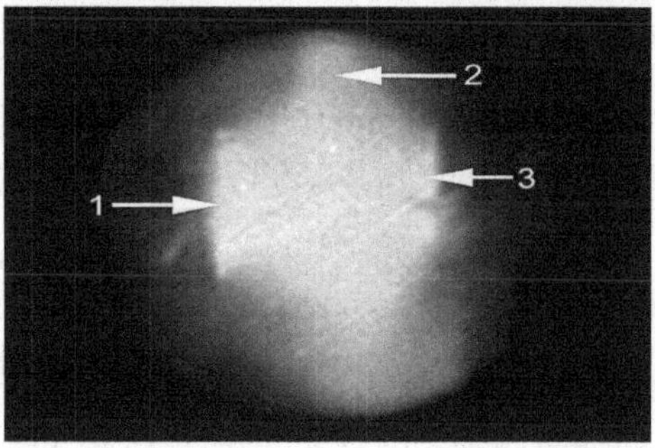

Fig. 4.9 Glowing zones (1, 3), shock-compressed area (2) formed at the collision of the anode and cathode jets [6]

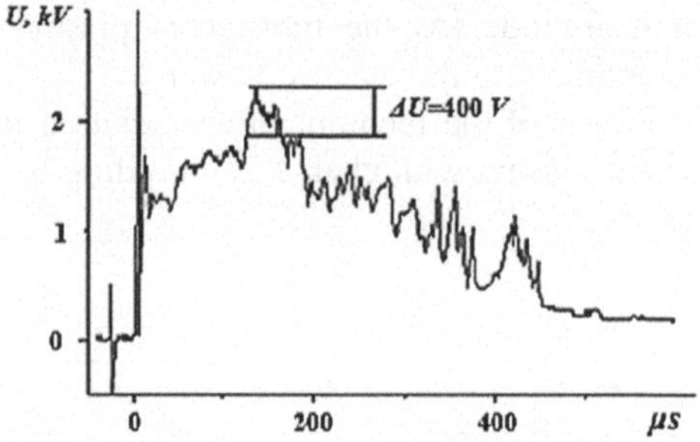

Fig. 4.10 Peak of voltage drop ΔU at the passing of the arc attachment spot from the side surface of the anode to its end [6]

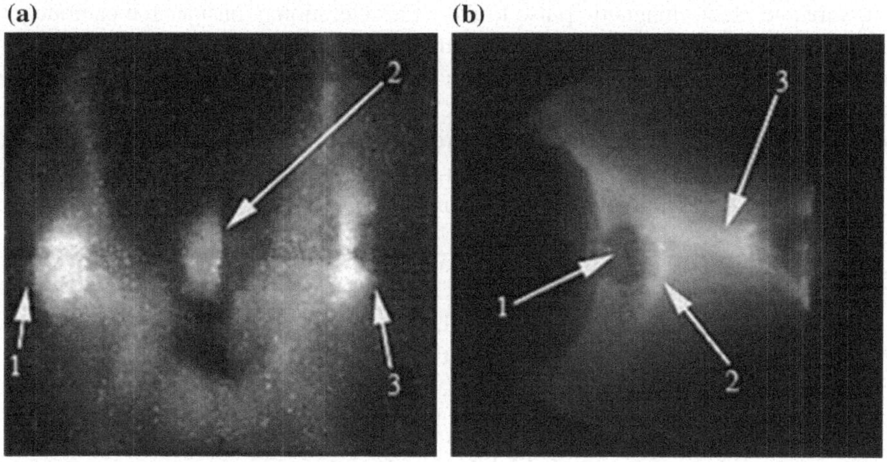

Fig. 4.11 Shock-compressed areas [6]

One of the causes of the abnormally high voltage drops on the arc areas can be the occurrence of the shock-compressed areas (2) at the interaction of the cathode (1) and anode (3) jets (Fig. 4.11a).

In the last case (Fig. 4.11b), interaction is occurred near the cathode (1). As the confirmation of it the dependences of the voltage drop across the arc and pressure in the discharge chamber on the interelectrode gap serve, obtained in the experiment with the copper electrodes of 10 mm diameter at the current of 300 kA, and initial pressure of 1 MPa (Fig. 4.12).

The graph shows that the voltage drop across the arc and steady pressure in the chamber are maximal when the interelectrode gap is ~15 mm; at this, the greatest impact of the anode jet on the cathode takes place (Fig. 4.11b). At increasing interelectrode gap, the shock-compressed area is not formed, and the voltage drop is reduced. Increase in the voltage drop can also be connected with the occurrence of the current's component directed perpendicular to the arc axis.

Using the results of the work [3], we can estimate the rate of the cathode jet in the area located behind the space charge, as well as the contributions of the dynamic

Fig. 4.12 Voltage drop across the arc (open circle) and pressure in the discharge chamber (open diamond) versus interelectrode gap [6]

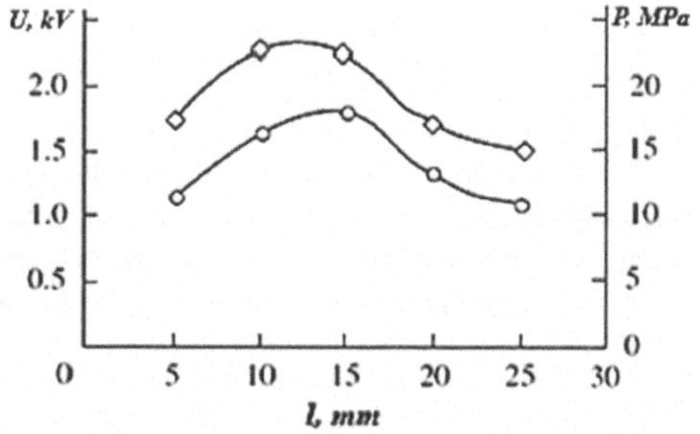

pressure and electromagnetic pulse to the jet acceleration. Consider the cathode jet at 70 μs, when the anode jet is not formed yet. For the estimation of the cross section for calculating, we select the area on the border with the space charge and denote it by index (1), and we assume that in this cross section there are temperature T and pressure P. We select the second cross section with index (2) before the area of the sharp rise of the voltage drop near the anode.

Considering $v_1 = 0$ and $P_2 << P_1$, then the law of momentum conservation is

$$Gv_2 = SP_1 + Q \tag{4.7}$$

where G is total erosion of cathode metal, S is cross-sectional area of the cathode, P_1 is pressure in the section (1), v_2 is jet velocity in section (2), and Q is contribution to the jet acceleration of electromagnetic pulse, which is defined as

$$Q = \left(\frac{I}{c}\right)^2 \ln\frac{r_a}{r} \sim \left(\frac{I}{c}\right)^2 \ln\frac{l}{r} \tag{4.8}$$

where I is total current of the arc, c is velocity of the light (3×10^8 m/s), r_a is arc radius in cross section (2), r is radius of electrode (0.3 cm), and l is interelectrode gap (1.7 cm).

Since the cathode jet has a cone shape, and generatrix of the cone has angle of $\sim 45°$ with the axis of the arc and $r < l$, then $r_a \sim l$.

According to data of the work [9], the specific erosion of the cathode alloy WNF is 6.5 mg/C; then, at the current of 300 kA, the rate of cathode metal erosion is 2.1 mg/s. Inserting the data into the formula (4.7), we obtain $v_2 = 10^4$ m/s.

Thus, at the current in the arc >300 kA, main role in accelerating the metal vapor in jet is played by the electromagnetic force, which by formula (4.8) is 1.7×10^4 N and more than in three times higher than gas-dynamic force, which by formula (4.7) is 4.8×10^3 N.

This estimation is approximate because pressure P_2 is different from zero, and the total consumption of the cathode metal according to the work [9] includes both the vapor and dropping phases. However, at the estimation of the velocity v_2, these errors partially compensate each other.

Another possible reason for the occurrence of high voltage drop across the arc near-electrode areas may be discrepancy between the vectors of the current direction and the expansion of jet (Fig. 4.13).

Additional increase in the electric field intensity in the arc E, due to the mutually perpendicular directions of the vector velocity expansion jet v and the magnetic field vector B, can be estimated as $E = v \times B$, where $B = \mu I/2\pi r$.

If the angle of the jet expansion is large, the transverse and longitudinal components of the speed are close in magnitude. According to the performed assessments and data of the works [10, 18–20], the longitudinal component of jet velocity is $\sim 10^4$ m/s. If the current I is $\sim 10^5$ A, radius of electrode r is 3 mm that

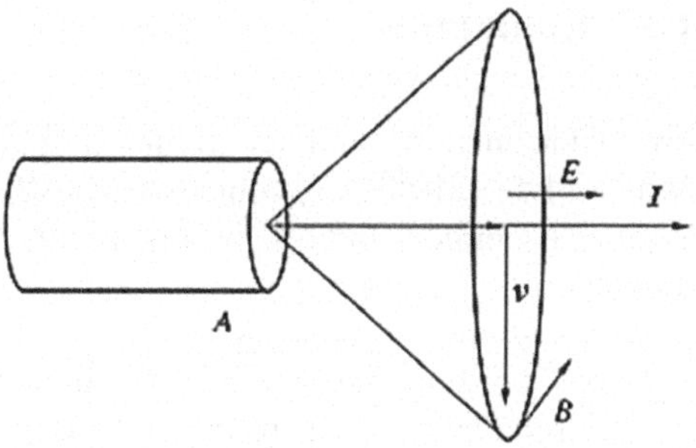

Fig. 4.13 Mutual location of the vector velocity expansion jet v and the magnetic field vector B [6]

corresponds to the experimental data, the electric field intensity $E \sim 10^3$ V/cm. For smaller angles of the jet expansion, such value of E should be matched to the larger discharge current.

4.3 Current Density

To estimate the current density in the arc, it is necessary to know the value of the current and cross-sectional area of the arc. The current is determined from the oscillograms, and diameter and cross-sectional area of the arc with a certain degree of confidence can be determined by the "autographs" method, i.e., by size of melted surface areas of the anode and cathode. Diameters of the melted spots depending on the distance from the point of ignition of the arc are varied from 2 to 5 cm, and the maximum value of the current corresponds to the maximum diameter of the spot. Thus, if we take the arc diameter equal to the diameter of the spot, then cross-sectional area of the arc is ~ 20 cm^2. When the average value of current is equal to the first maximum of 1 MA and the average cross-sectional of the arc is ~ 20 cm^2, the average current density j at experiments performed with the generator PPG-3 is $\sim 5 \times 10^4$ A/cm^2.

There is another method of estimating the current density at rise rate of 10^{10} A/s. In this case, it is assumed that under these conditions the self-regulating stabilization of the current density at (3–5) $\times$ 10^4 A/cm^2 takes place. At this, the electron emission sources are the thermal cathode spots of II type [19, 21]. As one can see, the obtained data are practically the same.

It should be remembered that when the current is ~ 1 MA, the arc contraction under the "pinch effect" takes place. The contraction greatly increases the current density in the arc. At some tests performed on the plasma generator PPG-5, current density reaches 10^7 A/cm^2.

4.4 Inductance

Arc inductance L_a and its variation ΔL_a have some effect on the arcing. At $dI/dt \sim 1.5 \times 10^{10}$ A/s, counter-emf (dIL_a/dt) at the experiments performed with plasma generator PPG-3 is up to several kilovolts. At these experiments, counter-emf is consisted of two components: one due to the inductance of the current lead and the other due to the inductance of the arc itself.

In order to eliminate the parasitic inductance and resistance of the current lead, the electrode system, where the voltage drop across the arc was recorded maximal near the binding spots of the arc to the electrodes, had been developed. Unfortunately, there was no reliable operation of this system, so this method of measuring the voltage directly across the interelectrode gap (with obvious advantages) had rejected.

To determine the current lead inductance, the short-circuit experiments were performed with the current lead in the discharge circuit and without it. Inductance of the current lead, which is defined as the difference in inductances of these two experiments 330 and 160 nH, was ~170 nH. The inductance of the arc itself is estimated by its characteristic dimensions, taking into account its specific resistivity ~20 nH [22].

Assessment of the total inductance of the current lead and arc can be made based on the oscillograms of the discharge current and voltage drop across the arc (Fig. 4.1a, b). In the time of 130–150 μs, the average rise rate dI/dt is $\sim 1.5 \times 10^{10}$ A/s and ΔU (due to the counter-emf) is ~3 kV. Thus, the inductance is ~200 nH that almost coincides with the data presented above.

As we can see, the arc inductance is ~6% of the inductance of the discharge circuit of the power supply CPS-10 and therefore cannot significantly affect the characteristics of the discharge.

Another method of inductance evaluation of the arc in the air at atmospheric pressure, based on the analysis of the programmable discharge of the power supply CPS-25, gives the value of arc inductance 20–70 nH [23], which is close to the value shown above.

The high inductance of the current lead (is mainly due to its material) gives the overvoltage peaks substantially exceeding the voltage at the power supply and decreases voltage drop across arc that negatively affects the efficiency of energy transfer from the power supply to the arc.

4.5 Temperature

To estimate the heat transfer from the arc to the gas, it is necessary to know the temperature of the arc. In order to determine the temperature of the arc, the arc plasma conductivity and ion concentration are necessary.

Arc plasma specific conductivity σ calculated by the formula $\sigma = j/E$ based on the current density $j = 5 \times 10^4$ A/cm^2 and electric field intensity $E = 260$ V/cm is $\sim 2 \times 10^4$ $(\Omega\ \text{m})^{-1}$.

The averaging for all experiments the concentration of evaporated from the electrodes metal atoms n_m allows to approve, that the mass of metal in the vapor state m_m less than 0.3 of the total mass of eroded metal [24] when its distribution is uniform in the discharge chamber by formula (4.9) is 7×10^{19} cm^{-3}

$$n_m = \frac{m_m}{\mu_m u V_c}. \tag{4.9}$$

where μ_m is average atomic mass of metal, V_c is volume of the discharge chamber, and u is atomic mass unit.

The averaging for the same number of experiments the concentration of hydrogen molecules n_0 at initial pressure $P = 31.5$ MPa and temperature 290 K by formula (4.10) is $\sim 8 \times 10^{21}$ cm^{-3}

$$n_0 = 10^{-6} N_l P. \tag{4.10}$$

where N_l is Loschmidt number.

Thus, the concentration of the metal atoms is $\leq 1\%$ of the hydrogen molecules' concentration.

It can be assumed that the ratio of the concentrations of hydrogen and metal atoms in the arc would be approximately the same that in the remaining volume of the discharge chamber. This assumption is based on the fact that arc moving at the speed of ~ 1500 m/s intensively mixes the metal atoms and hydrogen molecules. However, the metal atoms act on the arc conductance quite strongly, as shows the dependence of the arc resistance on weight of the metal eroded from electrodes (Fig. 4.5). The dependence can be explained by intense electrode jets, where the metal concentration is more than the average concentration.

On the other hand, this value can be considered as somewhat overestimated, since it corresponds to the total electric charge Q_0, whereas to the maximal current, i.e., to 100th μs, its current value does not exceed 30–50% of Q_0.

The pressure in the arc P_a in the maximum of the current is the sum of gas pressure P_g in the chamber and magnetic pressure P_m in the arc

$$P_a \geq P_g + P_m. \tag{4.11}$$

The average gas pressure P_g in the discharge chamber at this moment is 30–50 MPa. The magnetic pressure P_m in the arc at uniform current density distribution is determined by the formula (4.12) [25]

$$P_m = 10.6 \times 10^{-4} \frac{I^2}{r_a^2}. \tag{4.12}$$

where I is maximal current (A) and r_a is radius of the arc (cm).

When the current $I = 1$ MA and average arc radius $r_a = 2$ cm, the magnetic pressure $P_m = 40$ MPa. Thus, the pressure in the arc as the sum of these two pressures $P_a \approx 100$ MPa.

The arc in the hydrogen at the current 1 MA, particle concentration $>10^{19}$ cm^{-3}, and pressure ~ 100 MPa has temperature of (2, 3) $\times$ 10^4 K [26, 27]. At this in the near-electrode areas under similar conditions, the brightness temperature $>5 \times 10^4$ K had been registered [28]. At this temperature T and pressure P_a, the hydrogen dissociation degree is close to 1 and the degree of ionization $\alpha_i \sim 0.3$ [29].

Concentration of hydrogen atoms in the arc n_H calculated by the formula (4.13) and taking into account the increase in concentration due to ionization is 2.2×10^{20} cm^{-3}

$$n_H = \frac{P_a}{(1 + \alpha_i)kT}. \tag{4.13}$$

where k is Boltzmann's constant.

Electron density according to (4.14) is 7×10^{19} cm^{-3}

$$n_e = \alpha_i n_H. \tag{4.14}$$

At such particles' concentration, the plasma properties can differ substantially from the ideal. The criterion of distinction is the Debye coupling parameter D [30, 31]

$$D = \frac{Z_a e^2}{kT(r_D, r_i)}. \tag{4.15}$$

where $r_D = \sqrt{\frac{kT}{8\pi n_e e^2}}$, $r_i = \left(\frac{4\pi n_i}{3}\right)^{-\frac{1}{3}}$, e is electron charge, Z_a is atomic number, r_D is Debye radius, r_i is average distance between the ions, n_e is electron density, and n_i is ion concentration.

To estimate D, take the highest value of r_D and r_i. When $n_e \approx 10^{20}$ cm^{-3} and $T = 2.5 \times 10^4$ K, parameter $D = 0.5$. Calculation dependences for ideal gases and plasma are applicable at $D << 1$; therefore, for estimation of the temperature of the arc plasma based on its electrical conductivity, the Spitzer formula is not applicable.

Estimation of the arc plasma temperature can be made based on the data from the work [29], where the calculated values of the hydrogen plasma conductivity are presented. In accordance with these data at $\sigma = 250$ (Ω cm)$^{-1}$ and $n_e = 10^{20}$ cm^{-3}, the arc plasma temperature $T = 2.5 \times 10^4$ K.

Analogical data are presented in work [32], where at similar conductivity and concentration, the arc temperature T is 2.3×10^4 K.

Estimations based on the mathematical model of the discharge in hydrogen at initial pressure of 10 MPa and current of 10–30 kA [33, 34] also give similar $T = (2.4–2.7) \times 10^4$ K.

Evaluation of the transition layer temperature was made based on the spectrometric measurements data. Absorption lines are observed on the background of the continuous spectrum corresponding to radiation of the transition layer between the current channel and the cold gas. After determining the temperature of the layer, it is possible to estimate the minimal temperature of the arc plasma.

The temperature of the transition layer was calculated by the half-width of the absorption lines of the doublet Na 5890 E and 5896 E. It is known that at the concentration of the perturbing particles $n_d < 10^{22}$ cm^{-3} for determination of the line half-width, the impact approximation is applicable [35]

$$\Delta\gamma = 1.3\Delta c_6^{0.4} v_h^{0.6} n_d. \tag{4.16}$$

where v_h is average thermal velocity of the particles, Δc_6 is difference between the constants c_6 van der Waals potential of the initial and final states, and n_d is concentration of the perturbing particles.

The constants c_6 were calculated from the data of work [35], where they were defined for helium considering the difference in polarizability of the H_2 and He.

Experimental width of each line of the Na doublet is 4.5–5.0 Å and according to formula (4.16) is corresponded to the temperature of the transition layer T = (1.1–1.4) × 10^4 K. This result is in good agreement with the data of works [24, 25]; where at initial hydrogen pressure of 0.5–6.0 MPa and current of 60–70 kA, the surface temperature of 1.2 × 10^4 K was registered.

To determine the efficiency of energy transfer from the arc into the gas, the experiments at "manometric bomb" mode were carried out. Evaluation of the increment of internal energy of hydrogen was carried out based on the initial density of the gas and pulsed pressure in the chamber. Deviation of hydrogen properties from the ideal was taken into account based on the data of the work [36]. The efficiency of energy transfer from the arc into the gas at initial pressure of 4–5 MPa is 0.4–0.5.

Figure 4.14 shows the oscillogram of the pulsed pressure in the discharge volume, which is characterized by the acoustic oscillations.

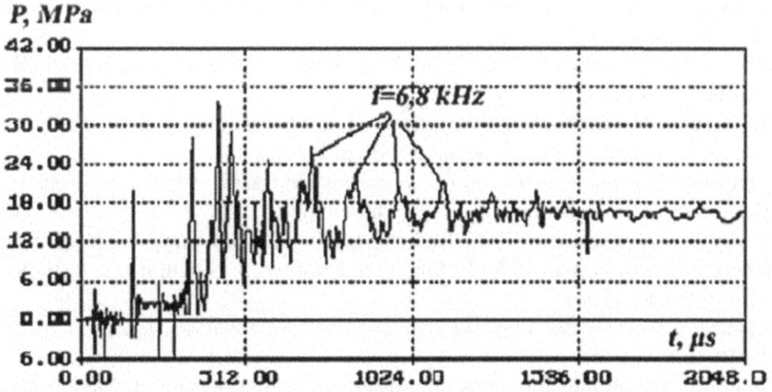

Fig. 4.14 Pulsed pressure at "manometric bomb" mode [6]

The propagation velocity of the oscillations is coincided with the speed of sound in hydrogen which had been determined by the final pressure. The frequency of the acoustic oscillations is 6.8 kHz at initial hydrogen density of 4.4×10^{-3} g/cm^3 and pressure of 22 MPa to 1100th μs.

To these conditions, the average temperature of hydrogen 1200 K and sound speed c of 2500 m/s are corresponded. The frequency of longitudinal acoustic oscillations f at chamber length l of 19 cm estimated by the formula

$$f = c/2l \qquad (4.17)$$

is 6.6 kHz, which is close to the value obtained from the pulsed pressure oscillogram (Fig. 4.14).

Based on these data, the temperature of the arc at our experiments is $(2.3\text{–}2.7) \times 10^4$ K and temperature of the transition layer is $(1.1\text{–}1.4) \times 10^4$ K.

References

1. V.A. Kolikov, Doctoral thesis, St.-Petersburg, 2005 (in Russian)
2. I.A. Glebov, F.G. Rutberg, *High-Power Plasma Generators* (Moscow, 1985) (in Russian)
3. G.A. Ljubimov, *Experimental Researches of Plasma Generators* (Novosibirsk, 1977) (in Russian)
4. Y.B. Zel'dovich and Y.P. Raizer, *Physics of Shock Waves and High-Temperature Hydrodynamic Phenomena* (New York, 1966)
5. S. Krist, P.M. Sherman, G.R. Glass, Sov. J. Aerosp. Sci. **1**, 87 (1966) (in Russian)
6. A.A. Bogomaz, Doctoral thesis, St.-Petersburg, 2012 (in Russian)
7. V.A. Nemchinskiy, Sov. J. Tech. Phys. **52**, 35 (1982) (in Russian)
8. V.A. Nemchinskiy, Sov. J. Tech. Phys. **52**, 235 (1983) (in Russian)
9. P.G. Rutberg, A.V. Budin, V.A. Kolikov, B.P. Levchenko, V.V. Leontiev, I.P. Makarevich, N.A. Shirokov, High Temp. **32**, 589 (1994)
10. O.B. Bron, L.K. Sushkov, *Plasma Flows in an Electric Arch of Switching Off Devices* (Leningrad, 1975) (in Russian)
11. B.V. Zamyshlyaev, E.L. Stupitskiy, A.G. Guz', V.G. Zhukov, *Composition and Thermodynamic Functions of Plasma, Handbook* (Moscow, 1984) (in Russian)
12. P.G. Rutberg, A.A. Bogomaz, A.V. Budin, V.A. Kolikov, M.E. Pinchuk, A.A. Pozubenkov, Tech. Phys. **47**, 26 (2002)
13. P.G. Rutberg, A.A. Bogomaz, A.V. Budin, V.A. Kolikov, A.G. Kuprin, Izvestia RAS. Energy **1**, 100 (1998) (in Russian)
14. A.A. Bogomaz, in *Materials of Seminars—Schools of Young Scientists, Students and Post-Graduate Students* (Petrozavodsk, 2004) (in Russian)
15. I.M. Podgornuiy, *Lectures on Plasma Diagnostics* (Moskow, 1968) (in Russian)
16. A.A. Bogomaz, A.V. Budin, S.V. Zakharenkov, V.A. Kolikov, A.I. Kulishevich, I.P. Makarevich, A.F. Savvateev, P.G. Rutberg, Izvestia RAS. Energy **1**, 64 (1998) (in Russian)
17. P.G. Rutberg, A.A. Bogomaz, A.V. Budin, V.A. Kolikov, A.G. Kuprin, Izvestia RAS. Energy **1**, 100 (1998) (in Russian)
18. G.A. Mesjats, *Ectons in the Vacuum Discharge: Breakdown, a Spark, an Arch* (Moskow, 2000) (in Russian)
19. G.A. Ljubimov, V.I. Rahovskiy, Sov. J. Adv. Phys. Sci. **125**, 665 (1978) (in Russian)
20. B. Juttner, IEEE Trans. Plasma Sci. **PS-15**, 474 (1987)

21. V.I. Rahovskiy, *Physical Bases of Switching of an Electric Current in Vacuum* (Moskow, 1970) (in Russian)
22. P.L. Kalantarov, L.A. Tseitlin, *Calculation of Inductances* (Leningrad, 1986) (in Russian)
23. B.E. Fridman, P.G. Rutberg, in *Conference Physics of Low-Temperature Plasmas* (Petrozavodsk, 1995) (in Russian)
24. G.S. Belkin, V.J. Kiselev, Sov. J. Tech. Phys. **36**, 384 (1966) (in Russian)
25. D.J. Sherklif, *Course of Magnetic Hydrodynamics* (Moskow, 1967) (in Russian)
26. J.P. Knjazev, E.S. Borovik, R.V. Mitin, V.I. Petrenko, Sov. J. Tech. Phys. **37**, 523 (1967) (in Russian)
27. V.I. Petrenko, R.V. Mitin, J.P. Knjazev, A.V. Zvjagintsev, Sov. J. Tech. Phys. **379**, 1827 (1969) (in Russian)
28. M.E. Pinchuk, PhD thesis, St. Petersburg, 2004 (in Russian)
29. N.B. Vargaftik, *Directory on Thermophysical Properties of Gases and Liquids* (Moskow, 1972) (in Russian)
30. Y.P. Raizer, *Physics of Gas Discharge* (Moskow, 1987) (in Russian)
31. V.E. Fortov, I.T. Jakubov, *Nonideal Plasma*, (Moskow, 1994) (in Russian)
32. W.B. Leung, N.M. March, Plasma Phys. **19**, 277 (1977)
33. H. Lee, J. Phys. D Appl. Phys. **18**, 425 (1985)
34. A.M. Voronov, PhD thesis, St. Petersburg, 1992 (in Russian)
35. S.E. Frish, *Optical Spectra of Atoms* (Moskow, 1963) (in Russian)
36. V.P. Kopushev, V.V. Hrustalev, Sov. J. Appl. Mech. Tech. Phys. **1**, 122 (1980) (in Russian)

Chapter 5
Erosion of Electrodes

Abstract It is important for understanding the processes in the electrodischarge chambers of the pulsed plasma generators are the basic mechanisms and quantitative characteristics of the electrothermal erosion of the electrodes. The fact is that the metal, which evaporated from the electrodes, affects both the characteristics of the arc (electric field intensity, emission characteristics, dynamic characteristics, etc.), and the heat transfer processes in the discharge chamber, changing the physical characteristics of the gas (heat capacity, thermal conductivity, density, and viscosity). In addition, the erosion of the electrodes is accompanied by the powerful electrode jets, which also impact the processes of heat transfer in the discharge chamber and the arc characteristics. In some cases, there was erosion in the form of the ejection of the surface layer of the electrode's metal. Besides, the change of physical characteristics, in particular an apparent molecular weight of the mixture gas + metal and associated directly with it the sound of speed, impact significantly the efficiency of heat transfer from the arc to the gas.

5.1 Specific Erosion of Electrodes

This section presents data on the specific erosion of different electrode materials used in research at electrodischarge chambers of coaxial type. The data were obtained at the discharge current of up to 1.4 MA, average gas temperature up to 4500 K, and electric charge up to 900 C.

Paper [1] presents the dependencies of the mass of eroded materials (copper, molybdenum, and titanium) on the electric charge Q, which defined by the formula

$$Q = \int_0^T |I| \mathrm{d}t. \tag{5.1}$$

where I is current and T is discharge time.

V. Kolikov et al., *Powerful Pulsed Plasma Generators*,
Springer Series on Atomic, Optical, and Plasma Physics 101,
https://doi.org/10.1007/978-3-319-95249-9_5

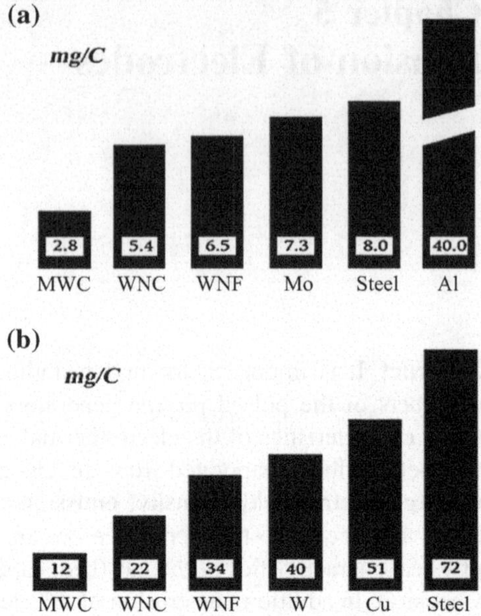

Fig. 5.1 Specific erosion of the cathode (**a**) and anode (**b**) materials [2]

This work indicates that the erosion is essentially independent on the electric charge at its values of up to 10 C for copper and 40 C for titanium. When the charge is above these values, erosion is increased abruptly and then is depended on the charge linearly.

Figure 5.1 shows the diagram of the specific erosion of electrode materials based on the data obtained at the average for all performed experiments discharge current of 800 kA, charge of 360 C, and average temperature of 2600 K.

Analysis of the data of specific erosion of the electrode materials showed the dependences on the discharge current for the cathodes (Fig. 5.2) and anodes (Fig. 5.3), and on the average gas temperature (Figs. 5.4 and 5.5).

Figure 5.2 shows the dependences of the specific erosion on discharge current for the cathodes of Al, steel, Mo, WNF, MWC, and WNC.

Graph shows that dependencies have general growing nature; at this highest erosion resistance has WNC and lowest the aluminum.

Figure 5.3 shows the dependences of the specific erosion on current of the anodes made of steel, Cu, W, WNF, and WNC. Here, as in previous case, dependences are growing and lowest erosion resistance has the steel. Using the steel is justified in those cases, where the average temperature of the gas, under the experimental conditions, does not exceed its boiling temperature of 3000 K.

In several experiments, the anodes that were made of brass, which is at low mechanical strength, have very high specific erosion ~180 mg/C, and therefore, its usage was abandoned, even in combination with WNF.

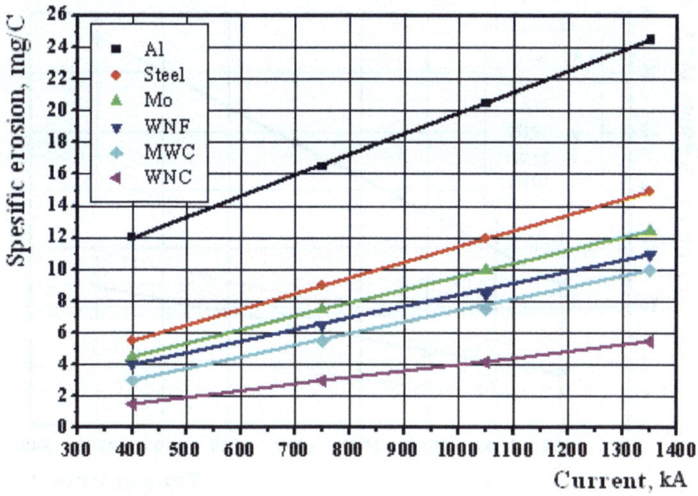

Fig. 5.2 Specific erosion of the cathode materials versus current [2]

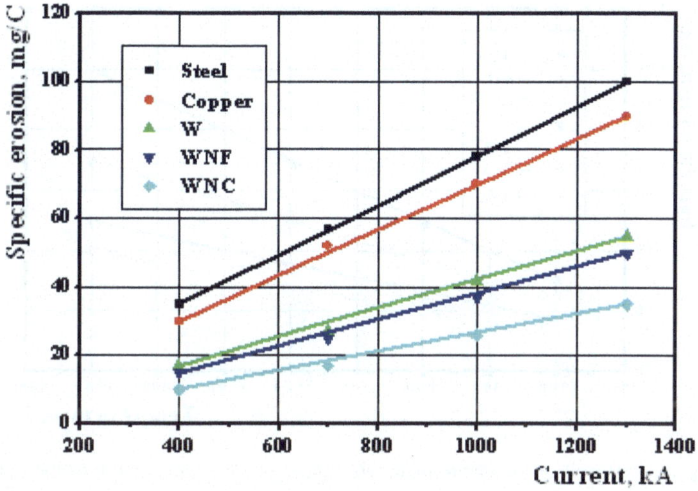

Fig. 5.3 Specific erosion of anode materials versus the current [2]

Figure 5.4 shows the dependence of the specific erosion of cathodes made of Al, steel, Mo, WNF, MWC, and WNC (dependence for aluminum was extrapolated to the high temperatures) on the average gas temperature.

The graph shows that the average value of the erosion of all the materials used at the temperatures from 1000 to 4500 K is changed by two to three times; at this, the character of the specific erosion change is linear.

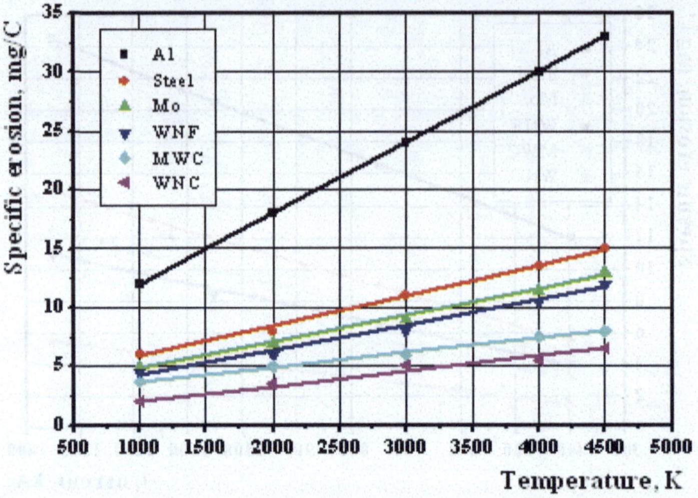

Fig. 5.4 Specific erosion of the cathode materials versus average temperature of the hydrogen [2]

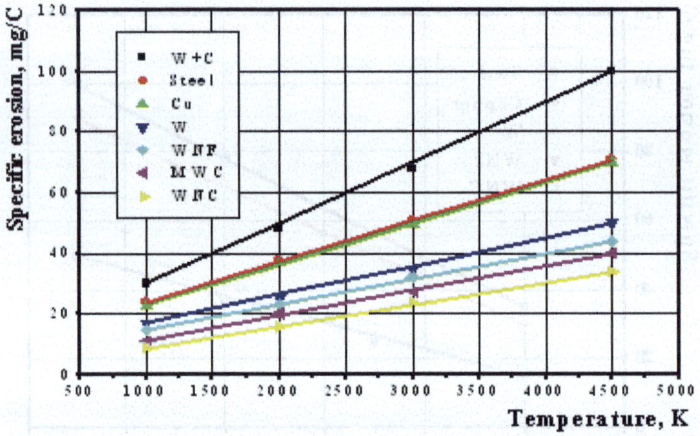

Fig. 5.5 Specific erosion of the anode materials versus average temperature of the hydrogen [2]

Figure 5.5 shows the dependence of the specific erosion of the anodes made of steel, Cu, WNF, WNC, and MWC, and plasma sprayed W with C mixtures and pure W on the average temperature of the gas.

For all four groups of the dependences is characteristic that the highest erosion resistance have the alloys based on tungsten, and lowest have Al, steel, and Cu.

Thus, these experimental data on the specific erosion of electrode materials should be considered as derivatives of the large number of additional factors, some of which have been mentioned above.

Some experiments to study the erosion of the electrodes were performed at installations with rod electrodes. At the current up to 400 kA and power of the heat

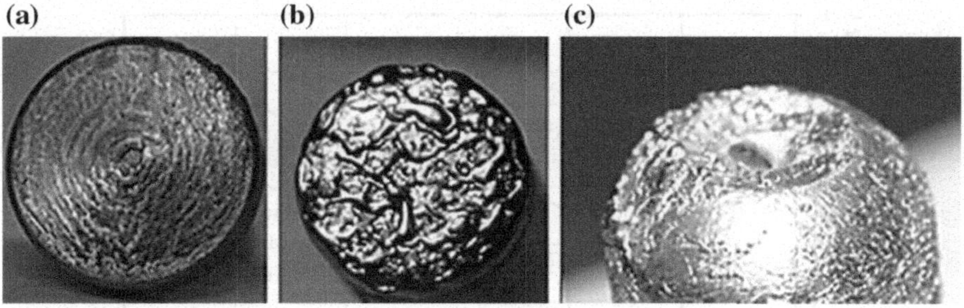

Fig. 5.6 Cathodes after the experiments at various heat flow power [3]

flow onto the electrodes $\sim(10^5$–$10^8)$ W/cm^2, the experiments at the diagnostic chamber (Fig. 3.16) were performed. At the current up to 1.6 MA and power of the heat flow onto the electrodes up to 10^9 W/cm^2, a study was performed at the generator PPG-5 (Fig. 3.10).

Figure 5.6 shows the photograph of the cathodes' end surfaces after experiments at various power of heat flow and at current of $\sim$200 kA. Differences in the surface structure of the cathode (alloy CuCr) at power of the heat flow of 10^5 W/cm^2 (Fig. 5.6a) and cathode (Cu) at power of 10^6 W/cm^2 (Fig. 5.6b) are due to thermophysical characteristics of their materials.

Figure 5.6c shows the surface of the steel electrode at current of $\sim$1.6 MA and power of heat flow of 109 W/cm2. Diameter of the crater on the surface of the cathode is corresponded to the diameter of the arc defined by a photographic of its glow in stable phase of arc burning by duration of $\sim$20 μs.

Figure 5.7 shows the dependence of the total specific erosion on the current of the steel rod electrodes of 10 mm diameter in hydrogen at current density of $\sim 2 \times 10^6$ A/cm^2.

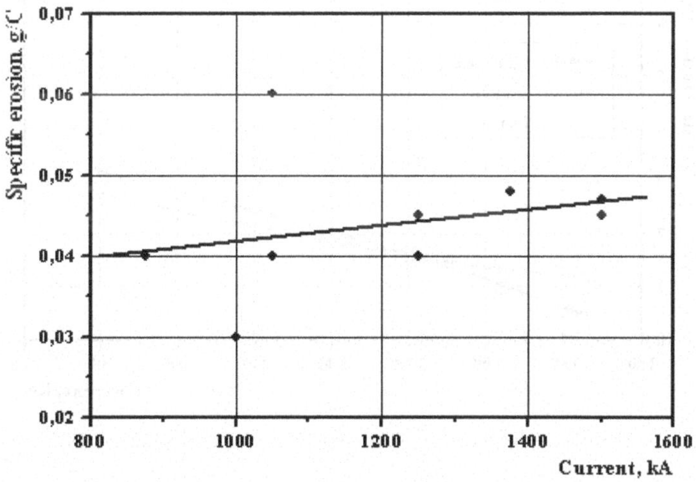

Fig. 5.7 Specific erosion of the steel electrodes versus current in hydrogen [3]

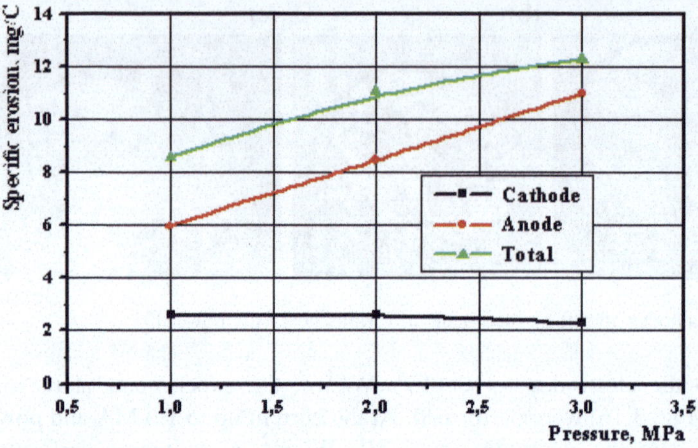

Fig. 5.8 Specific erosion of the tungsten electrodes versus initial hydrogen pressure [3]

Figure 5.8 shows the dependence of the specific erosion of the rod tungsten electrodes diameter of 6 mm on the initial hydrogen pressure at the current of ~220 kA and the interelectrode gap of 13 mm.

Figure 5.9 shows the dependence of the specific erosion of the rod tungsten electrodes diameter of 6 mm versus the current at interelectrode distance of 13 mm and air initial pressure of 0.1 MPa.

The total erosion of the electrodes of WNF in the air at the discharge current of 220 kA is 2.8 mg/C and copper electrodes is 2.1 mg/C.

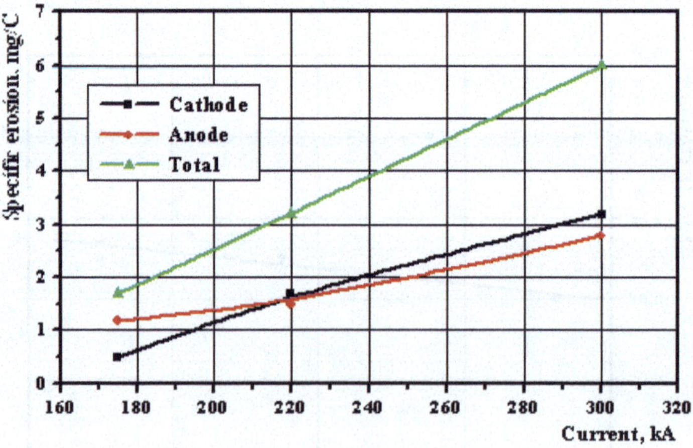

Fig. 5.9 Specific erosion of the tungsten electrodes in the air versus current [3]

5.2 Liquid and Vapor Erosions

Besides the above data, which characterize the total erosion of the electrodes' metal in the vapor and liquid phases, no less interest for the understanding of the processes in the discharge chambers of the plasma generators is the ratio between these phases or the emission factor *k*. The work [4] states that the emission factor *k* at the range from 70 to 800 kA is independent on charge *Q* and is 40% for copper.

The work [1] shows that for Mo emission, factor *k* is 40% for Ti—20%. This means that under such scatter of the emission factor, the differences in the erosion resistance Re [C/mg] for various materials is:

$$\mathrm{Re} = \frac{cT_{\mathrm{mt}}}{kU_{\mathrm{ne}}} \tag{5.2}$$

where c is heat capacity, T_{mt} is melting temperature, k is emission factor, and U_{ne} is near-electrode voltage drop.

Besides when replacing of the electrode material pair on other at small inter-electrode gap, the variation of the erosion may occur due to changes in the intensity of the electrode jet' impact onto the opposite electrode, and due to the transfer of the molten metal from one electrode to another.

This fact is confirmed by the data of the work [5], where it established a connection between erosion and interelectrode distance and experimentally shown that:

- The smaller the electrode diameter, the greater the energy of the jet
- The larger the area of the opposite electrode, the more energy it receives from the jets
- The smaller the atomic mass of the electrode material, the smaller the jet energy
- The greater the interelectrode distance, the less energy the electrode receives from jet.

Furthermore, this paper presents the erosion data for large number of copper alloys. Some distinctive features in erosion have the pseudo-alloys, which repeatedly had been used at our experiments due to their high resistance to erosion.

Authors of the work [6] investigated the nature of the electrode melting in places of the arc binding. It was found that the energy input into the pseudo-alloys is less than into homogenous materials. The reason for this is the heat flows in the surface layers of the material fix the cathode spots in places with facilitated vaporization and with high electron emission. At this does not form the continuous melt pool that makes it difficult the ejection of the melt, and hence, the erosion of the material is decreased.

In the works [7, 8], the erosion and destruction of the electrode materials at current from 230 to 500 kA were investigated. It is indicated that the removal of molten metal from the melt pool is due to the interaction of the magnetic field with a current flowing through the liquid metal.

Estimations of the ratio of the liquid and vapor phases of eroded metal versus power of the heat flow q at our experiments were performed at discharge duration $\tau = 10^{-5}$–10^{-4} s

$$q = jU_{\text{ne}} \tag{5.3}$$

where j is current density and U_{ne} is near-electrode voltage drop.

At transition from the liquid phase erosion to the vapor erosion that is due to the electrode jets appearance, the near-electrode voltage drop U_{ne} is increased from tens to thousands volts that at current density $j \sim 10^5$–10^6 A/cm^2 corresponds to an increase in the heat flow power from 10^6 W/cm^2 to the critical value of 10^9 W/cm^2.

Assuming $\overline{q}$ is an average value between the power of the heat flow q_1 required to achieve the temperature of the metal evaporation on the surface of the electrode and the power of the heat flow q_2, at which the evaporation rate is equal to the rate of heating the metal to the melting point then according to data of the work [9]

$$q_1 = \frac{0.085T\lambda}{\sqrt{a\tau}}, \quad q_2 = L\rho\sqrt{\frac{a}{\tau}}, \quad \overline{q} = \frac{q_1 + q_2}{2}. \tag{5.4}$$

where T is boiling temperature, λ is thermal conductivity, a is thermal diffusivity, L is specific heat of vaporization, ρ is density of the electrode metal, and τ is duration of the heat flow.

Let us compare the data of two experiments at currents of 380 kA [9] and 1.5 MA [11], where time τ of the electrode jet occurrence is known.

At the first experiment, the time of the electrode jet occurrence and its speed is determined from the photographs. Let us estimate $\overline{q}$ for the discharge in hydrogen with tungsten electrodes of 6 mm diameter and with time of the electrode jets occurrence of 70 μs. At this time, the discharge current I = 133 kA. Boiling temperature of tungsten T = 5680 K, thermal conductivity λ = 1.69 W/(cm × deg), thermal diffusivity a = 0.65 cm^2/s, and $\rho L = 9.5 \times 10^4$ J/cm^3 [8]. Then $q_1 = 1.2 \times 10^6$ W/cm^2, $q_2 = 9.2 \times 10^6$ W/cm^2, and $a = 5.2 \times 10^6$ W/cm^2.

Near-electrode drop of voltage U_{ne} before the occurrence of the electrode jets is 22 V that is consistent with the data of the works [3, 12, 13]. Thus, the time of occurrence of the electrode jet coincides with the calculated value U_{ne} = 20 V, what characterizes the erosion with a predominance of the liquid phase.

In the second experiment At $I_{\max}$ = 1.5 MA, the occurrence of the electrode jet was determined by the time of the sharp rise in the near-electrode voltage drops. Growth of the near-electrode voltage drops was determined by the decrease in electric field intensity in the arc at constant total voltage across the discharge gap. Time of occurrence of the electrode jet T is 7-14 μs. Let us estimate $\overline{q}$ in this experiment for steel electrodes of 20 mm diameter.

The boiling temperature of steel T = 3050 K, thermal conductivity λ = 0.51 W/(cm × deg), thermal diffusivity a = 0.15 cm^2/s, and $\rho L = 5.5 \times 10^4$ J/cm^3. Let us assume $\tau \sim 14$ μs, then $q_1 = 0.93 \times 10^6$ W/cm^2, $q_2 = 5.7 \times 10^6$ W/cm^2, and $a = 3.3 \times 10^6$ W/cm^2. If the average current density j for the 14 μs defined by a photographic of the arc glow and by current oscillogram is 4.5×10^5 A/cm^2, then $U_{ne} = 8.2$ V.

Ratio between vapor and liquid phases of the electrode's material after the intense evaporation (Fig. 5.10) is determined by the speed v of the boundary between the liquid (2) and vapor (1) phases, temperature T_0 of vapor on surface, and thickness of the melted layer (x_2) for various $\overline{q}$. Then the thickness of the vaporized layer $x_1 = vt$, where t is time of the heat flow.

Values of v and T_0 according to the data of the works [14, 15] were determined from the joint solution of equations:

$$v = \frac{q}{\rho\lambda + \frac{5}{2}\frac{RT_0}{A}} \tag{5.5}$$

$$v = c \times \exp - \frac{\lambda}{RT_0} \tag{5.6}$$

where R is universal gas constant, A is atomic weight of the metal, c is speed of sound in the electrode material, and λ is thermal conductivity.

Thickness of the layer of molten metal x_2 was determined by motion of the isotherm with temperature $T_{mt}(x, t)$, equal to the melting temperature of the electrode material, which is determined by the equation of thermal conductivity [14, 15]

$$T_n(x_2, t) = T_0 \cdot \exp - \frac{x_2 v}{a} \tag{5.7}$$

For a rough estimation, you can use the ratio $x_2 \sim a/v$.

Table 5.1 presents the results of the joint solution of equations for the thickness of the evaporated $x_1 = vt$, and molten $x_2 = a/v$ layers of the tungsten electrode at various powers of the heat flow onto the electrode during 160 μs.

Value of the $\overline{q}$ in Table 5.1 is corresponded to the average power of the heat flow during the electrode jet action.

For comparison with the estimated data were used the experimental data of the experiment at $I_{max} = 125$ kA. Power of the heat flow onto the cathode averaged over time interval t_1–t_2 in this experiment obtained from the formula (5.8) is 2.2×10^7 W/cm^2

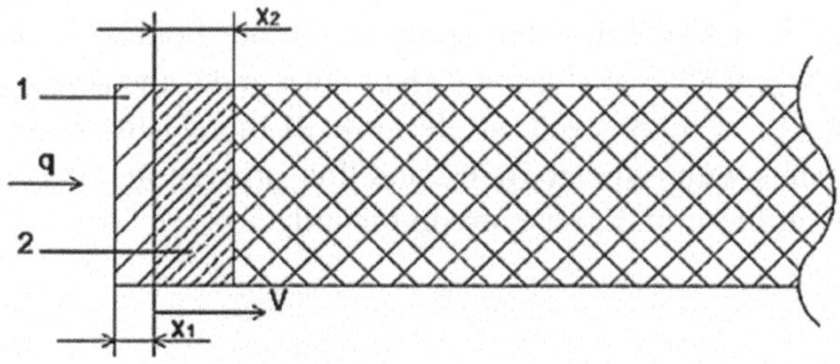

Fig. 5.10 Electrode under the heat flow action [3]

Table 5.1 Thicknesses of the evaporated x_1 and molten x_2 layers of the tungsten electrode

$\bar{q}$ (W/cm^2)	$T_0 \times 10^3$ (K)	v_1 (m/s)	x_1 (mm)	x_2 (mm)
10^9	30	79	13	8.2×10^{-4}
10^8	19	10	1.6	9.2×10^{-3}
2.2×10^7	14	1.2	0.19	5.2×10^{-2}
10^7	13	1.0	0.16	6.5×10^{-2}
5×10^6	12	3.2	0.051	0.2
10^6	10	1.0	0.016	0.58

$$\bar{q} = \frac{\int_{t_1}^{t_2} j_0 \sin \omega t \cdot U_{ne} \cdot \sin \omega t dt}{t_2 - t_1}. \tag{5.8}$$

where $j = I_{max}/S$ (4.4×10^5 A/cm^2), S is area of the electrode end, U_{ne} is near-electrode voltage drops (100 V), $T/2$ is current half-period (380 μs), and ω is angular frequency ($2\pi/T$).

Duration of the heat flow causing the evaporation of the electrode was determined from the photographs as the difference between the disappearance t_2 = 260 μs and occurrence t_1 = 100 μs of the electrode jet.

At the moment of maximal current, near-electrode voltage drop U_{ne} was determined from experimental data of the work [9]. For other moments, as in the work [16] we assumed that the near-electrode voltage drop at $j \geq 10^5$ A/cm^2 is proportional to the current density.

At $\bar{q}$ = 2.2×10^7 W/cm^2 mass of evaporated tungsten calculated by the formula (5.9) is 0.11 g

$$m = \rho S x_1 \tag{5.9}$$

where S is cross-sectional area of the electrode and ρ is density of the electrode material.

Specific erosion G in this case is 3.8 mg/C.

Data of Table 5.1 show that the weight of the vaporized electrode material is almost by four times more than of the liquid phase. Experimental value of specific erosion of the cathode for the same conditions is 3–4 mg/C, which is close to the calculated value.

Figure 5.1 shows that the data of the specific electrode erosion in the plasma pulsed generator PPG-3 [17] correspond to the second experiment at the current ~1.5 MA, current density of ~5×10^4 A/cm^2, and energy input into the gas up to 1 MJ. Since the spots of the arc binding in the discharge chamber of this generator move along the electrodes under the action of magnetic and gas dynamic forces at high speed, so the time of the electrodes heating less than in the experiments with rod electrodes, and thus erosion in this case, as in the papers [3, 18, 19] is in the liquid phase predominantly.

Table 5.2 Ratio of specific erosion of copper G_{Cu} to specific erosion of tungsten G_W for the liquid and vapor phases

Liquid phase ($j = 5 \times 10^4$ A/cm^2, $G \sim U_{ne}/cT_{mt}$)			Vapor phase ($j = 2.8 \times 10^5$ A/cm^2, $G \sim U_{ne}/r_{st}T_{st}$)		
W		cT_{mt} = 567 J/g	W		$R_{st}T_{st}$= 5.0 × 10^3 J/g
Cu		cT_{mt} = 400 J/g	Cu		$R_{st}T_{st}$= 5.4 × 10^3 J/g
$\frac{G_{Cu}}{G_W}$	Calculation	Experiment	$\frac{G_{Cu}}{G_W}$	Calculation	Experiment
	1.42	1.27		0.93	0.77

In the extreme case, for the rough estimation of the specific erosion G in the liquid phase the ratio can be used

$$G \sim U_{ne}/cT_{mt} \tag{5.10}$$

where c is specific heat of melting, T_{mt} is melting temperature.

For vapor phase

$$G \sim U_{ne}/r_{st}T_{st} \tag{5.11}$$

where r_{st} is latent heat of vaporization and T_{st} is evaporating temperature.

Data of Table 5.2 show that at lower current density the liquid phase is dominated and at higher current density the vaporous phase prevail.

At transition from the liquid to vapor phase, the average value of the specific erosion G is varied little and is 0.01–0.1 g/C. Table 5.2 shows that the product $r_{st}T_{st}$ approximately by an order of magnitude greater than cT_{mt}, then for small changing of the G is necessary to increase the U_{ne} at least on an order of magnitude.

Figure 5.11 shows the dependence of the voltage drop near the electrodes on the interelectrode distance at dI/dt ~ 10^{10} A/s.

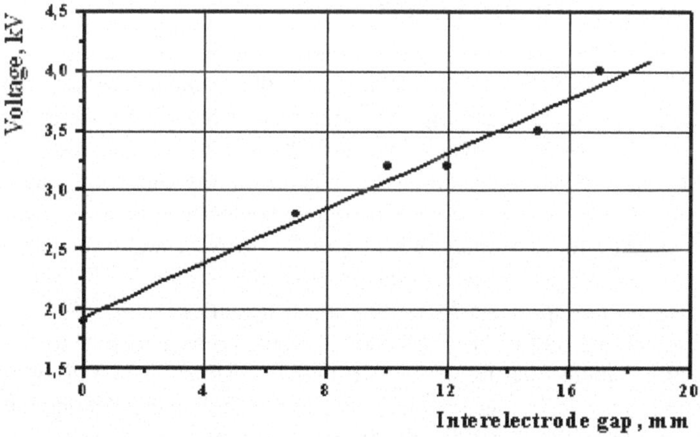

Fig. 5.11 Voltage drop across the arc versus interelectrode distance [3]

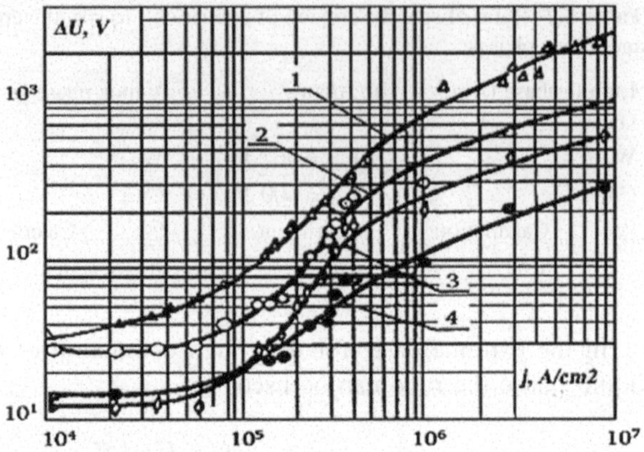

Fig. 5.12 Near-electrode voltage drop versus current density [16]

The dependence was obtained at the maximal current, thus after the appearance of the electrode jets. In the graph, the sum of the anode and cathode near-electrode voltage drops determined by extrapolation to zero of the interelectrode gap is ~ 2 kV.

Figure 5.12 shows the change in the nature of erosion and the corresponding sharp rise of the near-electrode voltage drops caused by the electrode jets at $j > 10^5$ A/cm^2 [16]. In the graph curve, (1) is total near-electrode voltage drop for tungsten electrodes. (2) is the total near-electrode voltage drop, (3) is near-anode voltage drop, and (4) is near-cathode voltage drop for copper electrodes.

Before the formation of powerful electrode jets at 16–24 μs, were observed the jet of the lower intensity, which are the sum of the microjets from rapidly moving and group electrode spots [20]. Specific erosion at this stage more than an order of magnitude less $\leq 10^{-4}$ g/C [13, 21].

5.3 Erosion as Electrode Jets

As noted above, the erosion of the electrodes in the megaampere arcs significantly affects both the arcing, and the improving the heat transfer from the arc to the gas. At high (10^8–10^9 W/cm^2) power of the heat flow onto the electrode, arc mainly burns in the vapor of the electrode metal. In addition, the anode and cathode jets at the collision with each other and with the opposite electrode generate shock waves, which also increase the efficiency of heat transfer from the arc to the gas, and affect the properties of the arc. This section presents the results of the study of these processes at currents up to 1.5 MA and current density of 2×10^6 A/cm^2.

Evolution and moving of the electrode jets in the hydrogen are shown in shadow pictures at incomplete filtering of its own radiation (Fig. 5.13). Photographs were done at experiments with tungsten electrodes of 6 mm diameter (anode left, cathode right) at interelectrode distance of 10 mm, current of ~ 125 kA, and initial hydrogen pressure of 1 MPa.

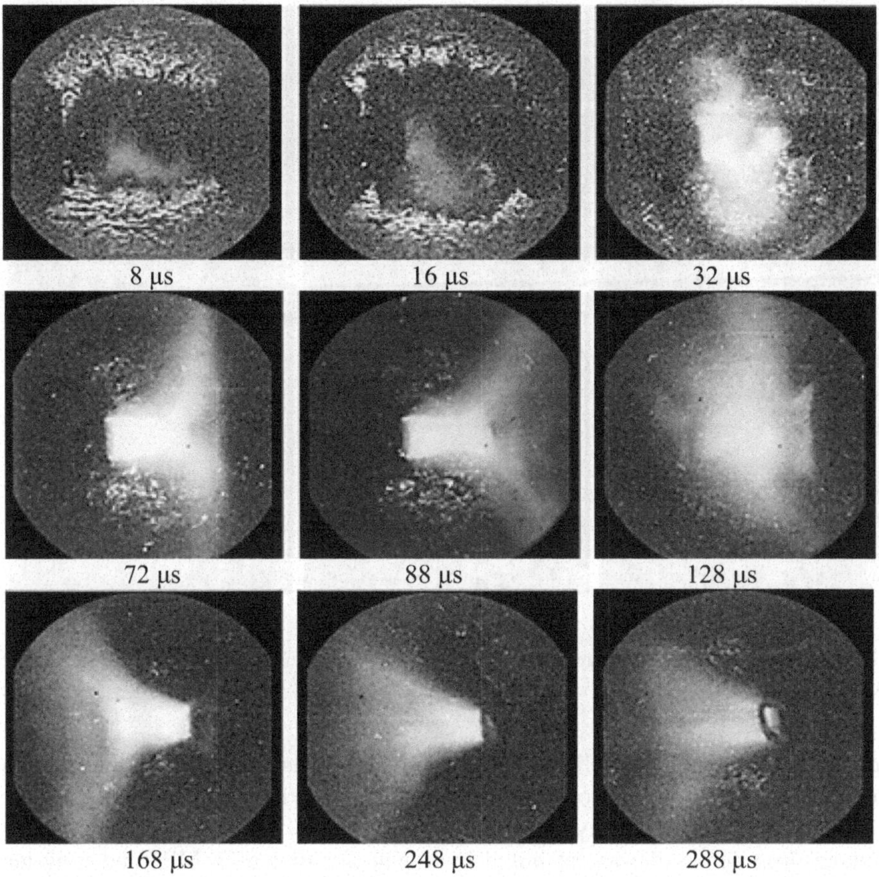

Fig. 5.13 Arc evolution and moving of the electrode jets in the hydrogen [3]

Similar photographs were made in the visible range of the spectrum (Fig. 5.14). Here the tungsten electrodes of 6 mm diameter (cathode left, anode right) have also been used at interelectrode distance of 17 mm, current of ~380 kA, and initial pressure of the hydrogen ~1 MPa.

Figure 5.14 shows that after the explosion of the wire at 4 μs, the arc is formed while the breakdown of the interelectrode gap. Arc is heterogeneous, due to the occurrence of hydrodynamic instabilities (necks), which are similar to those observed in fast Z-pinches. At this in points of the arc contraction, plasma ejections under high pressure are observed.

At 16-th μs, the cathode jet and then at 24-th μs anode jet are appeared. The anode and cathode jets are visible through the translucent expanding arc. By 8 μs, the electrode jets darker, i.e., the intensity of their glow is lower than of the arc, and by 16 μs is converse. At 24 μs intensely luminous area is appeared, as a result of the

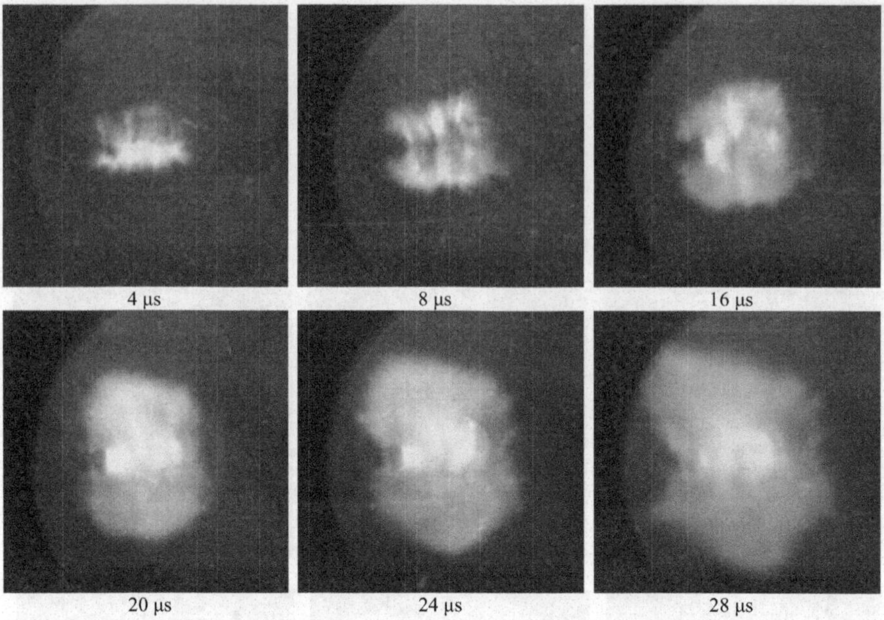

Fig. 5.14 Formation and moving of the cathode and anode jets at expanding of the translucent arc in the hydrogen [3]

anode and cathode jets collision. These so-called "jets of the first type" are by themselves the sum of microjets generated by fast-moving electrode spots.

Figure 5.15 shows the typical oscillograms of the discharge current and the voltage drop across the arc at initial hydrogen pressure of 1 MPa and tungsten electrodes of 6 mm diameter.

Figure 5.16 shows the dependence of the expansion speed of the translucent arc in the hydrogen versus initial pressure [22]. The data are consistent with those presented earlier in the work [23].

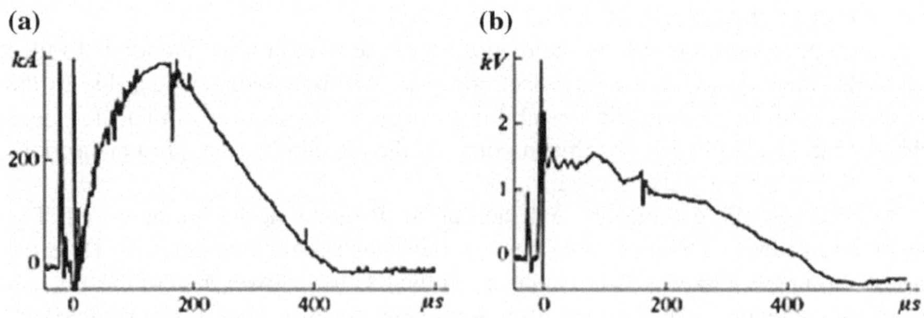

Fig. 5.15 Oscillograms of the current (**a**) and voltage drop across the arc (**b**) [3]

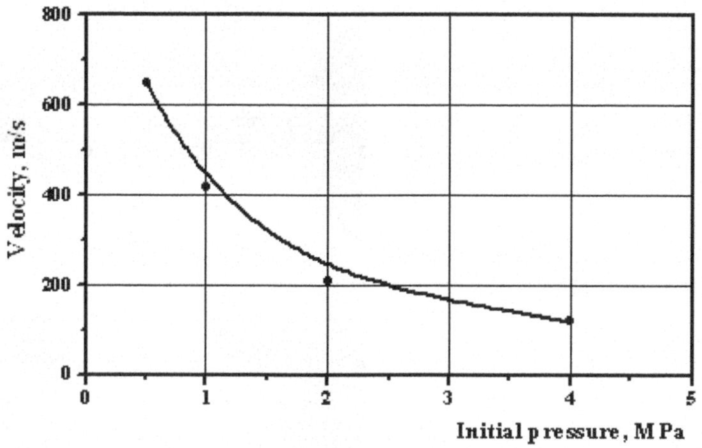

Fig. 5.16 Speed of the translucent arc expansion versus initial hydrogen pressure [3]

More complete picture of the evolution of the arc is obtained by joint analysis of shadow photographs with incomplete filtering of the arc radiation and photographs in the visible part of the spectrum with the use of the neutral filters their optical density varies from 0.5 to 2 that gives a 100-fold decrease in the intensity of the arc radiation [10].

Initial stage of the process at initial hydrogen pressure of 0.1 MPa, tungsten electrodes of 6 mm diameter, and current of 100 kA is clearly visible in Fig. 5.17, because the arc in the air is more optically transparent.

Later occurrence of the electrode jets in Fig. 5.13 due to the lower discharge current of 125 kA in comparison with 380 kA in Fig. 5.14.

The radiograph in Fig. 5.18 shows that the symmetrical cathode jet is appeared from the center of the left electrode and in the case when initiating wire is partially vaporized, and breaks down on separate fragments.

Powerful electrode "jets of the second type" are appeared from the common melt pool. Formation time of the melt pool depends on the properties of the electrode material, its diameter, and current density.

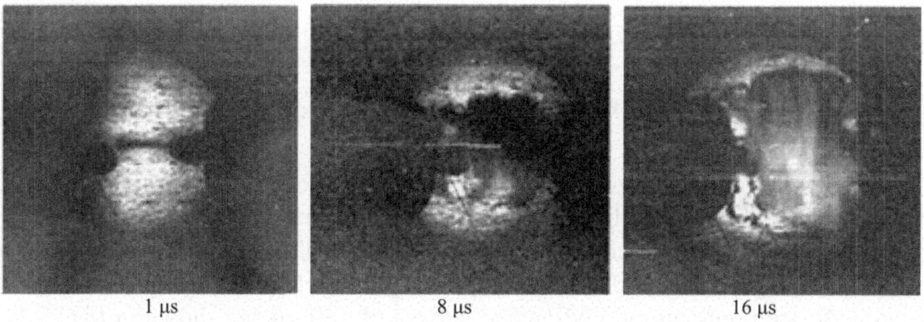

Fig. 5.17 Shadow photograph of the expanding arc in the air [3]

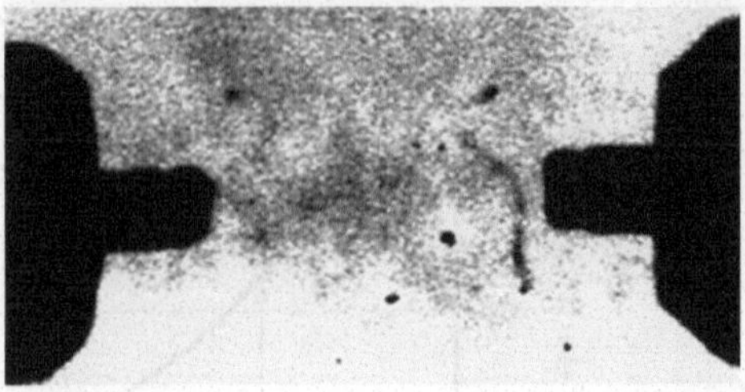

Fig. 5.18 Radiograph of the cathode jet and nonevaporated fragments of initiating wire [3]

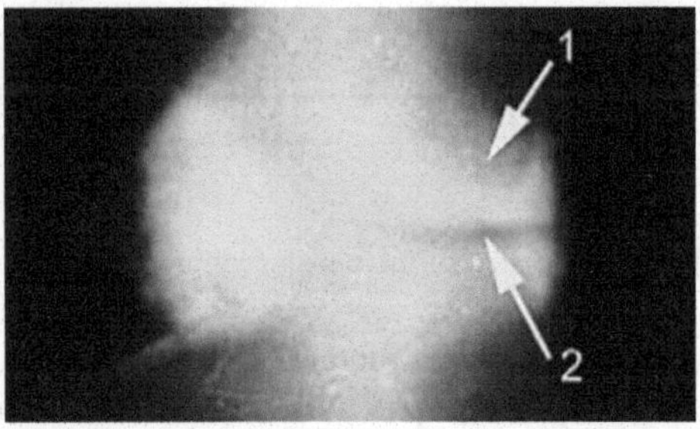

Fig. 5.19 Jet of the second type (2) from the anode on the brighter background of the first type jet (1) [3]

Formation of these jets is shown in Fig. 5.13, where at 72-nd μs and then at 168-th μs the powerful anode and cathode jets are formed. Time of 128 μs is corresponded to the same their intensities and appearance of the luminous shock-compressed area of the disk form.

Figure 5.19 shows the simultaneous existence of two types of the jets. On the anode the long, narrow and darker "jet of the second type" is visible on the brightness background of the broad conical "jet of the first type".

According to data of the works [24, 25], the cause of the "jets of the second type" on the anode may be "pinch effect". At current of ≥ 10 kA, the magnetic pressure becomes close to the gas-kinetic pressure that leads to contraction of the arc spot banding to the electrode and its (spot) further heating.

Central jet by its form and conditions of the appearance is similar to the "plasma focus" that at compressing of the arc by the magnetic field of the current takes place [26, 27].

5.4 Erosion as Electrode Surface Layer Ejection

At some experiments, the particular type of the electrode erosion was registered. This erosion is the ejection from the surface of the electrode the molten metal layer in the form of a hollow hemisphere. Photographs of this phenomenon at experiments with tungsten electrodes of 6 mm diameter, interelectrode distance of 10 mm, initial hydrogen pressure of 1 MPa, and current of 180 kA are shown in Fig. 5.20.

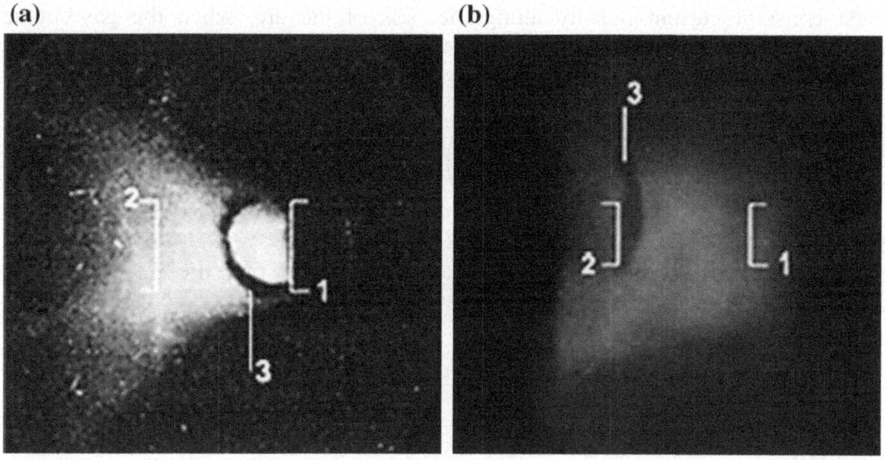

Fig. 5.20 Ejection of the surface layers from the tungsten cathode (**a**) and anode (**b**) [3]. 1 is cathode contour, 2 is anode contour, 3 are metal ejections

These ejections are caused by the uniform melting of the surface metal layer at the ends of the electrodes. By the metal ejection, the formation of the powerful electrode jet is preceded.

Besides, at the experiment with the same tungsten electrodes, interelectrode distance of 17 mm, current of 280 kA, and initial hydrogen pressure of 1 MPa after arc extinction, ejections of the metal particle clusters from the cathode were registered (Fig. 5.21).

It is obvious that the observed phenomenon is caused by exceeding of the gas-kinetic pressure near surface of the electrode over the magnetic pressure that (exceeding) is occurred till the arc extinction.

On the balance of these pressures at the surface of the anode spot near the current maximum (10–20 kA) is indicated in the works [24, 25].

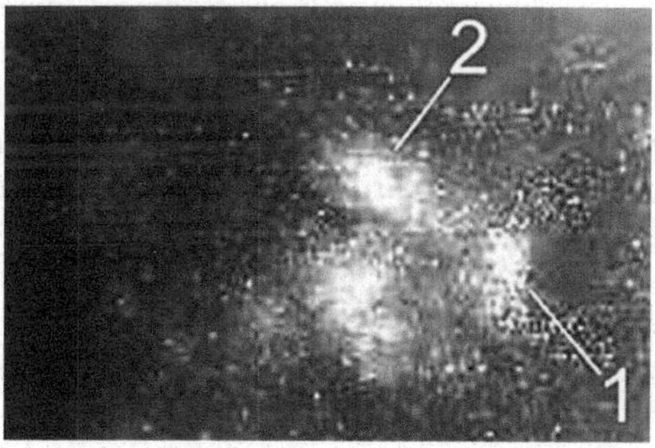

Fig. 5.21 Ejection of clusters of the metal particles from cathode [3]. 1 is cathode end, 2 are clusters of metal particles

At constant current density along the axis of the arc, when the gas-kinetic pressure is balanced by the magnetic pressure, the profile of the gas-kinetic pressure is determined by the formula [28]

$$P(r) = P_{\max}\left(1 - \frac{r^2}{r_0^2}\right) = \mu\frac{I^2}{4\pi^2 r_0^2}\left(1 - \frac{r^2}{r_0^2}\right) \tag{5.12}$$

where r_0 is arc radius equal to the radius of the electrode.

Average over the cross section, the maximal magnetic pressure $P_{\max}$ at I = 120 kA and r_0 = 3 mm estimated by the formula

$$P_{\max} = \mu\frac{I^2}{4\pi^2 r_0^2} \tag{5.13}$$

is 51 MPa.

At our experiments, due to the high power of the heat flows onto the electrodes through few microseconds, the common melt pool is formed of diameter close to the diameter of the electrode. On a similar phenomenon reported in the work [29], where on the surface of the electrodes with similar sizes and at same currents, the melt pools with a diameter close to the diameter of the electrodes and the depth of ~1 mm emitting the X-ray radiation with a continuous spectrum were resisted.

It was found that in common melt pool, the center hole is not formed, as the melt is not squeezed out by the gas-kinetic pressure toward the periphery. From this, it follows that within the melt pool, there is pressure, which balances the pressure above the melt surface. About this phenomenon for the single cathode spot is mentioned in the work [30].

Thus, when the current is decreased, the gas-kinetic and magnetic pressures over the surface of the molten pool are reduced faster than the pressure inside the melt. The result of this is ejection of the molten metal, which during expansion is partially converted to vapor. Since ejection of the molten metal from the electrode surfaces due to, in particular, high energy output in the near-electrode zones of the arc the molten metal is ejected both from the cathode and anode (Fig. 5.20).

At experiments with electrodes of more than 10 mm diameter in the center of anode by 20–50 μs is formed jet of 1–2 mm diameter, which indicates to higher current density in the electrode center. At this, more current density must comply more magnetic pressure than by the formula (5.13). In this case, in the center of electrode after current maximum, the metal jet of 1–2 mm diameter is formed (Fig. 5.19).

Let us consider in greater detail the phenomenon of ejection of molten metal from the electrode end in the form of hemisphere. Figure 5.22 shows the shadow pictures of the initial moment of the ejection of metal from the cathode at initial hydrogen pressure of 1 MPa and current of 125 kA. The metal is ejected by the powerful electrode jet with diameter equal to the diameter of the cathode, by 280 μs

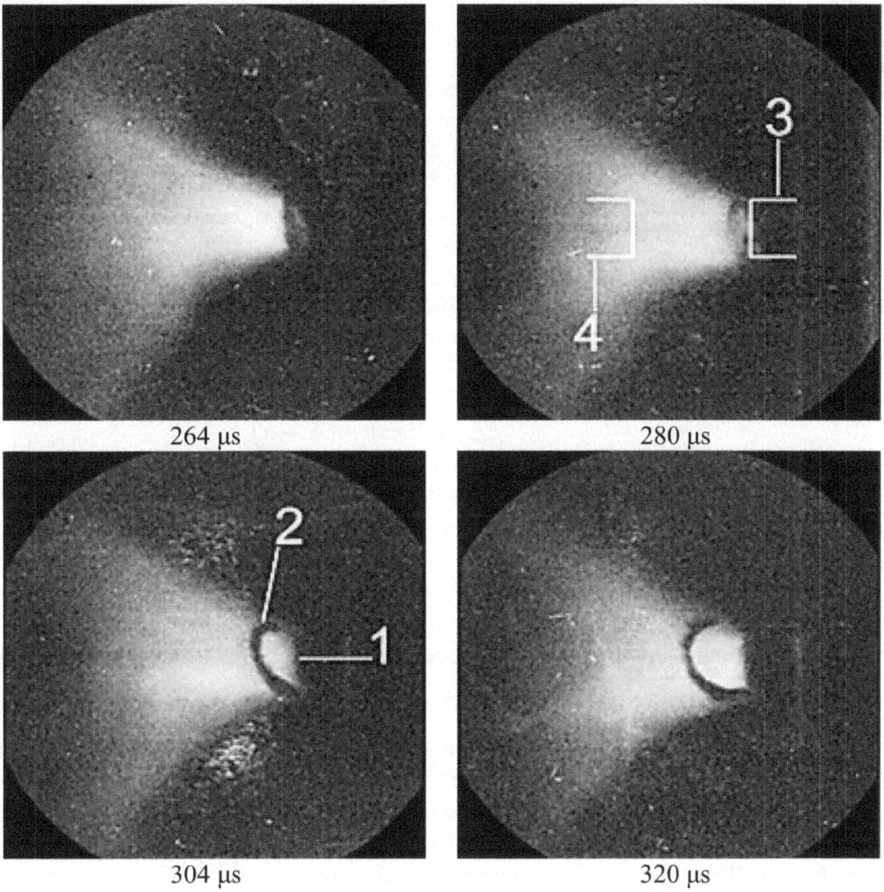

Fig. 5.22 Shadow photographs of tungsten ejection from the cathode surface [3]. 1 is cathode end, 2 is ejection of metal, 3 is contour of cathode, 4 is contour of anode

when the discharge current is decreased. The speed of movement of the metal jet determined from photographs in Fig. 5.22 is ~90 m/s.

At this experiment, when the power of the heat flow on the end surface of the electrode is 2.2×10^7 W/cm^2, the ejection of the molten metal in the form of hollow hemisphere is occurred. The thickness of the layer in according to Table 5.1 is 0.052 mm. However, in order to the ejection is looked like semicircle, but not as hemisphere, it is necessary that the metal is in the vapor state with the layer thickness according to optical measurements of 0.5 mm. In this case, the vapor concentration in the hemispherical layer should be 3×10^{21} cm^{-3} at temperature close to the boiling temperature of the tungsten ~6500 K.

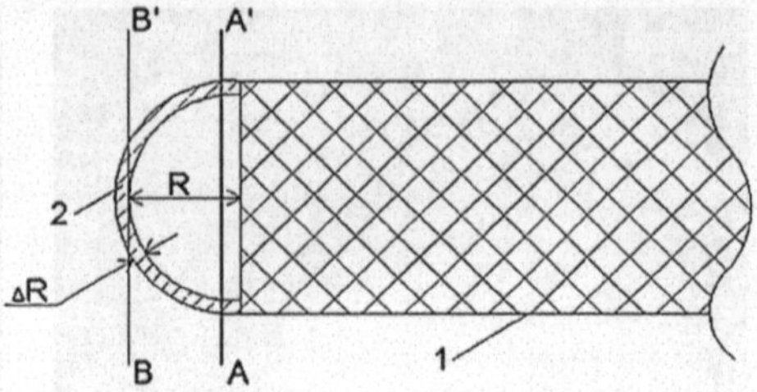

Fig. 5.23 Ejection of metal in form of hemisphere thickness ΔR from electrode end [3]

Absorption coefficient κ_0 of the radiation with a wavelength of 514.5 nm corresponding to the wavelength of light transmission defined by the Kramers' formula [31] is $\sim$4 cm^{-1}

$$\kappa_0 = 0.96 \times 10^{-7} \frac{NZ^2}{T_0^2} \frac{\exp-(x_1 - x)}{x^3}. \tag{5.14}$$

where $x_1 = I/kT$; $x = h\nu/kT$, I is the first tungsten ionization potential (8 eV), kT is temperature of metal in hemisphere ($\sim$6000 K), N is concentration of vapor $\sim 3 \times 10^{21}$ cm^{-3}, $h\nu$ is energy of transmission radiation photons with wavelength of 514.5 nm (2.45 eV).

The optical path length across the plasma hemisphere in section BB' (Fig. 5.23) is $2\sqrt{(R+\Delta R)^2 - R^2}$, the optical path length across the plasma hemisphere in section AA' (Fig. 5.23.) is $2\Delta R$, and their ratio is $\sqrt{2R/2\Delta R + 1}$. The smaller the relative thickness of the hemisphere, the ratio is greater.

At R = 0.3 cm and $\Delta R = 5 \times 10^{-2}$ cm, the ratio is $\sim$2.4. Then $\kappa_0 l_{AA}{'} = 0.4$, and $\kappa_0 l_{BB}{'} \sim 1$. It is known that $\kappa_0 l \sim 1$ is condition for the significant absorption of radiation. Therefore, the ejection of metal vapor is observed as semicircle, instead the hemisphere, as it really is.

Thus, the conclusion about the liquid phase predomination in the products of electrode erosion in high current arcs holds for $j \leq 10^5$ A/cm^2. For $j > 10^5$ A/cm^2 and times $\sim 10^{-4}$ s, the vapor phase is dominated.

5.5 Erosion and Molecular Weight of Gas

Erosion products of the electrodes being in a vapor state form a gas mixture with operating gases (hydrogen, helium, air), the speed of sound in which is lower the higher, the concentration of metal atoms, i.e., the higher molecular weight of the gas mixture. The apparent molecular weight of the gas mixture μ_m was calculated from Formula 5.15 assuming that $\sim$20% of the eroded metal is in the vapor phase [1]

$$\mu_{\rm m} = \frac{1}{\sum_1^n g_i/\mu_i}. \tag{5.15}$$

where g_i are weight fractions of gas mixture ($g_i = M_i/M_{\rm m}$, where M_i and $M_{\rm m}$ are weight of i-th component and total weight of mixture); μ_i is molecular weight of i-th component.

Dependence of the apparent molecular weight of the gas mixture on the ratio of the weight of pure hydrogen $M_{\rm g}$ to the weight of the metal $M_{\rm m}$ eroded from electrodes is shown in Fig. 5.24. Graph shows that the molecular weights of the gas mixture are from 2.05 amu (that corresponds to molecular weight of practically pure hydrogen) to 3.4 amu (that is somewhat lower than the atomic weight of helium). Average statistical value of the apparent molecular weight of the operating gas is 2.5 amu that is for an average statistical experiment this parameter by 25% higher than the molecular weight of pure hydrogen.

Dependences of the speed of sound in pure hydrogen and in operating gas on average temperature are shown in Fig. 5.25. It is noteworthy that the curve for pure hydrogen and most points of the real gas curve practically coincide up to 2000 K, and then the curve of the real gas is somewhat lower, at this the degree of its deviation increases with increasing of temperature.

This circumstance is due to the fact that at calculating of the molecular weight of the working gas (as a mixture of hydrogen and metal atoms of the electrodes), the ratio of the vapor fraction to the total weight of the eroded metal was assumed as constant (20%) and did not take into account its growth, depending on the gas temperature increase, although such dependence takes place, and at high gas temperatures it must be taken into account.

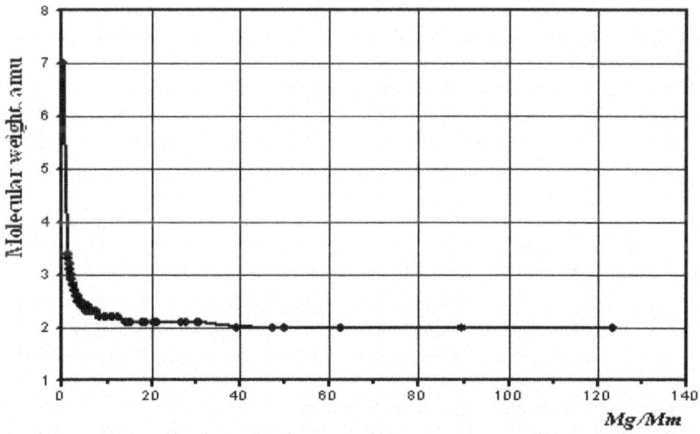

Fig. 5.24 Apparent molecular weight of the gas mixture versus the ratio of the weight of pure hydrogen to the weight of the eroded metal [2]

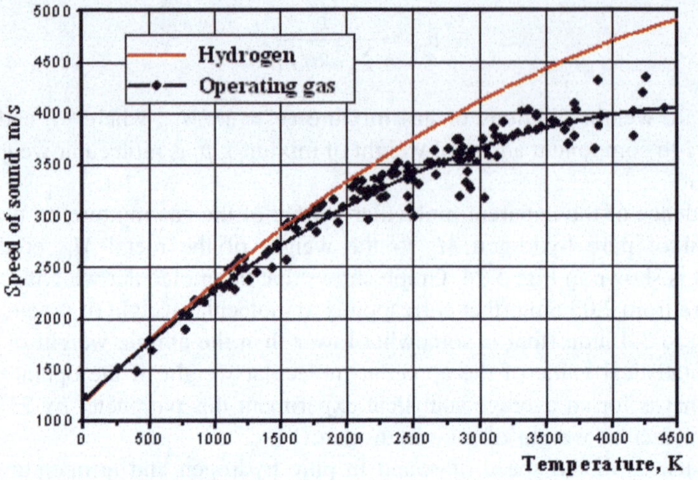

Fig. 5.25 Speed of sound in pure hydrogen and in operating gas versus temperature [2]

This dependence is due to an increase in the fraction of atomic metal in the gas at average temperatures exceeding the melting point of the metal of the electrodes.

References

1. G.S. Belkin, V.J. Kiselev, Sov. J. Tech. Phys. **37**, 977 (1967). (in Russian)
2. V.A. Kolikov, Doctoral Thesis, St. Petersburg, 2005 (in Russian)
3. A.A. Bogomaz, Doctoral Thesis, St. Petersburg, 2012 (in Russian)
4. G.S. Belkin, V.J. Kiselev, Sov. J. Tech. Phys. **36**, 384 (1966). (in Russian)
5. A.L. Donaldson, M. Kristiansen, A. Watson, K. Zinsmeyer, E. Kristiansen, R. Dethlefsen, IEEE Trans. Magn. **22**, 1441 (1986)
6. G.S. Belkin, M.E. Danilov, Sov. J. Electr. **8**, 45 (1972). (in Russian)
7. A. Watson, A.L. Donaldson, K. Ikuta, M. Kristiansen, IEEE Trans. Magn. **22**, 1799 (1986)
8. A.L. Donaldson, T.G. Engel, M. Kristiansen, IEEE Trans. Magn. **25**, 138 (1989)
9. N.N. Ruikalin, A.A. Uglov, A.N. Kokora, in *Laser Processing of Materials* (Moskow, 1975) (in Russian)
10. P.G. Rutberg, A.A. Bogomaz, A.V. Budin, V.A. Kolikov, M.E. Pinchuk, A.A. Pozubenkov, Tech. Phys. **47**, 26 (2002)
11. A.A. Bogomaz, A.V. Budin, M.E. Pinchuk, P.G. Rutberg, A.F. Savvateev, *Physics of Exstreme States of Matter—2005* (IPKhF RAN, Chernogolovka, 2005). (in Russian)
12. D.F. Alferov, V.A. Nevrovskiy, V.A. Sidorov, High Temp. **40**, 15 (2002)
13. B.A. Urjukov, in *Experimental Researches of Plasma Generators* (Novosibirsk, 1977) (in Russian)
14. S.I. Anisimov, Sov. J. High Temp. **6**, 116 (1968). (in Russian)
15. S.I. Anisimov, J.A. Imas, G.S. Romanov, J.V. Hoduiko, in *Action of High-Power Radiation on Metals* (Moskow, 1970) (in Russian)
16. V.P. Ignatko, G.M. Chernjavskiy, in *I All-Union Seminar on Dynamics of the High-Current Arc Discharge* (Novosibirsk, 1990) (in Russian)

17. P.G. Rutberg, A.V. Budin, V.A. Kolikov, B.P. Levchenko, V.V. Leontiev, I.P. Makarevich, N.A. Shirokov, High Temp. **32**, 589 (1994)
18. I.V. Tsvetkov, Izvestia RAS **58**, 156 (1994). (in Russian)
19. I.V. Tsvetkov, Rus. J. Plasma Phys. Rep. **19**, 1281 (1993). (in Russian)
20. I.A. Krinberg, E.A. Zverev, Plasma Phys. Rep. **25**, 82 (1999)
21. G.A. Ljubimov, in *Experimental Researches of Plasma Generators* (Novosibirsk, 1977) (in Russian)
22. A.A. Bogomaz, A.V. Budin, V.A. Kolikov, P.G. Rutberg, in *Conference PLTP-98* (Petrozavodsk, 1998) (in Russian)
23. A.A. Bogomaz, V.S. Borodin, B.P. Levchenko, P.G. Rutberg, Sov. J. Tech. Phys. **47**, 121 (1977). (in Russian)
24. V.A. Nemchinskiy, Sov. J. Tech. Phys. **52**, 35 (1982). (in Russian)
25. V.A. Nemchinskiy, Sov. J. Tech. Phys. **52**, 235 (1983). (in Russian)
26. V.I. Krauz, M.G. Levashova, M.A. Karakin, V.S. Lisitsa, A.N. Mokeev, V.V. Myalton, V.P. Smirnov, O.N. Krokhin, V.Y. Nikulin, A.V. Oginov, V.E. Fortov, Plasma Phys. Rep. **34**, 43 (2008)
27. W. Kies, G. Decker, U. Berntien, YuV Sidel'nikov, D.A. Glushkov, K.N. Koshelev, D.M. Simanovskiĭ, S.V. Bobashev, Tech. Phys. Lett. **25**, 802 (1999)
28. D.J. Sherklif, in *Course of Magnetic Hydrodynamics* (Moskow, 1967) (in Russian)
29. H. Chuaqui, M. Favre, R. Saavedra, E.S. Wyndham, P. Choi, C. Dumitrescu-Zoita, L. Soto, R. Aliaga Rossel, I.H. Mitchell, Phys. Plasmas **4**, 3696 (1997)
30. A.V. Zharinov, J.V. Sanochkin, Sov. J Tech. Phys. Lett. **9**, 1465 (1983). (in Russian)
31. Y.B. Zel'dovich, Y.P. Raizer, in *Physics of Shock Waves and High-Temperature Hydrodynamic Phenomena* (New York, 1966)

Chapter 6
Oscillations of Arc's Diameter

Abstract At the study of megaampere pulsed electric arcs, a number of phenomena have been established that have a significant effect on the arc parameters and processes in the discharge chambers of pulsed plasma generators. Such phenomena include transverse oscillations of the arc. At investigations of this phenomenon, the types of acoustic oscillations appeared in the discharge chamber were determined. The estimation was made of the arc diameter changing associated with the oscillation of the voltage drop. To the jump of the voltage drop across the arc of ~3 kV a change in the arc diameter of 0.4–0.7 cm is corresponded. It was established that periodic changes in arc parameters are associated with the impact of magnetic and gas-kinetic pressures. It was found that the resonance of acoustic oscillations and oscillations caused by magnetic and gas-kinetic pressures leads to a sharp increase in the electric field strength and energy density in the arc. It is shown that the fluctuation in the intensity of soft X-ray radiation at initial hydrogen pressure of 5–7 MPa is caused by a periodic change in the diameter of the arc. Analysis of the frequency and amplitude of the oscillations and the absorption of soft X-ray radiation indicates to the high-temperature zone in the arc of ~1 mm diameter and temperature 40–60 eV.

6.1 Acoustic Oscillations

Acoustic oscillations in the discharge chamber appeared as a result of multiple reflections of shock waves from the cylindrical wall and their cumulation in the center of the discharge chamber. Since the interelectrode distance in these experiments was 1–2 cm, diameter of the arc at the discharge current maximum is 0.7–1.0 cm, and diameter of the discharge chamber is 4.2 cm, the ratio of these dimensions determines the geometry for determining the characteristics of acoustic oscillations as something average between the spherical and cylindrical.

For spherical geometry, shock waves in hydrogen can be considered according to the data of the work [1] as a set of an infinite number of the own harmonic oscillations with period T_K

V. Kolikov et al., *Powerful Pulsed Plasma Generators*,
Springer Series on Atomic, Optical, and Plasma Physics 101,
https://doi.org/10.1007/978-3-319-95249-9_6

$$T_K = \frac{2\pi R}{\mu_K}\sqrt{\frac{\rho_0}{\gamma P_0}}, \tag{6.1}$$

where R is radius of discharge chamber, μ_K is set of roots of a special function similar to the Bessel function, ρ_0 is gas density, γ is adiabatic index, P_0 is gas pressure, and K is root number of the special function.

To the basic tone of acoustic oscillations $K = 1$ and μ_1, the smallest positive root of the equation $tg\mu = \mu$, equal to 4.49 corresponds. Thus, the period of the basic tone of the oscillations $T_1 = 1.39\ R/c$ (where c is speed of sound).

For cylindrical geometry, formula (6.1) is also applicable, but the roots μ_K correspond to the solution of the first order Bessel equation $I_1\ (\mu_K) = 0$. To the basic tone period value $\mu_K = 3.83$ corresponds, then $T_1 = 1.65\ R/c$. In the intermediate case corresponding to the conditions of our experiment, we will consider T_1 as the arithmetic mean of these values, i.e., 1.52 R/c (Table 6.1).

Let us also estimate the periods of acoustic oscillations, which are caused by the secondary breakdown of the interelectrode gap, according to formula (6.1). This means that the shock waves propagate in the heated gas and are accompanied by the expansion of the metal vapor of the electrodes.

Table 6.1 presents the results of these estimations for experiments, when the hydrogen pressure is 30–157 MPa, and the shock wave propagates through the "cold" gas and when it propagates in gas heated by the first discharge pulse. The speed of sound in hydrogen c was determined by its pressure and density after compression [2].

Figure 6.1 shows the oscillogram of pressure at preliminary adiabatic compression of hydrogen up to 60 MPa, where a slight decreasing of the period of acoustic oscillations due to heating of the gas takes place. Decreasing of the period is associated with increasing of the propagation speed of the shock wave. Up to 90 μs, the gas is heated and the period of acoustic oscillations is increased. The close values of the periods of the voltage drop oscillations across the arc and acoustic oscillations show that the shock waves impact the characteristics of the arc.

The amplitude of acoustic oscillations is increased at increasing of initial hydrogen pressure. As will be shown in the next section, the increasing of the electric field intensity in the arc corresponds to a decrease of the arc radius, which in its turn is caused by the impact of the compression wave. This is clearly seen in the oscillogram of the voltage drop of the experiment # 27 (Fig. 6.2) with a preliminary adiabatic compression of hydrogen up to the pressure of 157 MPa, which corresponds to hydrogen density of 5.6×10^{-2} g/cm^3.

The presented below estimations, based on the formulas (6.2–6.5), show that about value of the oscillation amplitude of the arc radius can be concluded by the electric field intensity in the arc, and consequently, by the voltage drop across the arc. A comparison of the voltage oscillograms in Figs. 6.1 and 6.2 indicates not only the increase in amplitude of the acoustic oscillations, but also the grows of their number at increase of the initial pressure from 60 to 157 MPa.

Table 6.1 Experimental and calculated periods of acoustic oscillations at the initial time and at highest temperature

Test number	N (GW)	P_i (MPa)	ρ (g/cm^3 × 10^{-2})	c_i (cm/s × 10^5)	T_1 calc. (μs)	T_1 exp. (μs)	P_{max} (MPa)	c_{max} (cm/s × 10^5)	T_2 calc. (μs)	T_2 exp. (μs)	Parameters to determine
№ 5	–	30	2.3	1.58	15.6	16	180	3.76	8.5	5.9	For voltage
№ 7	–	132	4.85	–	–	–	500	5.29	6.0	4.7	For voltage
# 23 (Fig. 6.3)	1.71	104	4.22	2.19	15	14	360	4.07	7.8	6.9	For voltage
# 27 (Fig. 6.2)	1.65	157	5.65	–	–	–	360	4.16	7.7	7.0	For voltage
# 28 (Fig. 6.4)	–	84	3.47	2.12	15	15	–	3.67	8.7	7.7	For voltage and pressure
# 34 (Fig. 6.5)	1.69	110	4.16	2.68	12	–	310	4.49	7.1	6.7	For pressure

[a]T_1 is period of acoustic oscillations at the initial time
[b]T_2 is period of acoustic oscillations at the maximum temperature

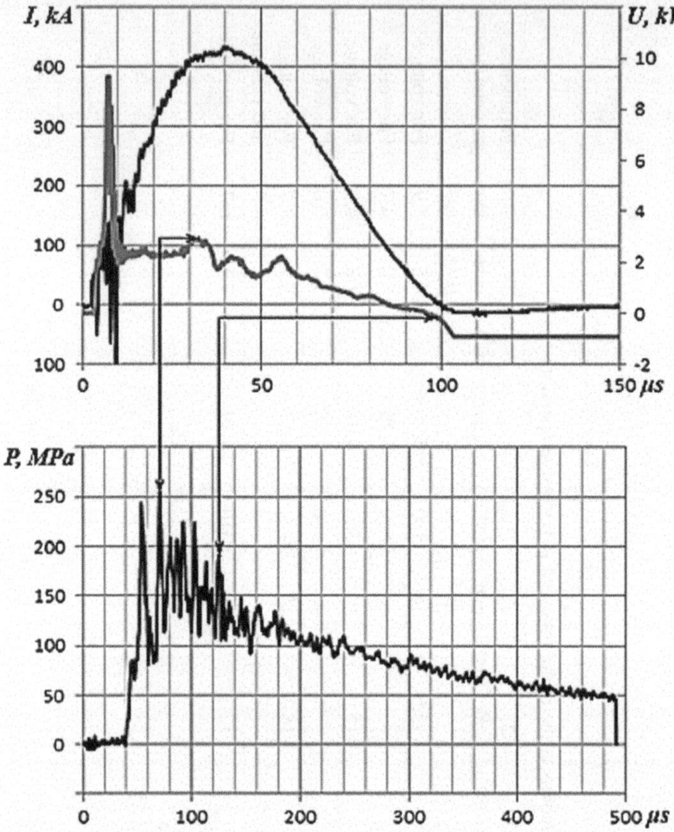

Fig. 6.1 Decreasing of the period of acoustic oscillations at the hydrogen heating [3]

6.2 Shock Wave Compression of Arc

Oscillations of the voltage drop across the arc ΔU can be caused by changes of the arc diameter, conductivity of the plasma in the arc, and inductance of the arc at its contractions. If we assume that the oscillations of the voltage drop across the arc are caused by a change of its inductance, then the voltage estimated by the formula (6.2) is $\sim$100 V, which is much less than the recorded amplitude of voltage oscillations reaching of several kilovolts

$$\Delta U = -I \frac{\mu_0 l}{2\pi r(t)} \frac{\mathrm{d}r}{\mathrm{d}t} \tag{6.2}$$

where I is discharge current, μ_0 is magnetic permeability of the vacuum, r is average radius of the arc, and l is arc length.

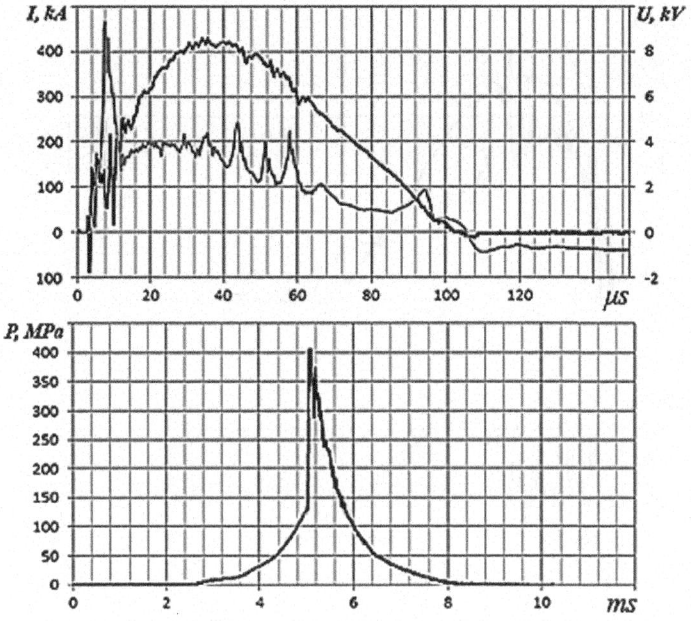

Fig. 6.2 Discharge in the hydrogen at initial pressure of 157 MPa (experiment # 27) [3]

Let us estimate the changes of the arc diameter at experiments # 23 (Fig. 6.3) and # 28 (Fig. 6.4) by the amplitudes of the voltage drop oscillations across the arc and the amplitudes of the pressure oscillations.

Since the conductivity σ for a multiply charged plasma at Spitzer conductivity with an average charge of the ion Z is proportional to $T^{3/2}/Z$, and Z can be approximated at the ion concentrations of 10^{18}–10^{20} cm^{-3} and temperature $\sim 10^5$ K according to the work [4] as $\sim \sqrt{T}$, then

$$E = \frac{j}{\sigma} \sim \frac{I}{r^2 \cdot T}, \tag{6.3}$$

where j is current density, r is radius of the arc, and T is temperature.

Assuming the compression of the arc is adiabatic, and magnitudes of the electric field intensity in the arc E_0, an arc radius r_0, and an arc temperature T_0 are the same as an arc parameters before compression, and, respectively, E_1, r_1, T_1—parameters after compression, and putting $\sigma \sim T$ in accordance with the data of the paper [4]

$$T_1 = T_0 \left(\frac{r_0^2}{r_1^2}\right)^{\gamma-1}, \tag{6.4}$$

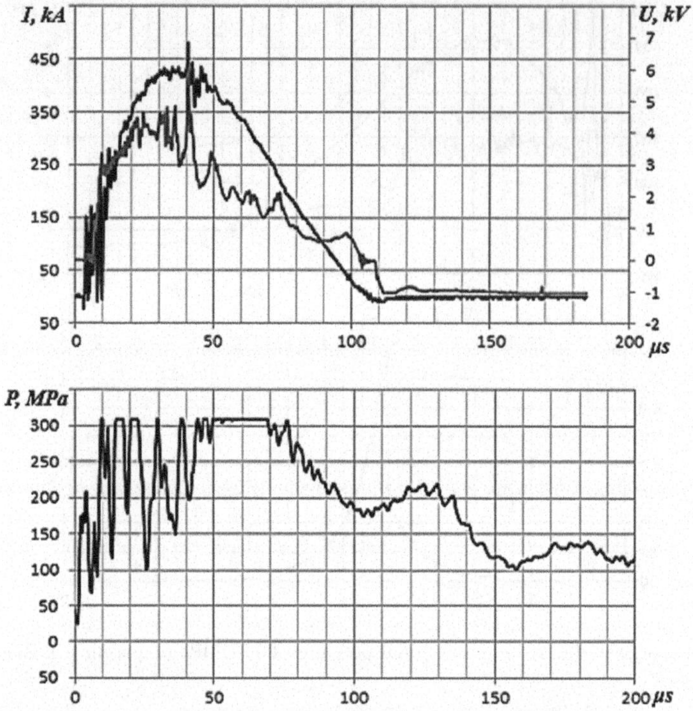

Fig. 6.3 Discharge in hydrogen at initial pressure of 104 MPa (experiment # 23) [3]

where γ is adiabatic index.

Since the oscillations of the arc occurred near the maximum of the current, the current can be considered as constant and then

$$\frac{E_1}{E_0} = \frac{r_0^2 T_0}{r_1^2 T_1} = \left(\frac{r_0}{r_1}\right)^{4-2\gamma}. \tag{6.5}$$

Since for a metal arc plasma at a particle concentration of $\sim 10^{20}$ cm^{-3} and temperature of $\sim 10^5$ K, the adiabatic index $\gamma \sim 1.1$ then

$$\frac{E_1}{E_0} = \left(\frac{r_0}{r_1}\right)^{1.8}. \tag{6.6}$$

At the experiment # 23 (Fig. 6.3) E_0 = 1.9 kV/cm, E_1 = 5.1 kV/cm, so using the formula (6.5), and assuming that, as at the experiment # 34 (Fig. 6.5) r_0 = 0.35 cm, r_1 = 0.2 cm, so this values of the electric field intensity corresponds to reduction of the arc radius by 0.15 cm.

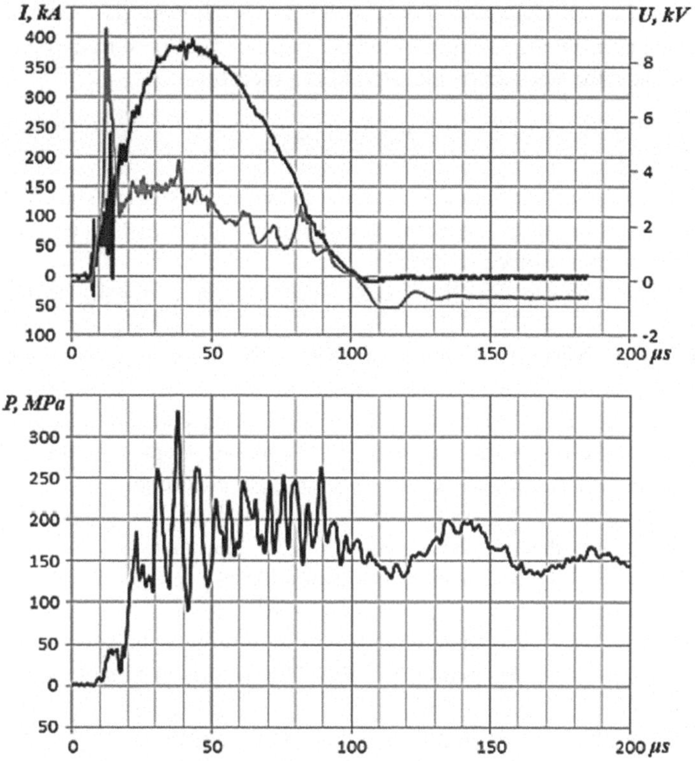

Fig. 6.4 Discharge in hydrogen at initial pressure of 84 MPa (experiment # 28) [3]

Fig. 6.5 Discharge in hydrogen at initial pressure of 110 MPa (experiment # 34) [3]. Photograph scan of the arc luminosity (above), graphs of arc diameter (D), brightness (B), and pressure (P) (bottom)

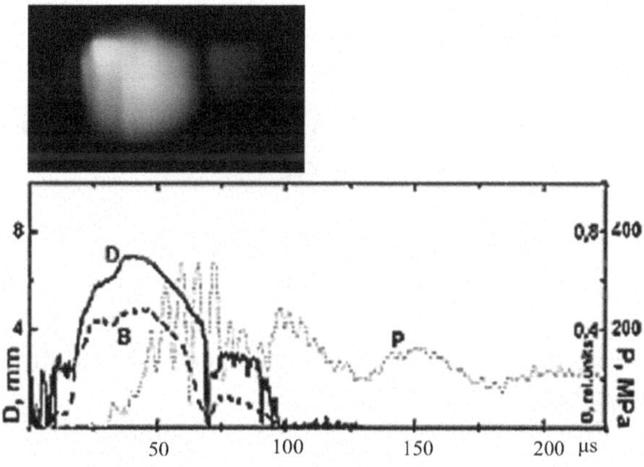

Graphs of the arc diameter (D) and the arc brightness (B) were receive by analysis of the photometric section of the arc luminosity. Pressure in the front of the sound wave near the surface of the arc [5] is given by

$$P_m = A\omega\rho c, \tag{6.7}$$

where A is decrease of arc radius, ω is circular frequency of acoustic oscillations, ρ is hydrogen density after adiabatic compression, and c is speed of sound in hydrogen at maximum of current.

At the experiment # 23 (Fig. 6.3), $A = 0.15$ cm, $\rho = 4.2 \times 10^{-2}$ g/cm^3, $\omega = 0.9 \times 10^6$ s^{-1}, $c = 4.1 \times 10^5$ cm/s, at this $P_m = 230$ MPa and pulse pressure magnitude on the chamber wall is of ~70 MPa. Assuming that the pressure decrease in the front of the cylindrical shock wave is inversely proportional to the square root of the radius of its coordinate, and then, calculated pressure on the wall at radius of the discharge chamber of 2.1 cm and average radius of the arc of 0.26 cm is ~80 MPa, which is close to the registered pressure.

At the experiment # 27 (Fig. 6.2), $E_0 = 1.7$ kV/cm, $E_1 = 2.1$ kV/cm, if we assume that, as in other experiments, $r_0 = 0.35$ cm, then by the formula (6.5) $r_1 = 0.89$ cm, $r_0 = 0.31$ cm, what is corresponded to arc contraction by 0.04 cm. For this experiment (Table 6.1), $\rho = 5.65 \times 10^{-2}$ g/cm^3, $c = 4.16 \times 10^5$ cm/s, and $\omega = 0.9 \times 10^6$ s^{-1}, and then, calculated pressure near the surface of the arc is 84 MPa, and on the chamber wall is 33 MPa. The recorded pressure on the chamber wall is ~40 MPa.

According to the presented estimations, the fluctuations of the electric field intensity in the arc are due to the impact of a shock wave repeatedly reflected from the wall of the discharge chamber.

However, oscillations of the arc can also be caused by the impact of magnetic and gas-kinetic pressures.

6.3 Oscillations of Arc's Diameter Caused by Magnetic and Gas Pressures

Let us consider the case, where the gas-dynamic pressure inside the arc P_g is balanced by the magnetic pressure P_m appeared due to the current in the arc. Suppose that under the action of a shock wave formed during a secondary breakdown of the interelectrode gap and reflected from the wall of the discharge chamber, the arc had contracted.

Since the contraction process is assumed to be adiabatic, then

$$P_g V^{\gamma} = \text{const}, \tag{6.8}$$

where V is volume of the arc, and γ is adiabatic index.

Since $V \sim r^2$, then $P_g \sim 1/r^2\gamma$ and $P_m \sim I^2/r^2$. If the arc had compressed, then at $\gamma > 1$ and I is constant, $P_g > P_m$, and the arc is expanded, and vice versa. As a result, radial oscillations of the arc are occurred, and their frequency is depended on the speed of sound in the gas, radius of the arc, and current. Theoretically, such oscillations are considered in the works [6–8].

In the work [7], the results of studies of radial MHD-oscillations of the plasma cylinder at the adiabatic approximation are presented. Solution is obtained for the frequency of nonlinear radial oscillations, where the initial current distribution can be specified arbitrarily. In the case of a homogeneous current density, the oscillations of the arc have period T

$$T = \frac{56r^2}{I}\sqrt{\frac{mn}{\gamma - 1}}. \tag{6.9}$$

where r is radius of the arc, I is current, m is mass of the metal atom in the arc, n is concentration of metal particles in the arc, and γ is adiabatic index.

In the work [6], it is indicated that the interaction between the pressure of the arc plasma and the magnetic pressure causes the radial oscillations of the Z-pinch. The solution of the system of MHD-equations presented in this paper describes the adiabatic process of radial oscillations of Z-pinch ideally conducting current. If $I \geq I_B$, where I_B is the Bennett current, at which the magnetic pressure is balanced by the gas-kinetic pressure, then oscillations occurred with a period equal to the time of several ranges of sound. The analysis of stability shows that the period of these oscillations can be less than the time constant of the constriction instability, which allows observing such oscillations at the experiment.

In the work [8], the solution was found that describes the evolution of the pinch oscillations with allowance of energy output by radiation. The case of a collapsing pinch system had considered, when at high current the energy loss by radiation exceeds the Joule heating. As an example, the Pisa-Braginskii model is used, when the output of energy is due to the bremsstrahlung accompanying electron-ion collisions. It is shown that of all dissipative processes, the largest is the emission of energy Q. For small Q, the system performs a large number of oscillations near the Bennett equilibrium. With increasing Q, the number of oscillations, their period, and amplitude are decreased.

At the experiment # 34 (Fig. 6.5) (Table 6.1), at initial hydrogen pressure of 110 MPa, tungsten electrodes of 6 mm diameter and interelectrode distance of 12 mm: the arc temperature $T = 10^5$ K and ion concentration $n_i = 1.6 \times 10^{20}$ cm^{-3} were determined on the pressure and conductivity with an error of ~20%.

The oscillation period T for this case, determined by formula (6.8) for $r = 0.35$ cm, $I = 4.9 \times 10^5$ A, $n = 1.6 \times 10^{20}$ cm^{-3}, $m = 10$–22 g, is 5.8 μs, which is close to the experimental value of 6.7 μs. However, the external pressure on the arc is not taken into account in formula (6.8), so it is more correct to apply this formula to experiments at initial hydrogen pressure of 5–7 MPa and current up to 1.5 MA, since under these conditions the external pressure is negligible compared to the magnetic pressure.

The close values of the frequencies of the two types of arc oscillations make it possible to explain the sharp increase of the electric field intensity oscillations in the arc, which in experiment # 23 (Fig. 6.3) increases from 1.9 to 5.1 kV/cm, what according to formula (6.8) at initial arc diameter of 0.7 cm is corresponded to contraction of the arc diameter to 0.4 cm.

At equal times of the arrival the compression wave reflected from the wall of the discharge chamber into the center of the chamber and of the moment of arc compression by the own magnetic field, diameter of the arc is decreased by the higher extent, which leads to additional increase of the electric field intensity and the energy density in the arc.

6.4 Oscillations of Soft X-rays Irradiation Intensity

Additional heating of the metal plasma near the arc axis can occur as a result of a collision of a shock wave reflected from the chamber wall with perturbation wave caused by the magnetic pressure. In this case, near the arc axis, a high-temperature zone is formed, which is the source of soft X-ray radiation (SXR).

It was found that the frequency of the SXR intensity fluctuation recorded in experiments with steel electrodes of 20 mm diameter, interelectrode distance of 10 mm, initial hydrogen pressure of 5–7 MPa and current of 1.0–1.5 MA varies in phase with the electric field intensity in the arc. As the SXR filter, the aluminum foil of 10–20 μm thickness was used. The X-ray diode that registered the radiation was located from the axis of the arc at distance of 16 cm.

At these experiments, the contraction of the arc was recorded at the discharge current equal the critical current of the Pisa-Braginskii. To the most contraction of the arc at 20–25 μs the maximum output of the SXR, feature on the oscillogram of the current, and peak of the voltage oscillogramm corresponds.

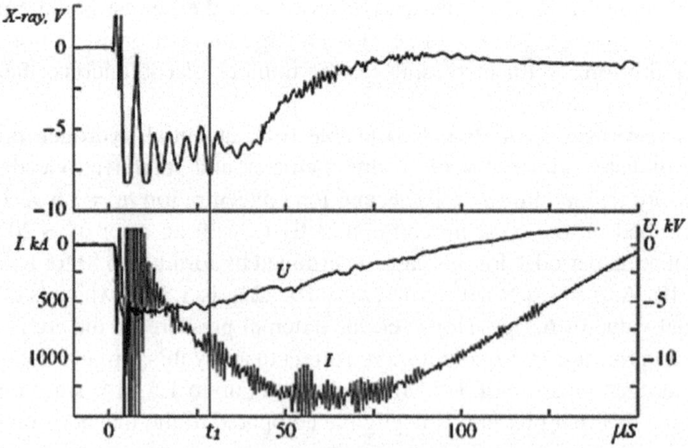

Fig. 6.6 X-ray signal before the maximal contraction of the arc (t_1), current (I), voltage (U) [3]

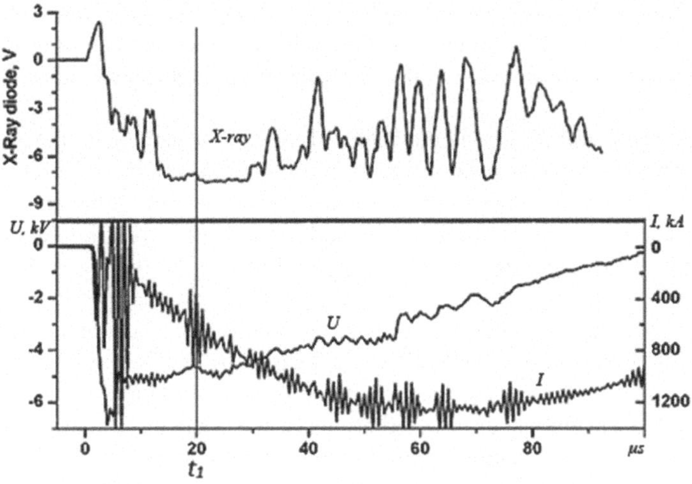

Fig. 6.7 X-ray signal after the maximal contraction of the arc (t_1), current (I), voltage (U) [3]

SXR intensity fluctuations with a frequency of 150–250 kHz were registered both before the arc contraction (Fig. 6.6), and after that (Fig. 6.7).

At initial hydrogen pressure of 30 MPa, the stable arcing phase of ~20 μs duration was observed before the contraction of the arc, while the diameter of the arc during this time is constant and equal to the diameter of the erosion crater on the electrode. When the initial hydrogen pressure changed from 20 mm Hg to 35 MPa, the diameter of the erosion crater changed insignificantly and is ~10 mm. The non-uniform distribution of temperature and pressure at the cross section of the arc is reflected on the shape of the erosion crater.

Figure 6.8 shows the end face of the steel electrode with the erosion crater formed after single discharge in hydrogen at current of 1.5 MA and half-period of 50 μs.

Fig. 6.8 Melt column in the center of the erosion crater [3]

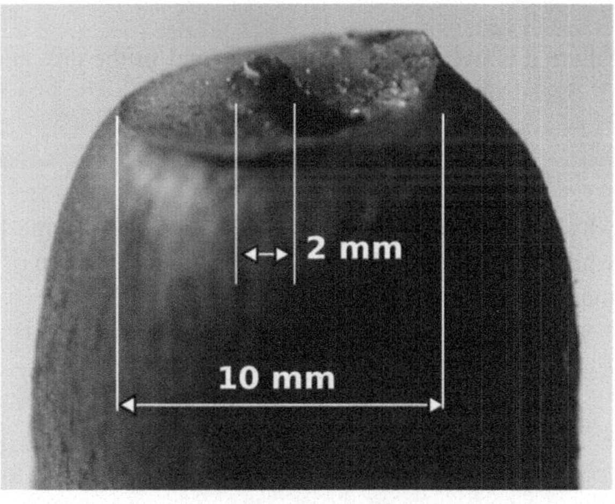

As a result of the action of arc binding spot to the electrode, the erosion crater with the melt column in its center is formed. The diameter and height of the column are ~2 mm that is correlated with the diameter of the high-temperature zone of the arc.

The high-temperature zone in the center of the arc formed due to the shielding of radiation from the arc was also observed in a number of studies at high initial gas pressure [9, 10]. The average temperature at the cross section of the arc is about ten electron volts, and the temperature of the high-temperature zone at later estimations is 30–40 eV. In our opinion, this high-temperature zone of the arc is the source of the SXR.

It was shown in the work [8] that as the intensity of radiation from the arc increases, the frequency of arc oscillations caused by the action of magnetic and gas-kinetic forces is increased, and amplitude of arc oscillations is decreased.

In this case, the frequency of oscillations of the SXR at the current maximum (50–75 μs) is decreased, and their amplitude is increased (Fig. 6.7). This indirectly confirms that the increasing heating of the transition layer of the arc leads to the decrease in the energy output from the arc by radiation. This is explained by the decrease of the arc transition layer transparency and, as a consequence, by increase of its shielding degree.

Using experimental data, we shall estimate the arc parameters and show that shock waves can cause noticeable fluctuations in the SXR intensity.

The concentration of metal ions ni near the maximum of current I = 1.2 MA we shall estimate by the average frequency of the SXR oscillation intensity equal to 200 kHz, assuming that frequency is due to oscillations of the arc.

The arc oscillation frequency we estimate when the radiation from the arc plays an important role, which is typical for arcs burning in gases at low (5–7 MPa) initial pressure, in contrast to arcs at ultrahigh initial pressure obtained by adiabatic compression.

According to the data of the work [8], the period of such oscillations T is

$$T = 3r/c. \tag{6.10}$$

where c is average speed of the sound in the arc, and r is radius of the arc.

Speed of the sound in the arc

$$c = \sqrt{\gamma P/\rho}, \tag{6.11}$$

where γ is adiabatic index (~1.1), and ρ is plasma-forming metal density ($n_i\ m_i$), where m_i is mass of the iron ion, n_i is concentration of iron ions, and P is pressure in the arc.

Assuming that the gas-kinetic pressure P_g is close to magnetic pressure $P_m = 1.6 \times 10^{-3} \times I^2/r^2$, then

$$n_{\mathrm{i}} = \frac{2 \times 10^{-4} I^2 T^2}{r^4 m_{\mathrm{i}}}. \tag{6.12}$$

Substituting the values we obtain $n_{\mathrm{i}} = 1.2 \times 10^{21}\ \mathrm{cm}^{-3}$. At such n_{i} and gas-kinetic pressure P_{g} equal to magnetic pressure P_{m}, pressure in the arc $P = 920$ MPa.

According to the data of the work [11], to these parameters the average temperature in the arc of ~8 eV and plasma conductivity of $1.3 \times 10^3\ (\Omega \times \mathrm{cm})^{-1}$ corresponds, what is agreed to the measured maximum brightness temperature of 6.5 eV at initial pressure $P_0 = 5$ MPa.

The electric field intensity in the arc E estimated by the conductivity is 1.2 kV/cm, which is close to measured earlier.

The average density of acoustic energy ρ_{a} by the period of oscillations is

$$\rho_{\mathrm{a}} = \rho A^2 \omega^2 / 2, \tag{6.13}$$

where A is amplitude of acoustic oscillations, ω is circular frequency of oscillations, and ρ is density of plasma-forming metal.

The value of A we estimate by the increment of the electric field intensity in the arc ΔE at the instant in time of the in-phase increase of the SXR intensity (Fig. 6.7).

According to Ohm's law $r = \sqrt{I/\pi\sigma E^{1/2}}$, hence

$$A \equiv \Delta r = \frac{1}{2}\sqrt{\frac{I}{\pi\sigma}} E^{-3/2} \Delta E. \tag{6.14}$$

Substituting the values in the formula (6.14), we obtain $A \equiv \Delta r = 6.5 \times 10^{-2}$ cm. If we use formula (6.5), which takes into account the change of conductivity under adiabatic contraction, then $A \equiv \Delta r = 6.2 \times 10^{-2}$ cm, and acoustic energy density $\rho_{\mathrm{a}} = 34\ \mathrm{J/cm^3}$.

Oscillations of the arc with the same frequency are recorded in the arc glow of the arc at initial pressure of 30 MPa. For the radius of the conduction zone, the radius of the glowing zone with density of blackening 0.7 was assumed. In this case, the change of the arc diameter at such density of blackening is consistent with the value of the voltage peaks. If the density of acoustic energy during the period of oscillations at the arc radius of 0.5 cm is 34 J/cm^3, then at the radius arc contraction up to 0.05 cm, it will be 3.4 kJ/cm^3.

The equality of frequencies of the arc contraction and the SXR oscillations intensity leads to the conclusion that the energy of the acoustic oscillations is converted to the energy of the SXR. At this, the temperature in region of the maximal arc contraction is by an order of magnitude more than average temperature at the arc cross section.

The concentration of iron vapor n determined by the frequency of the SXR oscillations (Fig. 6.7) is $1.2 \times 10^{21}\ \mathrm{cm}^{-3}$, i.e., by an order of magnitude higher than at the experiment # 34 (Fig. 6.5) with adiabatic compression (Table 6.1). This is due to both the difference in the thermophysical characteristics of tungsten and steel electrodes and the current density. Estimations of the time of the erosion jet occurrence from the tungsten electrode at the experiment # 34 (Fig. 6.5) show that the arc plasma is formed from the vapor of the copper initiating wire.

Let us estimate the concentration of the metal vapor n by the data of the experiments, where the erosion of the electrode is ~ 2 g. We will assume that the entrainment of the erosion products from the electrode is occurred at the speed of sound (Formula 6.10), assuming that the gas-kinetic pressure P_g is equal to half the maximum magnetic pressure P_m at current of 600 kA.

At $P_g = P_m = 1.6 \times 10^{-3} \times I^2/r^2$, $r = 0.5$ cm, $\gamma = 1.1$, we have $c = 5.2\times 10^{15}/\sqrt{n}$ cm/s. According to the experimental data, the rate of entrainment of the erosion products from the steel cathode at the current of ~ 1 MA is 4×10^{-2} g/s. In this case, during the first half-period ~ 120 μs, electric charge Q is 72 C. Then, the weight of the metal M evaporated from the electrode is 2.9 g, and the concentration of metal vapor n

$$n = \frac{M\sqrt{n}}{5.2 \times 10^{15} m t \pi r^2}, \tag{6.15}$$

where m is metal atomic weight, and t is time of discharge.

Then,

$$n = \frac{M^2}{m^2 \cdot 5.2^2 \times 10^{30} \cdot t^2 \cdot \pi^2 \cdot r^4}. \tag{6.16}$$

Substituting $M = 2.1$ g, $m = 56 \times 1.67 \times 10^{-24}$ g, $t = 1.2 \times 10^{-4}$ s, $r = 0.5$ cm in the formula (6.16) we obtain $n = 4 \times 10^{21}$ cm^{-3}.

Thus, the concentration of metal vapors estimated by the frequency of the SXR oscillations is close to the concentration had determined by the wear of the steel electrode. The higher vapor concentration estimated by this way is due to that it was assumed that the metal vapors are scattered as a cylinder with a diameter equal to the diameter of the electrode. However, photographs of the arc at a current of ~ 500 kA show that the metal vapors are scattered as a truncated cone, the smaller base of which is equal to the diameter of the electrode. In this case, the average concentration of metal vapor in the arc will be lower.

References

1. N.S. Koshljakov, E.B. Gliner, M.M. Smirnov, *Basic Differential Equations of Mathematical Physics* (Moskow, 1962) (in Russian)
2. N.B. Vargaftik, *Directory on Thermophysical Properties of Gases and Liquids* (Moskow, 1972) (in Russian)
3. A.A. Bogomaz, Doctoral thesis (St.-Petersburg, 2012) (in Russian)
4. V.I. Afonin, Plasma Phys. Rep. **25**, 620 (1999)

5. F.S. Felber, Phys. Fluids **25**, 643 (1982)
6. L.S. Solov'ev, Sov. J. Plasma Phys. Rep. **10**, 1045 (1984) (in Russian)
7. B.E. Mejerovitch, Sov. J. Plasma Phys. Rep. **11**, 1446 (1985). (in Russian)
8. R.V. Mitin, *Properties of Low-Temperature Plasmas and Diagnostics Methods* (Novosibirsk, 1977) (in Russian)
9. N.N. Ogurtsova, I.V. Podmyshenskiy, V.L. Smirnov, Sov. J. High Temp. **14**, 1 (1976). (in Russian)
10. B.V. Zamyshlyaev, E.L. Stupitskiy, A.G. Guz', V.G. Zhukov, *Composition and Thermodynamic Functions of Plasma, Handbook* (Moscow, 1984) (in Russian)
11. E.V. Aglickiy, V.V. Vikhrov, A.V. Gulov, *Multicharge Spectroscopy in Hightemperature Plasma* (Moscow, 1991) (in Russian)

Chapter 7
Processes and Heat Transfer in Electrodischarge Chamber

Abstract Investigation of the processes in the discharge chambers of plasma generators is complicated by the absence of optical windows, which are necessary for determining the main characteristics of the arc such as dimensions, optical characteristics, and nature of motion. The point is that to obtain limiting plasma parameters, the ultimate strength of the discharge chamber is necessary, whereas the openings for optical windows are concentrators of mechanical stresses, which substantially reduce the strength of the discharge chamber of the plasma generator [1, 2]. Naturally, in the absence of data of optical measurements, such parameters as temperature, electric field intensity in the arc, current density, etc., can be estimated only using various assumptions. Therefore, numerical estimations of the above parameters at the large-scale experiments were performed based on indirect data obtained by measuring current, voltage and pulsed pressure, and by visual and instrumental monitoring of the state of the electrodes and other elements of the generator before and after experiment. Based on these data, it is possible with a high degree of reliability to make conclusions about such parameters as discharge current and voltage drop across the arc, discharge power, arc resistance, electric charge, inputted electric energy in the arc, integral specific erosion of electrodes. With somewhat lesser degree of reliability, one can make conclusions about concentration and aggregate state of the products of erosion of electrodes and their quantitative ratio, average temperature, and internal energy of the operating gas. With even less reliability, we can conclude about the electric field intensity, current density in the arc, optical, and dynamic characteristics of the arc.

7.1 Arcing in Coaxial Discharge Chamber

The study of heat transfer from the arc to the gas is fundamentally important, because the knowledge of the mechanisms of heat transfer and their quantitative characteristics can help at choosing the optimal operating modes for the electric discharge chambers of pulsed plasma generators.

V. Kolikov et al., *Powerful Pulsed Plasma Generators*,
Springer Series on Atomic, Optical, and Plasma Physics 101,
https://doi.org/10.1007/978-3-319-95249-9_7

The arcing process in the discharge chambers of the coaxial type is characterized by a number of phenomena and features inherent for both other discharge chambers, and for this type of discharge chambers. Among the distinguishing features, the main is pulling the arc under the action of electromagnetic and gas-dynamic forces resulting in a multiple increases of its length; repeated, and, in some cases, multiple breakdowns of the interelectrode gap, after which follows the appearance and formation of a new arc is parallel to the already burning one. In addition, reflected shock waves and electrode jets have an appreciable effect on the arc and its characteristics.

To illustrate the processes during the arcing, in Fig. 7.1 the diagram of coaxial discharge chamber with designations of its main dimensions, arc parameters and the direction of the forces action are presented.

Let us consider the features of arcing in the coaxial discharge chamber of length *Lc*, diameter *Dc* and with an interelectrode distance *Lie*. At this, we use oscillograms of two experiments performed under relatively close initial conditions, and differing by the material of the electrodes.

An arc of diameter *d*2 initiating at point (*a*), under the effect of the resultant *F* electrodynamic and gas dynamic forces, occupies positions (*b* and *c*) step by step. When the arc reaches position (*d*), its contact with the chamber wall occurs at point (*e*) takes place. These conclusions are made based on the analysis of the arc binding spot traces on the inner wall of the discharge chamber, i.e., by the method of "autographs".

Figure 7.2 shows the oscillograms of the discharge current and the voltage drop across the arc at the experiment with voltage of 8.5 kV, initial pressure of hydrogen 25 MPa, volume of the discharge chamber of 1.12 dm^3, and interelectrode distance of 10 mm. The cathode material is tungsten, anode is steel.

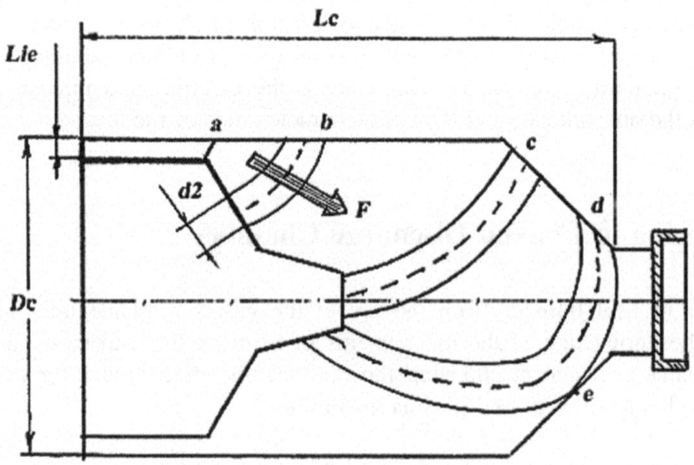

Fig. 7.1 Arc in the coaxial electric discharge chamber [26]

Fig. 7.2 Oscillograms of the current (**a**), voltage drop across the arc (**b**) [26]. Current—200 kA/div; voltage is 2 kV/div, time is 50 μs/div

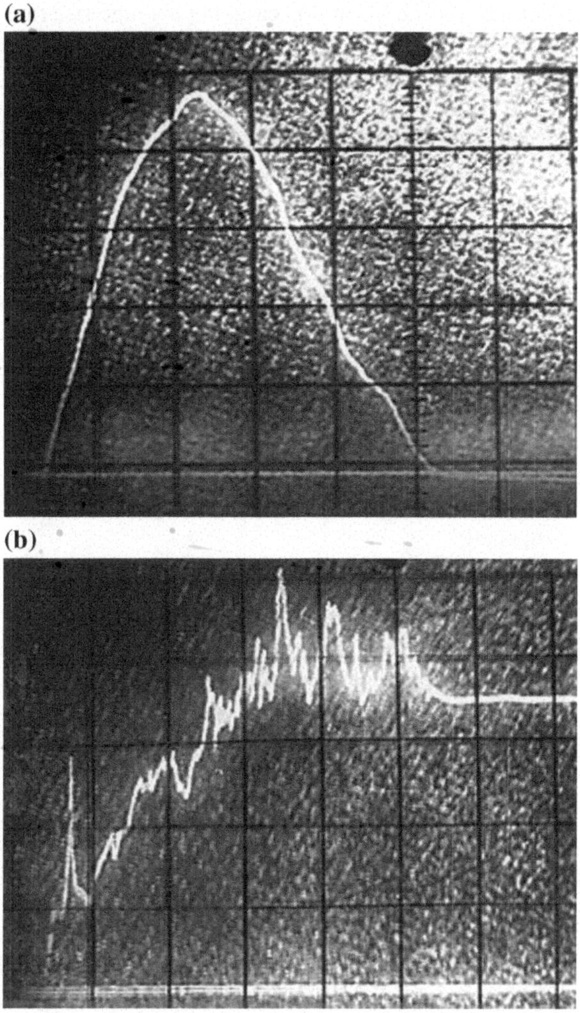

By the time before the current peak at the 86th μs, wire explosion (6th μs) takes place, to this moment the peak of voltage of 2 kV and duration of 4.5 μs is corresponded. Then within 6 μs, there is a sharp increase of the voltage across the discharge gap, which ends by the breakdown of the interelectrode gap through the vapor of the initiating wire at the 17th μs. After the breakdown during 14 μs, the arc channel is formed at the voltage drop across the gap from 6 to 2 kV. After this, beginning from 30th μs, under the action of electrodynamic and gas-dynamic forces, the arc is drawn into the volume of the discharge chamber, occupying successively the positions (b, c, and d). The drawing of the arc is accompanied by an increase in the voltage drop from 2 to 8 kV. The current to 96th μs reaches maximum (940 kA). In this case, the average rise rate of voltage d*U*/d*t* for a time of ~125 μs is 8×10^7 V/s, and the rise rate of current d*I*/d*t* at the linear region (50 μs) is 1.4×10^{10} A/s.

The voltage drop across the arc from 30 to 84 μs monotonically, but with small oscillations, is increased from 2 to 7.8 kV. At 84th μs, the feature is observed on the voltage oscillogram, which is expressed in a short-term decay and a subsequent increase of the voltage drop. The voltage drop of ~0.9 kV occurs for 9 μs at rise rate of 10^8 V/s, and the voltage peak of 2.8 kV with a rising front of ~14 μs and voltage rise rate of 2×10^8 V/s.

A similar feature, approximately at the same time, is also observed on other voltage oscillograms at experiments under similar initial conditions.

In our opinion, this feature can be due to four reasons:

- Repeated breakdown of the interelectrode gap
- Contact of the arc with the anode surface at point (c) position (d) (Fig. 6.1)
- Impact of a reflected shock wave on the arc
- Impact on the arc of the unloading wave that occurs when the diaphragm is opened.

Repeated breakdown of the interelectrode gap. The assumption that the singularity is caused by breakdown of the interelectrode gap is untenable, since the duration of the voltage drop after its maximum, fixed on the oscillogram, is about 15 μs, then, as is known, that in breakdown, it does not exceed fractions of a microsecond [3]. In addition, in coaxial discharge chambers with a large, as in this case, interelectrode distance and high initial gas pressure, a secondary breakdown of the interelectrode gap does not occur at all, as will be discussed later.

Contact of the arc with the surface of the anode. The possibility of such phenomenon itself raises doubts, but even if it is assumed that it had occurred, then in the place of contact, there should be new spot binding of the arc to the surface of the anode, somewhere near the point (*c*), when the arc occupies the position (*d*) (Fig. 7.1). In its turn, it should lead to a decrease of the voltage drop across the arc caused by a decrease of the arc length, i.e., by shunting. In this case, the voltage drop should be more than that recorded in the oscillogram. If we consider the relationships of the parts resistance of the arc at the moment of contact as linear, then the decrease of the voltage drop across the arc can be ~30%, whereas in our case, it is slightly more than 10%.

The impact on the arc of the reflected shock wave. Another factor that can affect the character of the arcing is the shock wave [4]. To estimate the time of encountering of the reflected shock wave with the moving arc, it is necessary to specify the speed of the arc motion and speed of the shock wave. The speed of arc motion can be determined by empirical relationship for the velocity of the plasma piston v_p in the air at atmospheric pressure [5]

$$v_p = 0.16\left(\frac{dI}{dt}\right)^{0.4}. \tag{7.1}$$

The calculated arc speed for the first 50 μs at dI/dt = 1.5 × 10^{10} A/s is 1800 m/s; then speed is decreased and for the next 40 μs at dI/dt = 5 × 10^9 A/s is 1200 m/s, which gives an average arc speed of ~1500 m/s. The data obtained are of an evaluation nature, since the formula (7.1) does not take into account the gas nature and its density.

The experimental data of our studies at current of 350 kA and initial hydrogen pressure of 5 MPa give the following values of the arc speed: At initial stage of 40 μs speed is 1600 m/s and then to the maximum of current, speed is decreased to 1000 m/s. Thus, the average speed of the arc motion up to the moment of maximal current is ~1300 m/s. If we take into account the difference in the discharge current, we can assume that this value will be somewhat higher. As we see, the values had obtained by calculation and experimentally, taking into account the difference of the initial conditions, are very close. Let us assume that the arc speed is ~1.400 m/s.

The speed of the shock wave was determined at the series of experiments were aimed to determine the time of the diaphragm opening. During the experiments, it was assumed that the cause of the diaphragm opening is the impact on it of the shock wave formed during the explosion of the wire and at the initial stage of expansion and motion of the arc. As a result, it was established that the opening of the diaphragm, i.e., the time of arrival of the shock wave to the diaphragm is 70–80 μs, while the time from the beginning to the full opening of the diaphragm is ~50 μs. Thus, at the distance from the initiating wire to the diaphragm in this experiment equal to 19 cm, the shock wave speed calculated on the moment of the diaphragm opening is ~2400 m/s that corresponds to the Mach number ~1.8.

The calculation of the coordinate, which the reflected shock wave from the opposite surface at the time of 84 μs is reached, indicates that it must be in 4–6 cm distance from the initiating wire, whereas the arc during this time passes the distance of 12–13 cm, and to the moment of its encounter with the arc is corresponded the time ~70 μs, which differs somewhat from the time of the voltage peak appearance.

The impact on the arc the unloading wave arising at the diaphragm opening. The time for which the unloading wave caused by the diaphragm opening reaches the arc is composed of the diaphragm opening time of 70–80 μs and the time of reaching by the unloading wave the arc of 20 μs. Thus, the total time is 90–100 μs, which approximately corresponds to the time when the peak is appeared in the voltage oscillogram.

Thus, only the last two factors: the impact on the arc of the shock and unloading waves are corresponded more or less to the time of the feature appearance.

At the time from 110 to 150 μs, the voltage is raised from 6.7 to 11 kV at the voltage rise rate of 10^8 V/s. The maximum voltage at the 150th μs is approximately by 2.5 kV higher than the voltage at the power supply, which is due to the counter-emf arising in the inductive components of the arc and current lead at the current decrease, at its rise rate of 1.5 × 10^{10} A/s.

At the 150th μs, there seems to be the self-breakdown of the arc section, as at this the voltage across the arc is dropped to 7 kV. In the oscillogram of the current, to the breakdown is corresponded the inflection point, after which the current rise rate is decreased halved (from 1.5 to 0.7) $\times$ 10^{10} A/s.

From the moment of self-breakdown, the current is redistributed between two parallel arcs—in the "new" arc, the current is increased, and in the "old" arc, the current is decreased, at this current rise rate d*I*/d*t* cannot exceed the limiting value of $\sim 3.3 \times 10^{10}$ A/s corresponding to the short circuit mode. The final current rise rate is the reason why the total current in two parallel arcs does not exceed the current at the time of breakdown in the "old" arc but leads only to somewhat slowing of the rate of its decrease. At the 180th and 225th μs, two self-breakdowns of the arc are appeared at 9.5 and 9 kV.

Another characteristic feature of this and a number of similar experiments is the large ($\sim$7 kV) residual voltage of the power supply. The reason for this is a large (10 mm) interelectrode distance, which makes it impossible the repeated breakdowns of the interelectrode gap with subsequent ignition of the arc, and consequently, more complete utilization of the energy accumulated in the battery (this problem will be discussed later).

The dependence of the residual voltage of the power supply on the interelectrode distance for refractory and fusible metals is shown in Fig. 7.3. For both types of electrode metals, the dependence is increasing; however, for refractory metals, the residual voltage is much higher than for fusible metals. In the shown range of the interelectrode distances, the optimal, in our opinion for various reasons, is the interelectrode distance of $\sim$2 mm. This interelectrode distance for fusible and refractory electrode metals is corresponded to residual voltages of 0.5 and 1.5 kV, respectively.

This interelectrode distance is optimal from the point of view of reducing the contamination of the operating gas by the products of electrode erosion at practically equal energy parameters. So, the increasing of the residual voltage reduces the electrodes erosion linearly and, for example, at initial voltage of 7 kV and residual voltage of 1.5 kV, the electrodes erosion is decreased by 20% in comparison with

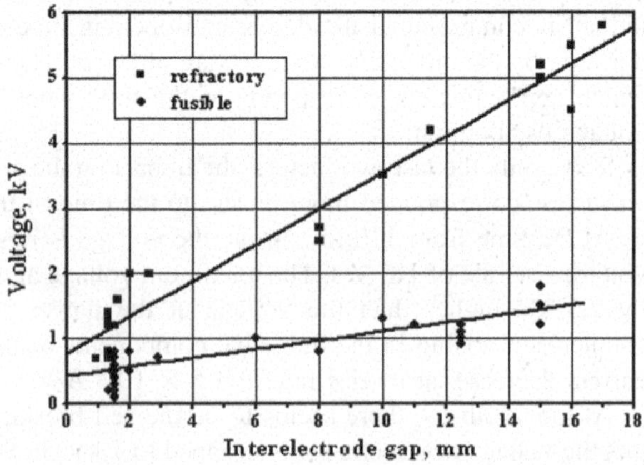

Fig. 7.3 Residual voltage of the power supply versus interelectrode distance [26]

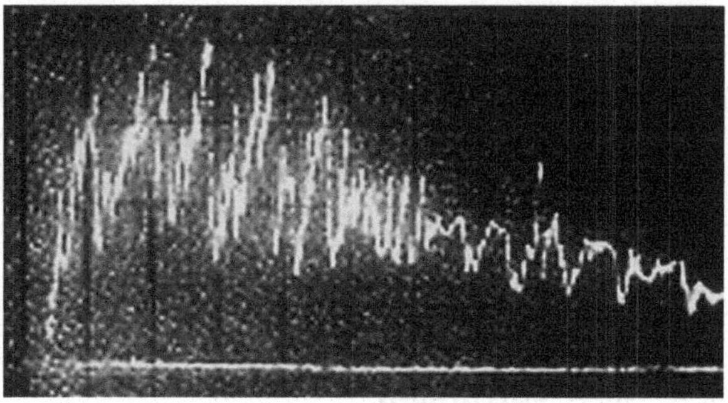

Fig. 7.4 Oscillogram of the voltage drop across the arc [26]. Voltage is 2 kV/div, and time is 100 μs/div

the case when the residual voltage is zero, at this case the unused energy is only ~5% of the initial energy.

At analyze of the oscillograms of the voltage drop across the arc at the most experiments, where the multiple breakdowns are occurred, a criterion is needed according to which it is possible to distinguish the *breakdown* of the interelectrode gap against the self-breakdown of the arc (both of them can occur during the one arcing process). Such a criterion can be the ratio of the post-breakdown voltage to the voltage before the breakdown, let us denote this ratio by the coefficient k_{bd}. If we refer to the oscillogram of the voltage drop across the arc (Fig. 7.2b), then at the breakdown at the 150th μs, the k_{bd} is 0.64. For comparison, let us present the oscillogram of the voltage drop at the experiment performed under similar initial conditions, but differing by that the metal of the electrodes is steel (Fig. 7.4).

The process of arcing in this experiment is accompanied by numerous breakdowns, while the k_{bd} is from 0.38 to 0.64. Statistical analysis of the experimental data shows that the boundary value k_{bd} for the two types of breakdowns is ~0.5. Thus, we assume that the $k_{bd} > 0.5$ corresponds to self-breakdown of the arc, and $k_{bd} < 0.5$ to breakdowns of the interelectrode gap.

Figure 7.5 shows the dependences of the voltage of the first breakdown of the interelectrode gap and of the post-breakdown voltage on the interelectrode distance.

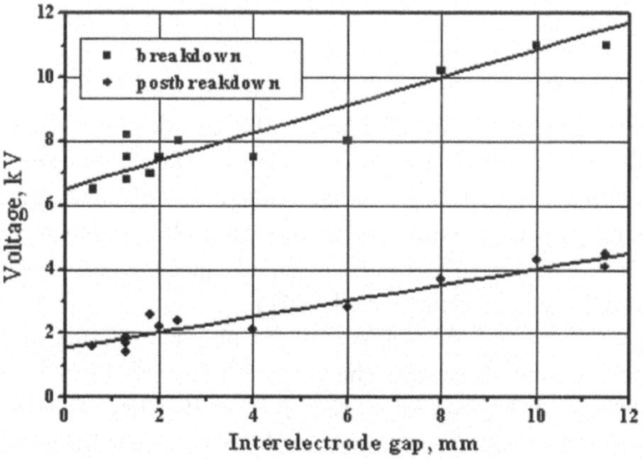

Fig. 7.5 Breakdown and post-breakdown voltages versus interelectrode distance [26]

At determination of those dependencies, the factors that, in one way or another, impact on the parameters of interest to us were identified, and rigid frameworks for them were given. Thus, the dependences are valid for refractory metals at initial pressure $P_0 = 30–40$ MPa, final pressure $P_f = 200–300$ MPa, and discharge current $I \leq 1$ MA.

Figure 7.5 shows that the value of k_{bd} is 0.3–0.4, what in agreement with the chosen criterion value.

7.2 Arc Movement

As mentioned in the introduction to this chapter, the lack of opportunities for optical measurements (because of the high requirements to the strength of the generator discharge chamber) significantly narrowed our understanding of the processes in the discharge chamber of plasma generators, and especially about the arcing process and size of the arc.

Partially, this gap was filled with the results obtained during the research at the diagnostic chamber (Fig. 7.6). The design of the diagnostic discharge chamber (Fig. 3.16) is similar to that of the electric discharge chambers of powerful plasma generators [6]. At simulating the operation of such generators, the camera housing (5) is the anode. The initiating wire was installed in the narrow gap between the conical cathode (7) and the camera housing.

The main task of these studies at the diagnostic chamber was to verify the assumptions about the nature of the arc movement in the discharge chamber of the powerful plasma generators, and about physical parameters and geometric characteristics of the arc.

The diagnostic chamber was powered by the one section of 0.018 F capacitance of the power supply CPS-10. The maximal electrical energy inputted into the arc was 200 kJ, discharge current ~500 kA.

At the diagnostic chamber for the first time by means of the high-speed camera, the images of the evolution and movement of the arc were obtained throughout the entire arcing process. Analysis of the arc motion and changes of its parameters was carried out by the juxtaposition of the oscillograms of the current, voltage drop across the arc, and pulsed pressure in the chamber with photos of the arc.

Figures 7.7 and 7.8 show the oscillograms of the discharge current, voltage drops across the arc and arc images at the experiment when the volume of discharge chamber was 1 dm^3, initial voltage 4.2 kV and initial hydrogen pressure 5 MPa. The pulsed pressure in the discharge chamber was 15 MPa, discharge current was 340 kA, electric energy inputted into the arc was 140 kJ, and pressure of the diaphragm opening was 12 MPa.

Obviously, the values of the above experimental parameters are significantly lower than those are characteristic for the PPG generators, where initial voltage $U_0 = 7–8$ kV, initial pressure $P_0 = 40$ MPa, electric energy inputted into the arc $W_e = 1–2$ MJ, pulsed pressure $P_p = 400–500$ MPa. However, the comparison of the

Fig. 7.6 Diagnostic discharge chamber

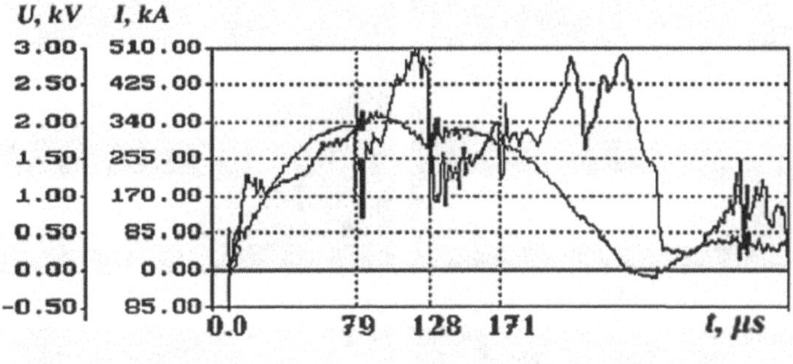

Fig. 7.7 Oscillograms of the current and voltage drop across the arc [26]

oscillograms of the discharge current and voltage drop across the arc lead us to the conclusion that in both cases there are common characteristic features that allow speaking about the similarity of the arcing processes.

Figure 7.8 shows that in the images corresponding to 2-nd and 16th μs, after the wire explosion, the arc is widened, whose diameter to the 16th μs is 6.5 cm, and at this, the front boundary of the arc is moved by the distance of 9.5 cm, which corresponds to the average speed at this time of 1200 m/s. While the speed is average and at the end of the time interval it is 250 m/s, it can be assumed that at the beginning, speed should significantly exceed the average value and can be 2000–2200 m/s and higher, what is consistent the earlier estimates of the speed of

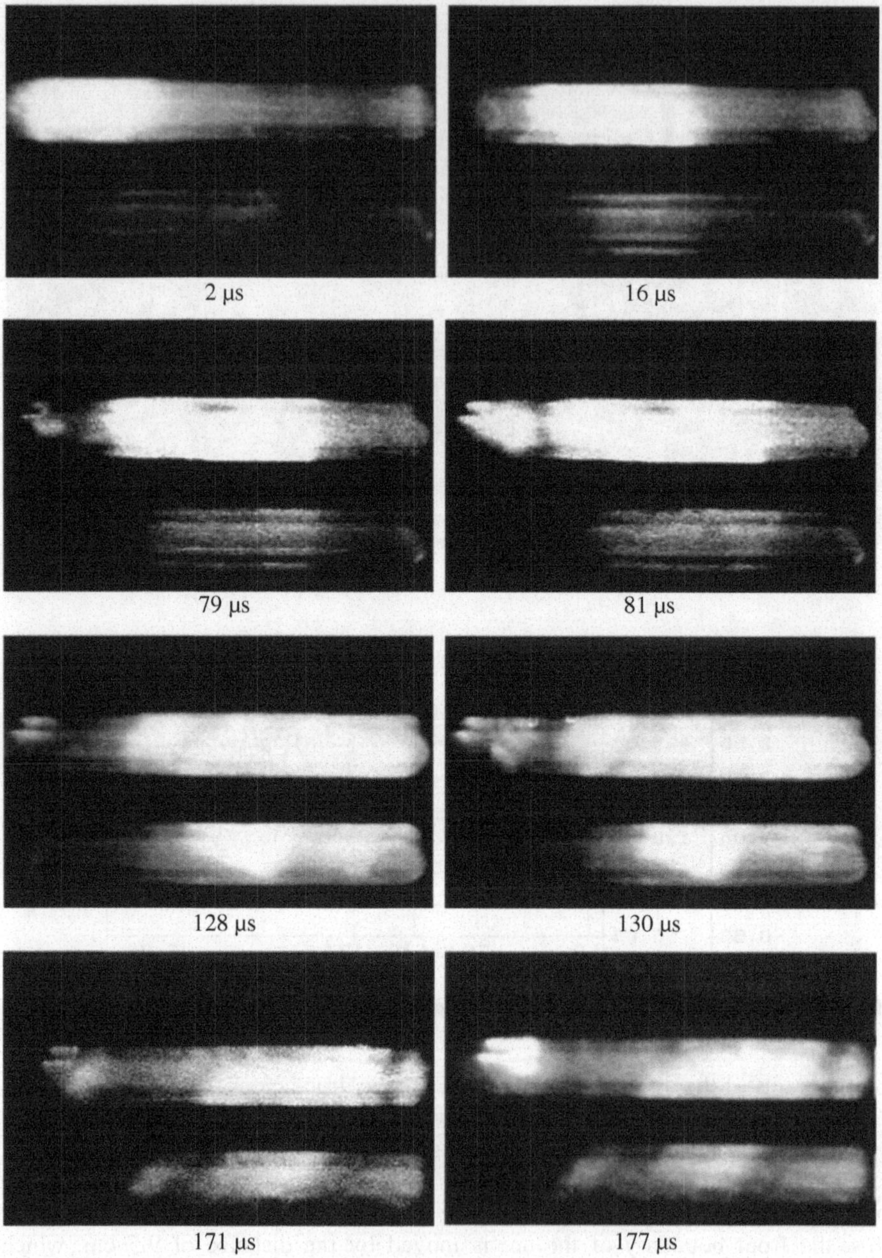

Fig. 7.8 Arcing process in the diagnostic chamber [26]

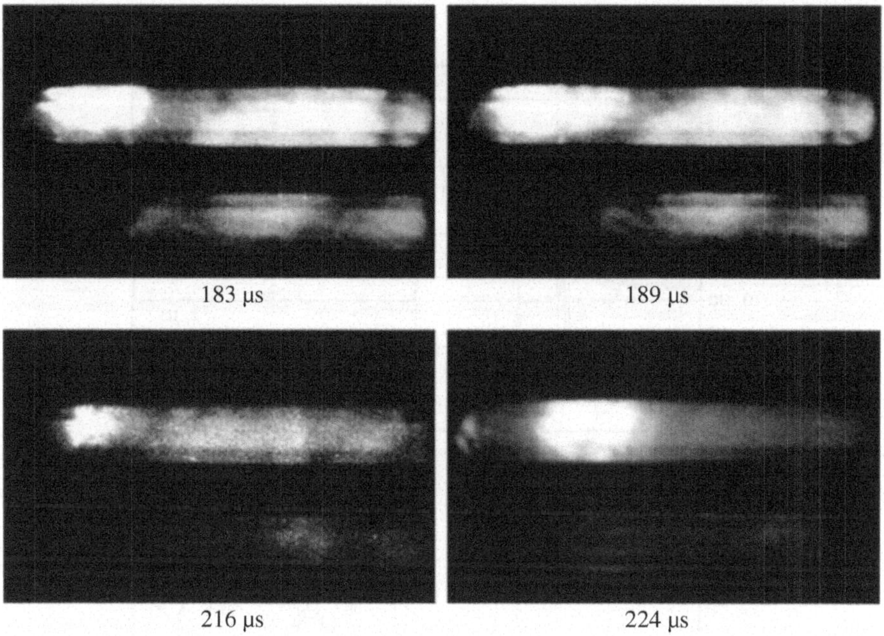

Fig. 7.8 (continued)

the shock wave ~2400 m/s, if we estimate it by the speed of the front boundary of the arc at the initial stage of its formation. The current in this time is ~100 kA, voltage drop across the arc is 1.2 kV.

In the images corresponding to 79, 128, 171, and 216th μs, the moments of breakdowns of the interelectrode gap are shown. The breakdowns are occurred when the arc is burning in the volume of the discharge chamber.

At the 79th μs, the first breakdown of the interelectrode gap is occurred at the current of 340 kA, to which is corresponded the decrease of the voltage drop across the arc from 2 to 0.7 kV and the current growth from 340 to 370 kA. The next breakdown is occurred at 128th μs when the arc occupies the maximal volume and has the longest length at the current of 280 kA. At this, the pre-breakdown voltage is 3 kV, and the post-breakdown voltage is 0.7 kV. At the 216th μs, the third breakdown of the interelectrode gap is observed at the current of 130 kA.

The speed of the front boundary of the arcs moving in the heated hydrogen is 3000–4000 m/s. For the discharge that occurred at the 216th μs, the arc speed is ~2000 m/s.

Figure 7.9 shows the oscillograms of the pulsed pressure in the diagnostic chamber of the same experiment. Two pressure transducers were used: one located just above the initiating wire (T1), and the second (T2) at the distance of 16 cm from the first.

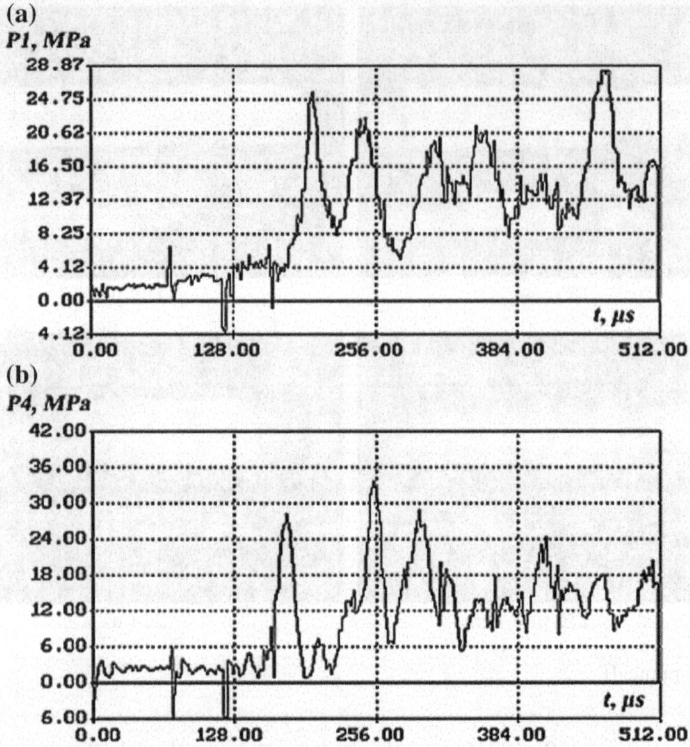

Fig. 7.9 Oscillograms of the pulsed pressure in the diagnostic chamber [26]

The first two peaks in the oscillogram from the T1 transducer (Fig. 7.9a) are corresponded to the compression waves from the discharges caused by repeated breakdowns, they are also observed in the oscillogram from the T2 transducer (Fig. 7.9b). The propagation speed of these waves is ~2700 m/s that correspond to the Mach number of ~2.3 for hydrogen at pressure of 5 MPa and temperature of 300 K.

Attention is drawn to the fact that on the oscillogram obtained from the T1 sensor (Fig. 7.9a), the time of appearance of the first pressure peak exceeds the time of the first peak registered by the sensor T2 (Fig. 7.9b). Naturally, this circumstance causes bewilderment. However, there was the perfectly logical explanation that this peak is corresponded to the first shock wave formed during the arc movement from the T1 sensor coordinate, and thus, it was not registered by this sensor.

7.3 Heat Transfer from Arc to Gas

At the hydrogen pressure >10 MPa and arc temperature >12,000 K, practically all energy transfer is by the radiation [7], and the coefficient of radiant thermal conductivity exceeds by three orders of magnitude the coefficients of electron and atomic thermal conductivities.

To determine the diameter of the arc, it is necessary to estimate the speed of its expansion v_e. In the work [8], the v_e = 102 m/s for the arc in the hydrogen at close parameters P_0 = 40 MPa and I = 30 kA had been registered. The same value of the expansion speed is presented in the work [9]. Based on the theoretical model of the discharge in hydrogen presented in this paper at P_0 = 10 MPa, I = 100 kA, and the discharge duration of 200–250 μs, the arc radius is 2.0–2.5 cm.

In the work [10], where a "quasi-stationary" model of a high-current pulsed discharge in dense gases at current of 1 MA and rise rate of 10^{10} A/s is presented, the radius of the arc r_a = 3 cm. This estimation is in good agreement with r_a = 2.5 cm presented above and obtained under the assumption that the current density is constant.

7.3.1 *Irradiation*

The obtained parameters of the arc make it possible to estimate the contribution of different types of heat transfer from the arc to the gas. To do this, let us set the next initial conditions:

- Volume of the discharge chamber 1 dm^3
- Initial hydrogen pressure 40 MPa
- Discharge current 1 MA
- Electric energy inputted in the arc 1 MJ
- Arc temperature 2.5×10^4 K
- Radius of the arc 2 cm
- Length of the arc 8 cm
- Surface area of the arc 100 cm^2
- Duration of discharge 1 ms.

At a pressure of 100 MPa and temperature of 2 eV < T < 8 eV, the hydrogen plasma radiates as the absolutely black body and energy in the radiation W_r according to the formula (7.2) is 200 kJ

$$W_r = \sigma_B T^4 S_a \tau, \tag{7.2}$$

where σ_B is the Stefan–Boltzmann constant, T is temperature of the arc, S_a is surface area of the arc, τ is duration of discharge.

Consequently, by the radiation ~20% of the electric energy inputted in the arc is transferred. In order to estimate the fraction of the radiation energy consumed on heating of the gas, we use the dependence of the absorption coefficient of the radiation kv by hydrogen on the energy of quantum (Fig. 7.10).

The graph shows the dependence of kv on the energy of quantum hv in the continuous spectrum at hydrogen pressure of 100 MPa and temperature of 2000 K (curve 1), 3000 K (curve 2), and 4000 K (curve 3).

The estimation was performed under the assumption of local thermodynamic equilibrium in accordance with the data of the work [11]. Figure 7.10 shows that kv has two ranges of values: $kv < 0$ for $hv < I_H$ (where $I_H = 13.6$ eV is the ionization potential of hydrogen) and $kv > 1$ for $hv \geq I_H$.

Thus, the low-frequency radiation from the arc almost unimpeded reaches the wall of the discharge chamber, while the high-frequency radiation is absorbed by the hydrogen near the arc. The fraction of absorbed energy is

$$\Omega(T) = \frac{1}{\sigma_\Lambda T^4} \int_{I_{H/h}}^{\infty} J_{pv}(T)\mathrm{d}v, \quad (7.3)$$

where $J_{pv}(T)$ is the Planck frequency (v) distribution of energy of the absolutely black body at temperature T.

Figure 7.11 shows the dependence of the fraction of energy absorbed by hydrogen $\Omega(T)$ on the arc temperature.

At arc temperature of 25,000 K, ~40% of the radiated energy is heated the gas, the rest is absorbed by the wall of the discharge chamber. The value of $\Omega(T)$ is somewhat lowed since the energy absorbed in hydrogen and metal lines is not taken

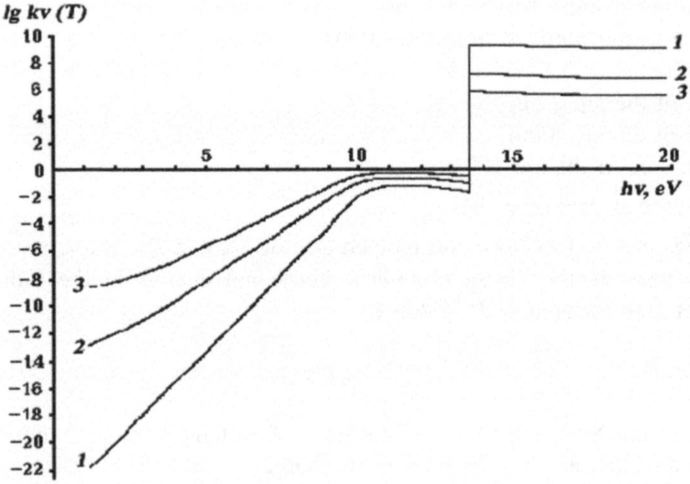

Fig. 7.10 Coefficient of absorption of radiation by hydrogen versus energy of quantum [26]

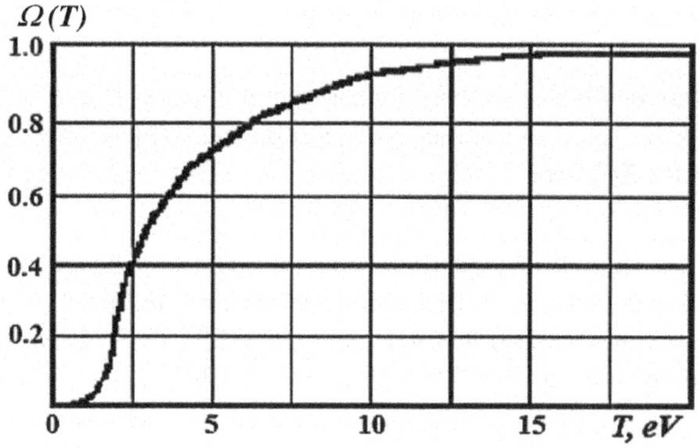

Fig. 7.11 Dependence of fraction of the arc radiation energy absorbed by the hydrogen versus arc temperature [26]

into account. Estimation of this energy shows that up to 50% of the radiated energy can be absorbed in the chamber volume, however, and in this case the fraction of arc energy transferred to the gas by radiation $\lesssim 10\%$.

7.3.2 *Thermal Conductivity*

The density of the energy flux from the arc to the gas by means of thermal conductivity

$$I_0 = \lambda_{\sum} \mathrm{grad} T. \tag{7.4}$$

where $\lambda_{\sum}$ is sum of the coefficients of the atomic λ_a, electron λ_e, and dissociative λ_d thermal conductivities (the ionization thermal conductivity at such concentrations can be neglected).

Assuming $T = 25{,}000$ K and $n = 3 \times 10^{20}$ cm^{-3}, the coefficients of thermal conductivity can be calculated by the formulas from the work [12]:

$$\lambda_a = l_a v_a \rho c_v / 3, \tag{7.5}$$

$$\lambda_e = 1.83 \times 10^{-12} \frac{T^{5/2}}{\ln \Lambda}. \tag{7.6}$$

where $l_a = (n\sigma_a)^{-1}$ is mean free path of atom, $v_a = \sqrt{\frac{8kT}{\pi m_a}}$ is average thermal speed of atom, ρc_v is specific heat capacity, $\ln\Lambda$ is the Coulomb logarithm (~ 1.5).

The coefficient of thermal conductivity due to dissociation and recombination of molecules in the transition layer, according to the data of the work [12], is connected with the coefficient λ_a by the relation

$$\lambda_{\mathrm{d}} = \lambda_{\mathrm{a}} \times 0.1(D/kT_{\mathrm{D}}) = 0.9\lambda_{\mathrm{a}}. \tag{7.7}$$

where D is dissociation energy of the hydrogen molecule (4.48 eV), T_D is temperature at which the hydrogen dissociation degree $\alpha_d \sim 0.5$ (~ 0.5 eV) at pressure of 100 MPa.

Thus, the sum of the coefficients $\lambda_{\sum} = 0.14$ W/(cm × K). The maximal value of the temperature gradient can be estimated based on the data of high-current discharge studies in the helium, where the thickness of the transition layer between the arc and the cold gas was determined by the optical method $\Delta x \sim 3 \times 10^{-3}$ cm [9]. Considering that the temperature difference between the arc and the gas is $\Delta T \sim 23{,}000$ K, we obtain grad$T = 8 \times 10^6$ K/cm.

Energy transferred to the gas by heat conduction

$$W_{\mathrm{tc}} = I_0 S_{\mathrm{a}} \tau = 100\,\mathrm{kJ}. \tag{7.8}$$

That is 10% of the energy inputted in the arc.

7.3.3 *Turbulent Heat Transfer*

The main mechanism of energy transfer from the high-current arc to the gas, when the arc temperature does not exceed the temperature of the hydrogen more than by 3×10^4 K, is turbulent heat transfer [13]. It is known that for large Rayleigh numbers, the laminar motion of the gas becomes the turbulent. In this case, the Rayleigh and Grashof numbers turn out to be of the same order of magnitude [14]. The Grashof criterion Gr

$$\mathrm{Gr} = \frac{g d^3 \rho^2 (T_{\mathrm{a}} - T_{\mathrm{w}})}{\eta^2 T_{\mathrm{w}}}. \tag{7.9}$$

where g is acceleration of gravity, d is diameter of the arc, ρ is gas density, η is dynamic viscosity coefficient, T_a is surface temperature of the arc, and T_w is wall temperature of the discharge chamber.

Substituting in the formula (7.9), the average values: $d = 3$ cm, $\rho = 5 \times 10^{-3}$ g/cm^3, $\eta = 2.5 \times 10^{-4}$ g/(cm×s), $T_a = 25{,}000$ K, $T_w = 2000$ K, we obtain Gr $\sim 10^8$. To numbers of Gr $\geq 10^4$ the developed turbulent gas flow corresponds [14]. The energy transferred by turbulent heat transfer

$$W_{\mathrm{t}} = \lambda_{\mathrm{t}} \frac{\partial T}{\partial r} S_{\mathrm{a}} \tau. \tag{7.10}$$

where $\lambda_t = \rho c_V D_t$ is coefficient of turbulent thermal conductivity, $D_t = l_t v_t$ is coefficient of turbulent diffusion, l_t is scale of turbulence, v_t is average speed of inhomogeneities motion in direction of the temperature gradient [14].

The value of v_t cannot be less than the speed of the arc expansion, i.e., $v_t \geq 10^2$ m/s; as for l_t, the exact estimating seems difficult; however, it can be assumed that the scale of turbulence is of the same order of magnitude as the thickness of the transition layer at turbulent heat transfer, and the expression for W_t can be simplified by setting

$$l_t \frac{\partial T}{\partial r} \approx l_t \frac{\Delta T}{\Delta r} \approx \Delta T. \qquad (7.11)$$

Then for $\rho = 5 \times 10^{-3}$ g/cm^3, $v_t = 10^2$ m/s, $\Delta T = 23{,}000$ K, $S_a = 100$ cm^2, and $\tau = 1$ ms, we obtain $W_t = 700$ kJ. Such estimation of W_t is approximate, but it allows us to conclude that turbulent heat transfer is the main share of the energy inputted in the arc and transferred to the gas. The maximal electric power of the arc, at its surface area $S_a = 200$–300 cm^2, reaches 3–5 GW.

7.3.4 Shock Waves

Along with the heat transfer mechanisms presented above, a certain role is played the shock waves. Depending on the experimental conditions and the gas, the contribution of shock waves to heat transfer is varied significantly.

At increase of the energy input and the initial hydrogen pressure in the chamber of the PPG-3 plasma generator (Fig. 3.9) at current rise rate of 10^{10} A/s, the increase of the pressure in the front of the shock wave was observed. At discharge current of 900 kA and initial hydrogen pressure of 40 MPa, the shock wave with the pressure in the front of 100 MPa with Mach number 1.4–1.5 was registered. However, as will be shown below, the energy transferred by the shock waves in hydrogen does not exceed a few percent of the energy inputted in the arc, and, therefore, cannot be an essential factor for hydrogen heating. At the diagnostic chamber (Fig. 3.16) at initial hydrogen pressure of 5 MPa, discharge current up to 450 kA, and current rise rate of 10^{10} A/s, shock waves with the Mach number of 1.2 were registered.

The heat transfer by shock waves in the helium is significantly different than in the hydrogen. On the low-inductance electric discharge installation (Fig. 3.14) the pulsed arc in the helium at initial pressure up to 15 MPa and in the air at atmospheric pressure was studied [15]. At initial voltage of 50 kV and discharge circuit inductance of 80 nH, the current rise rate was 6×10^{11} A/s. At initial pressure up to 15 MPa, discharge current of 600 kA, and current rise rate of 6×10^{11} A/s, the shock waves are moved with Mach number of $\sim$5 that at radius of the discharge chamber 1–2 cm due to their multiple reflections from the wall provide the main heating of helium. At this, the pulsed pressure on the arc axis was 440 MPa [16].

7.3.5 Shock Waves in Hydrogen

The high-current arc in the high-density gas is the source of the shock waves. The shock waves are transferred some share of the electric arc energy, heating the gas in the electrodischarge chamber.

Figure 7.12 shows the oscillogram of the pulsed pressure in the electrodischarge chamber of the plasma generator PPG-3.

The front pressure jump (Fig. 7.12) is corresponded to the front of the shock wave, by the magnitude of which will estimate the fraction of energy transferred by the shock wave. The relative change of the temperature when the shock wave passing through the gas is expressed by the formula

$$\frac{\Delta T}{T_0} = \frac{2(\gamma - 1)}{(\gamma + 1)^2 M_1^2} \left(M_1^2 - 1\right)\left(1 + \gamma M_1^2\right). \qquad (7.12)$$

where T_0 is initial temperature of the gas; γ is adiabatic index; M_1 is ratio of the shock wave speed to speed of sound in unperturbed gas.

The typical speed of the shock wave in the plasma generator chamber is 1600 m/s; then $M_1 = 1.25$, and for $\gamma = 1.4$, we have: $\Delta T/T_0 \sim 0.16$; i.e., the gas temperature increase by $\sim 16\%$.

7.3.6 Shock Waves in Helium

The pulsed arc in helium was investigated at the low-inductance installation (Figs. 3.14 and 3.15). At initial pressure up to 15 MPa, the arc was initiated by the jet of helium plasma with radius of ~ 0.1 cm from the anode-injector (Fig. 3.15b). As the results of the calculations presented in the work [17] show, this method of

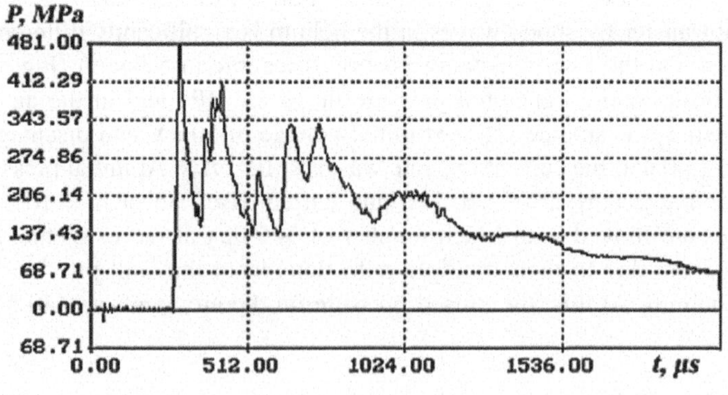

Fig. 7.12 Oscillogram of pulsed pressure in the electrodischarge chamber of the plasma generator PPG-3 [26]

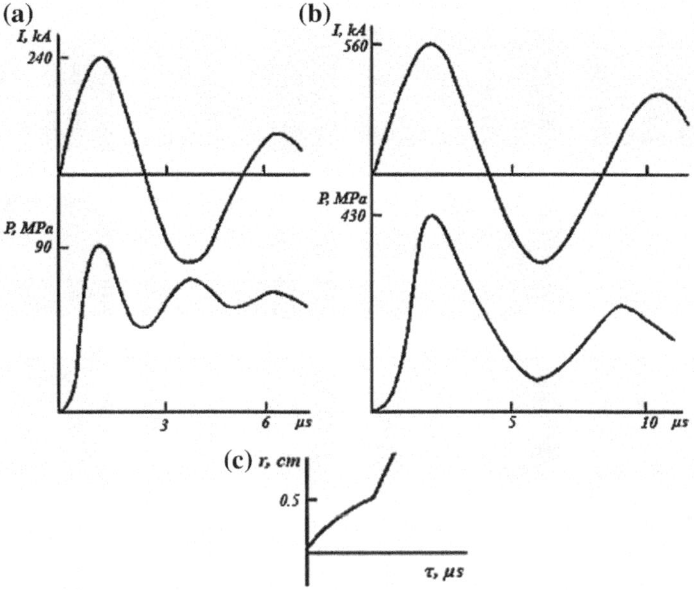

Fig. 7.13 Graphs of discharge current (**a**), pressure (**b**), and photostreak of the glow of the arc in helium (**c**) [27]

initiating of the interelectrode gap, except more stable breakdown due to the presence of metal vapors, increases the temperature and pressure in the arc. The radius of the discharge chamber varied from 0.9 to 2.0 cm.

Figure 7.13a shows the experimental data at initial helium pressure of 1 MPa, capacitance of the power supply is 7.4 μF, initial voltage is 35 kV, and interelectrode distance is 4.2 cm. The discharge was initiated by self-breakdown.

Figure 7.13b shows the experimental data at initial helium pressure of 10 MPa, capacitance of the power supply is 22.2 μF, initial voltage is 50 kV, and interelectrode distance is 3 cm. The discharge was initiated by the plasma injector (Fig. 3.14).

Figure 7.13a shows the graph of current and pressure at the arc axis, at initial helium pressure of 1 MPa, here the second maximum in the pressure plot is the synchronous with the current maximum in the second half-period. At initial helium pressure of 10 MPa (Fig. 7.13b), the second maximum of pressure comes later and corresponds to the front of the shock wave reflected from the wall of the discharge chamber.

In the graph of the arc glow (Fig. 7.13c) obtained at initial pressure of 10 MPa and current of 560 kA, there is the curve inflection point corresponding to the sudden change of the arc expansion speed (from 2 to 5.7) $\times$ 10^3 m/s that coincides with the beginning of the decrease of the discharge current after maximum.

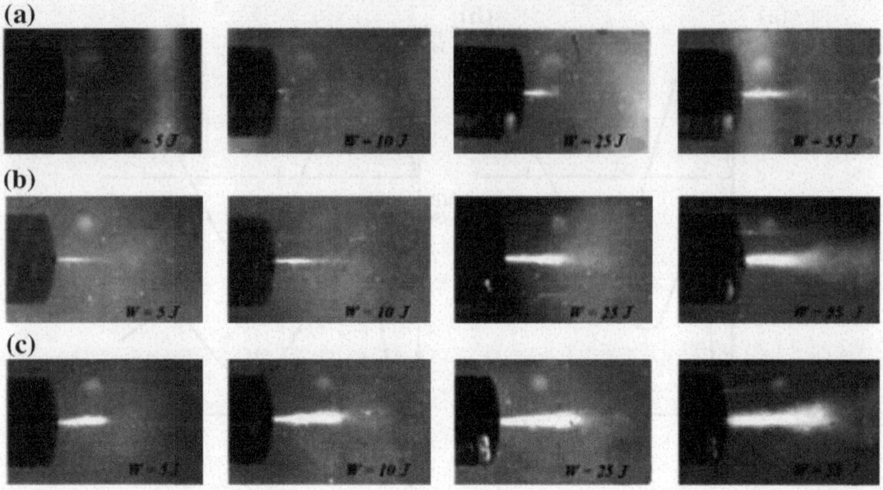

Fig. 7.14 Plasma jet at various inputted energy *W* and diameter of outlet hole a—diameter 0.4 mm, b—diameter 0.7 mm, c—diameter 1.0 mm [27]

The gas heating was registered by the pressure sensors located on the axis and wall of the discharge chamber. At the research, the diameter of the discharge chamber and the energy inputted in the arc were varied (Fig. 7.14).

The front of the pressure on the chamber wall lags by 4–7 μs relative to the pressure on the axis (Fig. 7.15), what corresponds to the expansion speed of the high-pressure zone (3–5) × 10^3 m/s (Fig. 7.16) and agrees with the expansion speed determined from the photo scanning. The capacitance of the power supply at these experiments was 7.4 μF. The curve (1) corresponds to initial voltage of 30 kV and discharge current of 240 kA, curve (2) to 50 kV, and 390 kA.

Fig. 7.15 Graphs of pressure on the axis P_0 and wall of the discharge chamber P_r [27]

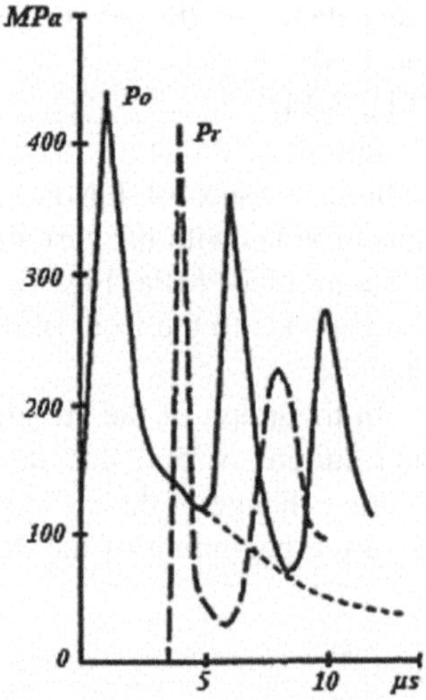

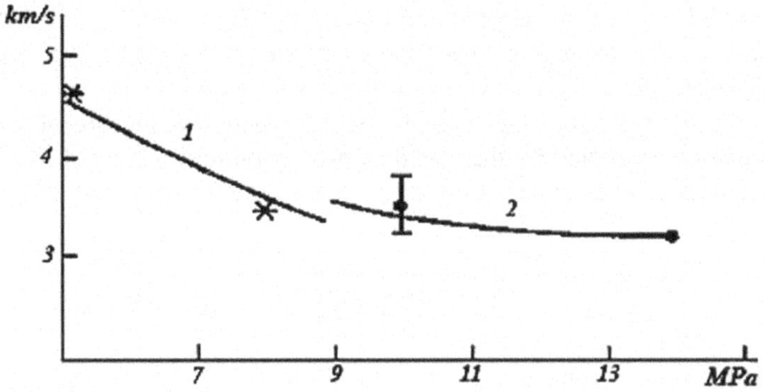

Fig. 7.16 Speed of the shock wave propagation versus initial pressure of helium [27]

At the radius of the discharge chamber of 1.5–2.0 cm, there is no noticeable equalization of the pressure during the discharge and shock waves are fixed on the chamber wall; their amplitude is comparable with the pressure at the arc axis. At radius of the discharge chamber of 0.9 cm, the pressure drop on the arc axis after the current pulse is negligible.

The change of the pressure in the chamber on the arc axis and wall of the chamber during the heating of the gas is in qualitative agreement with the results of the calculations presented in the work [16]. As is known, for a pinch in radius r_0 with uniform current density, the distribution of pressure on radius r is expressed as:

$$P(r) = P_{\max}\left(1 - \frac{r^2}{r_0^2}\right) \tag{7.13}$$

Then for average pressure on the square with radius r_1, it will be:

$$\bar{P} = \int_0^{r_1} \frac{2\pi r p(r) \mathrm{d}r}{\pi r_1^2} = 2\pi \int_0^{r_1} \frac{P_{\max}\left(r - \frac{r^3}{r_0^2}\right)\mathrm{d}r}{\pi r_1^2} = P_{\max}\left(1 - \frac{r_1^2}{2r_0^2}\right), \tag{7.14}$$

where $\bar{P}$—means value averaged over square of the sensor's rod.

$$\bar{P} = P_{\max}\left(1 - \frac{r_1^2}{2r_0^2}\right) = 3.2 \times 10^{-4}\frac{I^2}{r^2}\left(1 - \frac{r_1^2}{2r_0^2}\right). \tag{7.15}$$

The pressure determined by formula (7.15) is 380 MPa, The pressure registered by the sensor at the distance from the arc axis $r_1 = 0.2$ cm at the current $I = 570$ kA is 440 MPa, what is close to the calculated value for a pinch with uniform current density in the arc of 0.5 cm radius determined from the photo scanning.

The active resistance of the arc at discharge current of 200–600 kA is $(3\text{–}5) \times 10^{-2}$ Ω. In this case, the average electric field intensity in the arc is increased slightly at increasing of initial helium pressure from 2 to 15 MPa and is (4–7) kV/cm. As the distance between the electrodes is decreased, the electric field intensity is increased.

The average temperature and density of the plasma in the arc at the current maximum can be estimated by the speed of propagation of the shock wave and by the pressure at the current maximum. Using the formula from the work [16], we have

$$\frac{P_a}{P_i} = \frac{2\gamma_i M_i^2 - (\gamma_i - 1)}{(\gamma + 1)} \left\{ 1 - \frac{(\gamma_a - 1)a_a}{(\gamma + 1)a_a} \left(M_i - \frac{1}{M_i} \right) \right\}^{-\frac{2\gamma_a}{(\gamma_a - 1)}}. \tag{7.16}$$

where P_a is helium pressure in the arc (400 MPa), P_i is initial helium pressure (10 MPa), γ_i and γ_a are adiabatic indexes (1.67), a_i is speed of sound in the chamber ($\sim 10^3$ m/s), a_a is speed of sound in the arc.

According to the photo scanning, the maximum arc expansion speed is 5.7×10^3 m/s, what is corresponded to the Mach number $M_i = 5.38$. By the formula (7.16), we determine the speed of sound in the arc a_a, and then the arc temperature, assuming the helium is the ideal and completely ionized gas [16].

Speed of sound in the arc

$$a_a \equiv c = \sqrt{\frac{\gamma(1 + Z)kT}{A}}, \tag{7.17}$$

where Z is charge number, c is speed of sound in the arc, A is atomic weight of helium.

At $Z = 2$ and $c = 6.4 \times 10^4$ m/s, $A = 4 \times (1.67 \times 10^{-24})$ g, from the formula (7.15) we have $T = 3.4 \times 10^5$ K, and from the relation $P = (1 + Z)\ n_i kT$, $N_i = 2.4 \times 10^{19}$ cm^{-3}.

The specific conductivity of the helium plasma of 100–200 $(\Omega \times \text{cm})^{-1}$ obtained from the experiment is below then Spitzer plasma. At increasing of temperature, the number of Coulomb collisions is decreased and the magnetic field of the arc current begins to effect on the conductivity of the arc plasma.

7.3.7 *Shock Waves in Air*

In addition to data on the shock waves in the hydrogen and helium, we present here the results of investigations at discharges in the air. The discharge in the air at atmospheric pressure was initiated by the plasma injection from the anode and self-breakdown. The discharge chamber radius at these experiments was 1.5 cm, initial voltage 20 kV, capacitance of the power supply 7.4 μF, and interelectrode distance 1 cm.

On the photo scanning of the arc glow at the initial stage, the weak glow of the front of the shock wave propagating at speed of 10^4 m/s was registered. The shock wave occurs during the breakdown of the interelectrode gap. At the same time, the arc is expanded at speed of $\sim 10^3$ m/s.

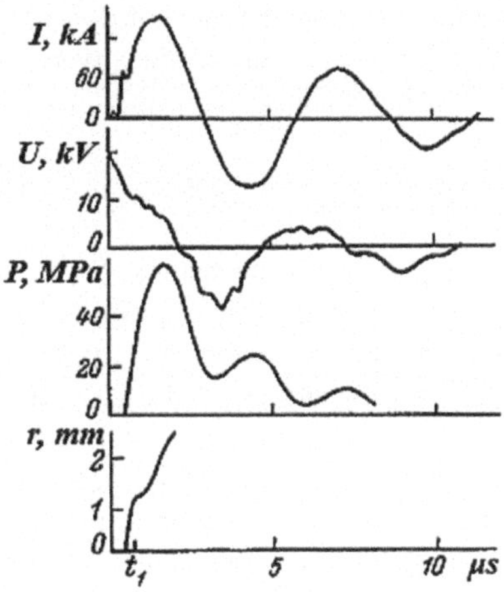

Fig. 7.17 Graphs of the current, voltage drop, pressure, and photo scanning of the discharge glow in the air at atmospheric pressure [27]

At the initial stage of arc expansion, the measured radii and pressure are consistent with those calculated from the theory of the thermal wave proposed in the work [18]. After the time t1 (Fig. 7.17), the rate of the intense glow zone expansion is changed due to the impact of the arc own magnetic field. Since the current rise rate in our case is higher than in the work [18], stage of the arc expansion is occurred earlier.

As is known, for a pinch with a uniform current density, if the magnetic and gas-kinetic pressures are equal, the average cross section pressure

$$\bar{P} = 1.6 \times 10^{-4} \frac{I^2}{r^2}, \tag{7.18}$$

where $\bar{P}$ is average pressure in cross section of the arc (Pa), I is current (A), and r is arc radius (cm).

At the time t_1, the arc radius $r = 0.12$ cm, discharge current $I = 60$ kA (Fig. 7.16), what corresponds to the pressure of 40 MPa by formula (7.18). Since the radius of the arc r at this time is smaller than the radius r_p of the probe measuring pressure placed into the electrode ($r = 0.12$ cm, $r_p = 0.165$ cm), the pressure from the pressure probe is understated by $(r_p/r)^2$ times. With this correction, the values of the calculated and measured magnetic pressure are close to each other. At the time corresponding to the maximum current (144 kA), the radius of the glowing zone $r_z = 0.25$ cm completely overlaps the end of the pressure probe.

It is easy to show that for the pinch of radius r with the uniform current density, the average pressure measured at the radius r_1 corresponding to the radius of the probe is expressed by the formula (7.15). Substituting the values of r and I, corresponding to the maximal current, we have $\bar{P} = 83$ MPa, at this measured pressure $\bar{P} \sim 60$ MPa.

Measurements of the pressure on the arc axis at radii r_1 and r_2 make it possible to estimate the distribution of the current density along the radius of the arc. From formula (7.15), the ratio of the two average pressures measured at radii r_1 and r_2 at constant radial current density is

$$\frac{\overline{P_1}}{\overline{P_2}} = \frac{2r_0^2 - r_1^2}{2r_0^2 - r_2^2}. \tag{7.19}$$

The ratio of pressures $\overline{P_1}/\overline{P_2}$ at the current of 144 kA, measured at radii $r_1 = 0.1$ cm and $r_2 = 0.2$ cm, is 1.6. Using formula (7.19), we determine the radius of the conduction zone $r_0 = 0.21$ cm, which is close to the radius determined from the photo scanning of the glow and equal to 0.23 cm.

Since for the current of 144 kA, its density over the cross section is constant and the current radius is close to the light one, then the conductivity, which is 200 $(\Omega \times \text{cm})^{-1}$, can be found from the measured arc resistance.

The radiation spectrum of the arc is continuous with absorption lines of the Fe, Mn, and Cr. For the current of 144 kA, the arc's active power is about 1 GW. The power balance is fulfilled if it is assumed that the arc radiates as the absolutely black body with the radius of 2 mm and surface temperature of $\sim 10^5$ K.

Let us define the parameters of the arcing in the air (Fig. 7.17), at the time corresponding to the maximal current by the conductivity σ and pressure P measured along the axis, using the data of the work [19].

At the current 144 kA, conductivity is 200 $(\Omega \times \text{cm})^{-1}$, and pressure is 77 MPa, the ion concentration $n_i = 1.1 \times 10^{19}$ cm^{-3}, temperature $T = 10^5$ K, and average ion charge $Z = 3.2$. According to the data of the work [20], $\ln\Lambda = 5.2$. The conductivity of the plasma σ in the magnetic field

$$\sigma = \frac{\sigma_0}{1 + \frac{\omega^2}{\nu^2}}. \tag{7.20}$$

where σ_0 is conductivity in absence of magnetic field, ω is Larmor frequency, and ν is frequency of Coulomb collisions of electrons with ions.

Since for this case, $\omega = 2.4 \times 10^{12}$ s^{-1} and $\nu = 4.8 \times 10^{13}$ s^{-1}, the plasma can be considered as unmagnetized. Indeed, for these parameters, the Spitzer conductivity is 290 $(\Omega \times \text{cm})^{-1}$, which is close to the experimental value of the conductivity.

The coincidence of the temperature calculated on the radiation of the arc as an absolutely black body with the temperature determined from the balance of the power and conductivity indicates the reliability of the measured voltage.

At the current of 290 kA based on the experimental data, we have $\frac{\overline{P_1}}{\overline{P_2}} > 2$, what does not correspond to the condition of constant current density over the cross section of the arc. This means that at the radius r_1 smaller than r_2, the current density is higher. At the same time, on the radius $r_1 = 0.1$ cm the average pressure is of 170 MPa, that under assuming the plasma density in the center of the arc be lower than at atmosphere pressure, corresponds to the plasma temperature of $\geq 10^5$ K.

7.3.8 *Impact of Electrode Jets on Heat Transfer*

As the result of the studies of the pulsed arcs in hydrogen, helium, nitrogen, and argon at d*I*/d*t* ~ (1–3) × 10^8 A/s and initial pressure of 0.1–4.0 MPa, the arc temperature, speed of the arc expansion, and the heat transfer efficiency from the arc to gas had been determined [21–23].

Figure 7.18 shows the photo scanning of the arcs and Fig. 7.19 shows the graphs of the arc temperature changing in the argon, helium, and hydrogen.

Data on the speed of the arc expansion in various gases determined by the intensity of the glow (*I*), and signal from the pressure transducer (*P*) are presented in Table 7.1.

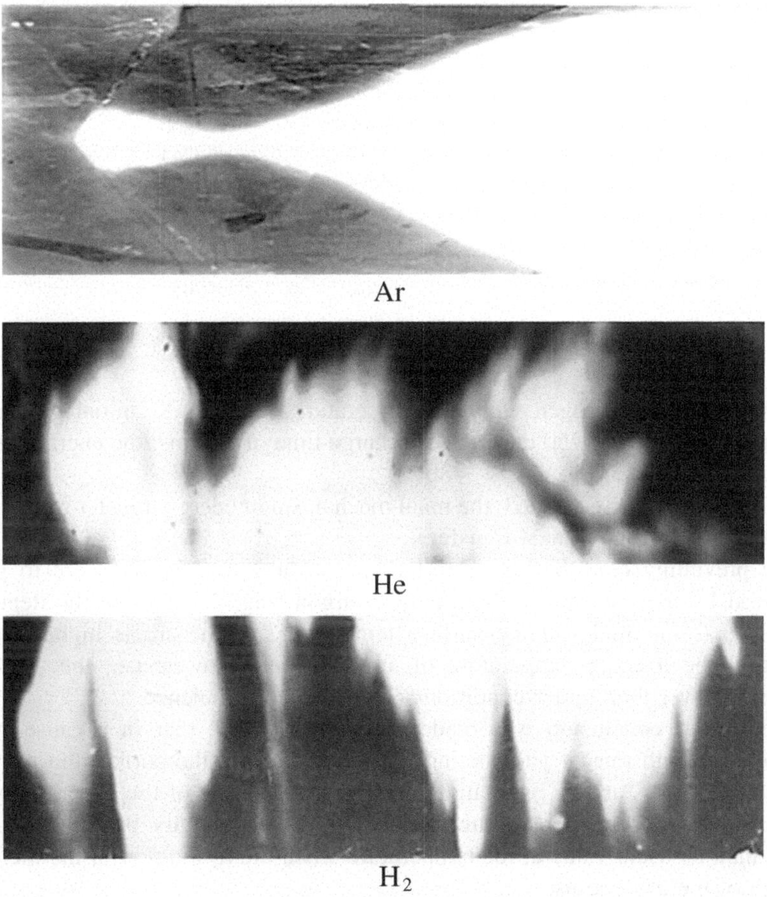

Fig. 7.18 Photo scanning of the arcs in argon, helium, and hydrogen [27]

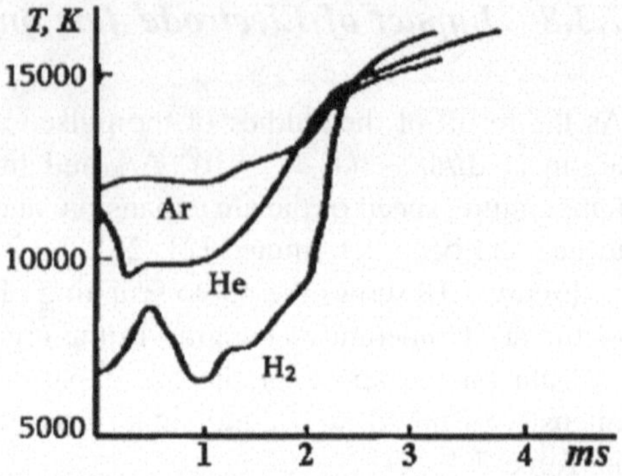

Fig. 7.19 Arc temperature changing in argon, helium, and hydrogen [27]

Table 7.1 Speed of the arc expansion

Gas	P_{in} (MPa)	V_w (m/s)	
		I	*P*
H_2	0.18	–	175
	0.6	152	150
	1.0	210	200
	1.25	–	250
He	0.13	600	–
	1.0	244	200
	1.5	164	–
	2.1	190	–
Ar	0.19	192	280
	0.54	161	–
	0.96	97	–
	2.0	76	–

For the arc in hydrogen with the temperature of $\sim 10^4$ K, initial pressure of 1 MPa, surface area of 250 cm^2, and discharge time of ~ 1 ms, the energy transfer efficiency from the arc to gas is $\sim 80\%$.

As was shown in Sect. 7.3.3, the main mechanism of energy transfer from the arc to hydrogen is turbulent heat transfer.

The prevailing contribution of turbulence to heat transfer from arc to gas is evidenced by the fact that at the arcing in argon (Fig. 7.18) when the degree of turbulence is minimal and the surface temperature of the arc is higher than in hydrogen, the average temperature of argon is much lower; i.e., the total heat transfer is lower than under conditions of developed turbulence.

The similar conclusion was made in the work [24], that in a cause of the anomalously high rate of gas cooling after extinction of the current the turbulent gas-dynamic flows arises. Estimations made by the results of the experiment give the value of the coefficient of turbulent thermal conductivity by hundred times higher than its usual value at the temperature when the restoration of the dielectric strength of the gas begins.

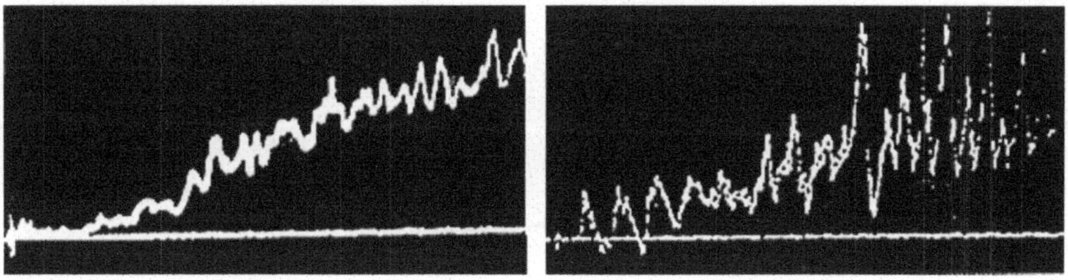

Fig. 7.20 Increase of the amplitude of pressure oscillations at the discharge in hydrogen at interelectrode distance decrease [27]

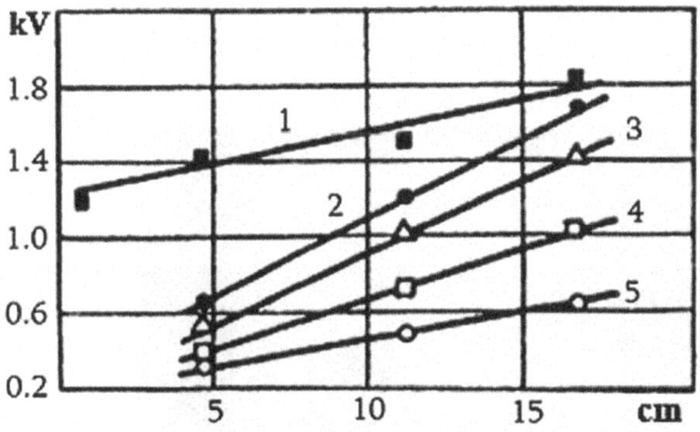

Fig. 7.21 Voltage drop across the arc versus interelectrode distance [27]. 1—is hydrogen, 2—is helium, 3—is nitrogen, 4—is argon, 5—is vacuum

To the turbulent regime of heat transfer at the arcing in the hydrogen is corresponded the voltage and pressure oscillations [21, 22], which increase the efficiency of gas heating at the interelectrode gap decreasing (Fig. 7.20). This phenomenon was detected at the PPG-1 generator at the discharge current of 80 kA, interelectrode distance from 5 to 30 cm, initial hydrogen pressure of 1.6 MPa, and duration of the first half-period of the current of 2 ms.

The arcing in hydrogen differs from that in other gases by significantly higher near-electrode voltage drops of ~ 1 kV (Fig. 7.21).

As the atomic number of the gas is increased, the degree of turbulence is decreased, and turbulence thus plays less important role in heat transfer from the arc to gas. Thus, on the image of the photo scanning of the glow of the arc in helium, turbulence is still existed, but in the argon the expansion of the arc is symmetrical (Fig. 7.18).

One of the reasons for the turbulence occurrence was established at the studies performed at the diagnostic chamber; this is interaction of anode and cathode jets. It can be assumed that electrode jets had being appeared at the generator PPG-1 at operation on hydrogen at initial pressure of 0.6 MPa, discharge current of 80 kA, and half-period of 1 ms (Fig. 7.20). The decrease of the interelectrode distance was accompanied by increase of the final hydrogen pressure (Fig. 7.22).

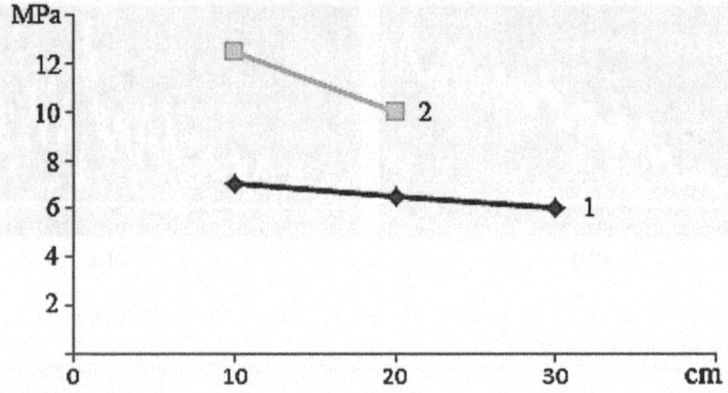

Fig. 7.22 Pressure in the chamber of the generator PPG-1 versus interelectrode distance [27]. 1—steady-state pressure, 2—pressure at maximal current

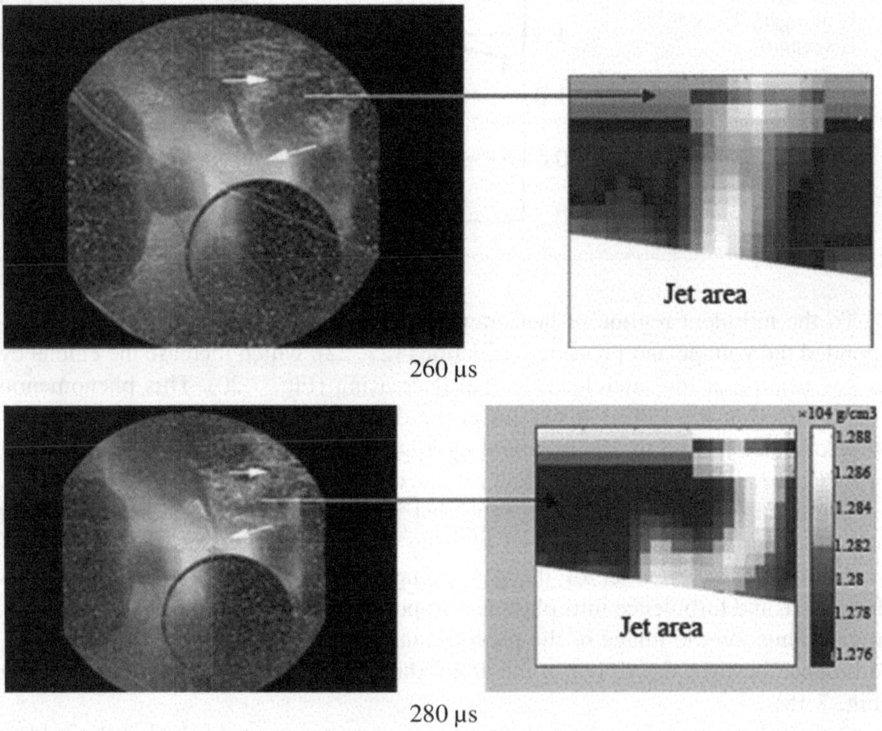

Fig. 7.23 Gas flow in the diagnostic electrodischarge chamber [27]

These curves are consistent with the nonmonotonic increase of the pressure and voltage drop across the arc due to the interaction of the anode and cathode jets (Fig. 5.13).

In addition, electrode jets are the some kind of "fan" pumping gas through itself, which also efficiently mix the gas in the discharge chamber.

This effect is demonstrated in Fig. 7.23. Experiment conditions are next: tungsten electrodes of 6-mm diameter, interelectrode distance 12 mm, gas is air, maximum current 110 kA, and anode on left, cathode on right.

The distribution of gas density was determined from the shadow photographs under the assumption of cylindrical symmetry. The gas surrounding the arc is sucked at the base of the jet, heated, ionized, and ejected into the discharge chamber. Gas speed in the jet at its periphery is 1000–1500 m/s.

In the discharge chamber, the turbulent flow is formed, which develops efficient heat transfer from the arc to the gas. The velocity of the reverse flow of gas near the walls of the chamber is ~300 m/s.

The most share of the gas is entered into the arc at the base of the jet, i.e., at the end of the electrode. This is one of the reasons that the zone at the end of the electrode is the zone of maximum energy output and increased electric field intensity.

7.4 Energy in Arc and Heat Transfer Components

Except the heating of the gas and walls of the chamber, some share of the power supply energy is consumed to maintain arc burning, heating, melting, and evaporation of the electrodes. The energy of the arc W_a is composed of the gas-kinetic, dissociative, and ionization components, and estimated by the formula (7.21) is of 150 kJ.

$$W_a = V_a n_a\left((1+\alpha)\frac{3}{2}kT + \frac{D}{2} + \alpha I\right). \tag{7.21}$$

where V_a is maximal volume of the arc (~300 cm^3), n_a is concentration of atoms in the arc (3×10^{20} cm^{-3}), α is degree of ionization of hydrogen in the arc (0.3), D is dissociation energy of the hydrogen molecule (4.48 eV), I is ionization potential of the hydrogen atom (13.6 eV).

After the extinction of the current, the arc energy is transferred to the gas by the heat conductivity.

At arcing in a mode of thermal cathode spots of the second kind, the share of vapor in the erosion products does not exceed 10% [25]. The average mass of the eroded metal at the experiments at the generator PPG-3 ~ 30 g, for melting and partial evaporation of this metal is consumed ~80 kJ. Part of this energy can also be transferred to the gas.

Based on the performed estimations, we present the heat transfer components and their fractions in the energy balance of the high-current arc in the high-density hydrogen.

Table 7.2 presents the approximate distribution of the energy inputted in the arc among the main types of the heat transfers for the energy of 1 MJ.

Table 7.2 Distribution of the electric energy inputted in the arc

Type of heat transfer	Transferred energy (kJ)	Gas heating (kJ)	Looses (kJ)
Radiation W_r	200	80	120
Thermal conductivity W_{tc}	100	100	–
Turbulence W_t	470	470	–
Arc energy W_a	150	150	–
Electrode erosion	80	40	40
Total	1000	840	160

The data in Table 7.2 show that the calculated coefficient of the energy transfer from the arc to the gas is ~85%, what is close to its experimental values.

At experiments of the "manometric bomb" mode, the higher energy transfer coefficients were recorded, which is explained by the relatively low energy input and, consequently, the smaller losses by radiation into the wall and for erosion of the electrodes.

Another conclusion based on the presented data is that during the discharge, the main share of the energy is transferred from the arc to the gas by the turbulent heat transfer.

References

1. *Experiment and Technics of High Gas and Solidphase Pressure* (Moscow, 1978) (in Russian)
2. D.S. Tsiklis, *Technics of High Pressure* (Moscow, 1976) (in Russian)
3. Artsimovich, L.A.: *Controlled Thermonuclear Reactions* (Moscow, 1961) (in Russian)
4. L.E. Belousova, *Generators of Low-Temperature Plasmas* (Moscow, 1969) (in Russian)
5. A.A. Bogomaz, A.V. Budin, S.V. Zakharenkov, V.A. Kolikov, A.I. Kulishevich, I.P. Makarevich, A.F. Savvateev, PhG Rutberg, Izvestia RAS. Energy **1**, 64 (1998). (in Russian)
6. A.T. Onufriev, V.G. Sevast'janov, Sov. J. Appl. Mech. Tech. Phys. **2**, 17 (1968). (in Russian)
7. V.I. Petrenko, R.V. Mitin, Ju.P. Knjazev, A.V. Zvjagintsev, Sov. J. Tech. Phys. **379**, 1827 (1969). (in Russian)
8. A.M. Voronov, PhD thesis (St. Petersburg, 1992) (in Russian)
9. I.P. Fedotov, V.V. Vikhrev, in *4th All-Union Scientific and Technical Conference* (Nikolaev, 1988) (in Russian)
10. Yu.P. Raizer, *Physics of Gas Discharge* (Moscow, 1987) (in Russian)
11. Ya.B. Zel'dovich, Yu.P. Raizer, *Physics of Shock Waves and High-Temperature Hydrodynamic Phenomena* (New York, 1966)
12. K.A. Naugolnuikh, N.A. Roi, *Electric Discharges in Water* (Moscow, 1971) (in Russian)
13. B.M. Smirnov, *Introduction in Plasma Physics* (Moscow, 1982) (in Russian)
14. PhG Rutberg, D.A. Andreev, A.A. Bogomaz, A.M. Shakirov, Tech. Phys. **37**, 485 (1992)
15. PhG Rutberg, D.A. Andreev, A.A. Bogomaz, A.M. Shakirov, Tech. Phys. **38**, 56 (1993)
16. K.V. Dubovenko, Rus. J. Tech. Phys. **62**, 83 (1992). (in Russian)
17. S.V. Shevkunov, Plasma Phys. Rep. **36**, 902 (2010)
18. A.F. Aleksandrov, V.V. Zosimov, A.A. Rukhadze, V.I. Savoskin, I.B. Timofeev, Br. Messag. Phys. **8**, 72 (1970). (in Russian)

19. N.M. Kuznetsov, *Thermodynamic Functions and Shock Adiabatic of Air at High Temperatures* (Moscow, 1965) (in Russian)
20. S.Ju. Luk'janov, *Hot Plasma and Controlled Nuclear Synthesis* (Moscow, 1975) (in Russian)
21. I.A. Glebov, F.G. Rutberg, *High-Power Plasma Generators* (Moscow, 1985) (in Russian)
22. A.A. Bogomaz, in *Materials of Seminars—Schools of Young Scientists, Students and Post-graduate Students* (Petrozavodsk, 2004) (in Russian)
23. A.A. Bogomaz, V.S. Borodin, B.P. Levchenko, PhG Rutberg, Rus. J. Tech. Phys. **47**, 121 (1977). (in Russian)
24. E.A. Azizov, S.A. Kravchenko, S.G. Solodovnikov, Plasma Phys. Rep. **31**, 616 (2005)
25. G.A. Ljubimov, V.I. Rakhovskiy, Sov. J. Adv. Phys. Sci. **125**, 665 (1978). (in Russian)
26. G. Reter, *Electronic Avalanches and Breakdown in Gases* (Moscow, 1968) (in Russian)
27. A.A. Bogomaz, *Doctoral thesis* (St.-Petersburg, 2012) (in Russian)

Chapter 8
Modes of Arcing

Abstract The searches for the ways to increase the efficiency of the plasma generator are leaded, among other things, to the determination of such modes of the arcing, at which the share of the energy transferred to the arc from the power supply is the maximal. This is realized under the condition when the active resistance of the arc repeatedly exceeds the resistance of the discharge circuit and internal resistance of the power supply. As has been repeatedly noted, the characteristic feature of the arcing in the discharge chambers of the coaxial type is the effect of the so-called magnetic blow, i.e., "pulling" the arc and moving of the anode and cathode spots along the surface of the electrodes in the direction of the Lorentz force. The term "magnetic blow" needs to be adjusted, because it unilaterally characterizes the mechanism of the occurrence of this phenomenon. The point is that at the initial stage of the arcing, the gas-kinetic pressure is much higher than the magnetic pressure, and consequently, it determines the mechanism of the arc motion [Lebedev, in I All-union seminar on dynamics of the high-current arc discharge in magnetic field, Novosibirsk, 1990, 1]. Similar conclusions were made in the work [Gurevich et al. in Rep. AS USSR 293, 1102 (1987), 2], where it was stated that in the "plasma focus" at the gas pressure higher than atmospheric, the motion of the current sheath is also due primarily to the gas-kinetic pressure. Thus, partially retaining the terminology of the works [Bianchetta and Sivier, in Second symposium on technic research at hyper sound speeds, Denver, (1962), [3, Rotert and Sivier in National symposium on hypervelocity techniques (1960), 4], it is more correct to call this effect as "gas-magnetic blow".We have determined the basic principles controlling the arcing regime:

- The interelectrode distance should be such that the final length of the arc is maximal
- There must be guaranteed repeated breakdown/breakdowns of the interelectrode gap, to maximize the transfer of the power supply energy into the arc.

The consequence of the not performing of the last condition is the unused energy in the capacitor bank and, consequently, the decrease of the parameters of the plasma generator.

V. Kolikov et al., *Powerful Pulsed Plasma Generators*,
Springer Series on Atomic, Optical, and Plasma Physics 101,
https://doi.org/10.1007/978-3-319-95249-9_8

8.1 Arc at Refractory Electrodes

In addition to the above principles for controlling the arcing regime, it is necessary to mention other factors that also affect the arcing regime, among them:

- Initial gas pressure
- Initial voltage of the power supply
- Metal of the electrodes.

The first experiments at the coaxial discharge chambers were carried out with tungsten electrodes at large interelectrode distances [5].

Figure 8.1 shows the oscillograms of the current and voltage drop across the arc of such experiment at volume of the discharge chamber 1.12 dm^3, initial pressure of the hydrogen of 29 MPa, voltage of 8.5 kV, and interelectrode distance of 8 mm.

The oscillograms show that the current has almost sinusoidal shape and the voltage drop across the arc is increased. The duration of the current pulse is 150–200 μs, which by 2–3 times shorter than the half-period of the battery discharge at the short-circuit mode. This indicates that the increase of the arc resistance, by increasing its length, intensive heat transfer to the cold gas, while reducing the voltage on the battery, and reducing the discharge power, lead to the extinction of the arc.

In this case, the interelectrode distance is such that the residual voltage (7 kV) at the source does not ensure breakdown of the interelectrode gap and reignition of the arc. Thus, under these conditions, is realized the single-pulse mode of the discharge that is characterized by underusing of the significant, in this case ~70% of the energy stored in the capacitive bank.

Naturally, such regime of arcing cannot be considered as acceptable. To solve this problem, the main requirements for regulating the discharge of the battery on the load were developed and the conditions for repeated breakdown of the inter-electrode gap and ignition of the arc were experimentally determined.

The comparative analysis of the data of the experiments has shown that the voltage necessary for the repeated breakdown of the interelectrode gap is

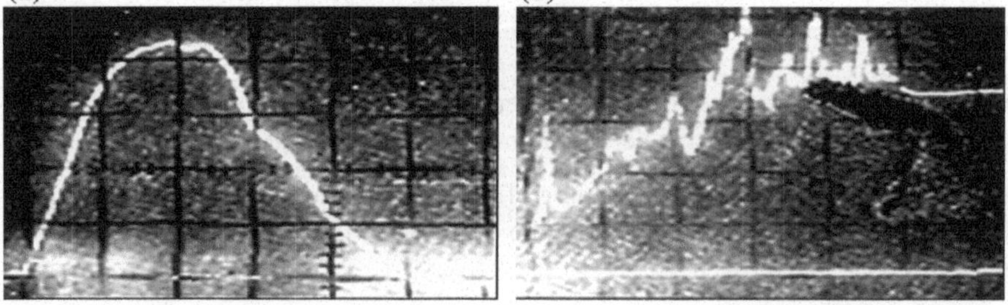

Fig. 8.1 Oscillograms of the current (**a**) and voltage drop across the arc (**b**) [5]. Current is145 kA/div, voltage is 2 kV/div, time is 50 μs/div

proportional to both its value of L_{ie} and to the pressure of the gas Pp in the discharge chamber at the time of extinction of the first current pulse. Thus, for given conditions of the experiment (initial gas pressure, metal of the electrodes, etc.), the solution to the problem was to determine the interelectrode distance L_{ie}, at which repeated breakdown of the interelectrode gap and arc ignition are guaranteed.

Changing the interelectrode distance, we thereby have changed the arcing regime in the discharge chamber obtaining firstly the separation through the time intervals current pulses at $L_{ie} \sim 2$ mm (Fig. 8.2), and then their continuous follow-up at $L_{ie} \sim 1$ mm (Fig. 8.3).

It was also established that at low interelectrode distances L_{ie} there is no single-valued and precise correspondence between L_{ie} and the arcing regime. Initially, it was assumed that as L_{ie} is decreased, the time intervals between the separated current pulses should be reduced up to their disappearance. However, in practice this assumption was not confirmed as fully.

The reason for this is that at small interelectrode gaps, there is then uncertainty about their actual value during the discharge process. This uncertainty was due to the transverse bending (in the direction of increasing the interelectrode gap) of the central current lead and cathode under the action of electrodynamic force and intensive erosion of the electrodes in the zone of the minimal interelectrode gap. This was confirmed by the significant residual bending of some current leads at currents exceeding 1 MA and considerable erosion of the electrodes.

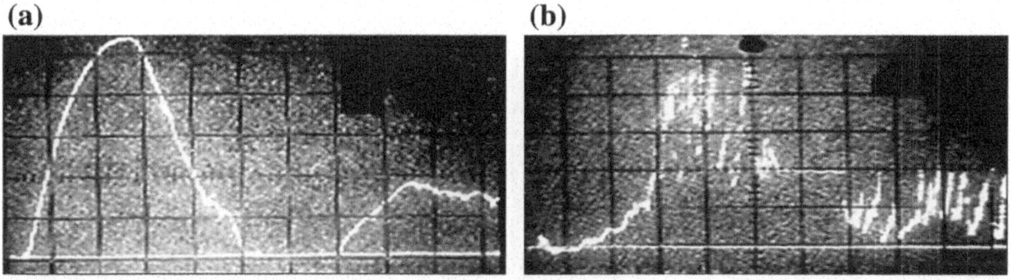

Fig. 8.2 Oscillograms of the current and voltage drop across the arc at regime of separate pulses [5]. Current is 300 kA/div, voltage is 2 kV/div, time is 50 μs/div

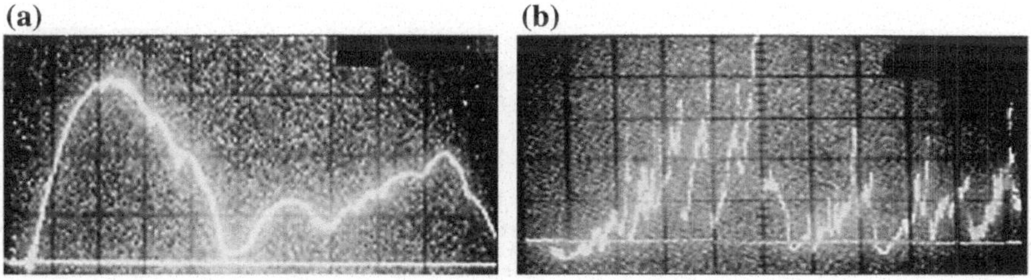

Fig. 8.3 Oscillograms of the current and voltage drop across the arc at the mode of continuously following pulses [6]. Current is 300 kA/div; voltage is 2 kV/div; time is 50 μs/div

Thus, it was established that the maximal interelectrode gaps for the coaxial discharge chambers with refractory electrodes at which the arcing occurs in the regime of continuously following current pulses and residual voltage on the capacitor bank are ≤ 2 kV. The interelectrode gaps for these conditions are 2–3 mm.

In addition to the maximum, the minimal value of L_{ie} was determined, below which there is no pulling out of the arc and restoring the dielectric strength of the interelectrode gap, and the arc burning mode corresponds to a short circuit (Fig. 8.4).

Such regime is characterized by lowest coefficients of the energy transfer from the power supply to the arc and from the arc to the gas (17%), because of the arc is localized in the point of the smallest interelectrode gap.

At this experiment, with interelectrode gap L_{ie} = 0.65 mm and initial voltage U_i = 6.5 kV, the maximal current was 1.6 MA, average voltage drop ~2 kV. However, if we take into account that the total voltage drop consists of the voltage drop across the arc and current lead, then the voltage drop only across the arc is ≤ 1.1 kV, and the active resistance of the arc at the maximal current is 0.7 mΩ.

The initial pressure of hydrogen has a significant effect on the arcing regime, which is confirmed by comparing the oscillograms of the experiments in Figs. 8.5 and 8.6. The experiments were carried out with the same interelectrode distance and electrodes of the similar erosion properties: WNF and WNC, respectively. At the experiment (Fig. 8.5), the initial pressure P_i was 42 MPa, and at the experiment (Fig. 8.6) 19 MPa.

(a) (b)

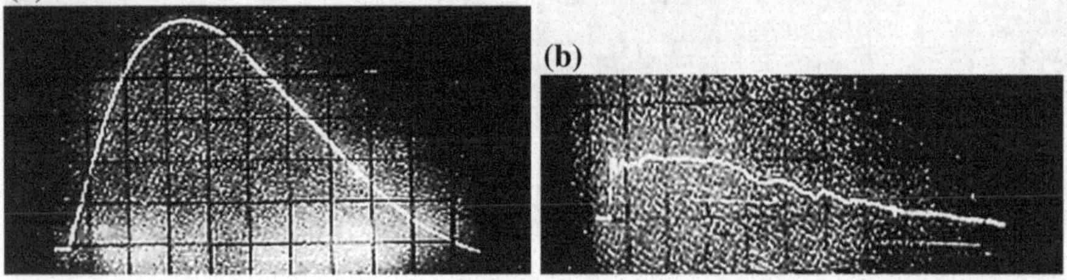

Fig. 8.4 Oscillograms of current and voltage drop across the arc at the regime of extremely small interelectrode gap [5]. Current is 300 kA/div; voltage is 2 kV/div; time is 50 μs/div

Fig. 8.5 Oscillogram of the current at P_i = 42 MPa [5]. Current is 300 kA/div, and time is 100 μs/div

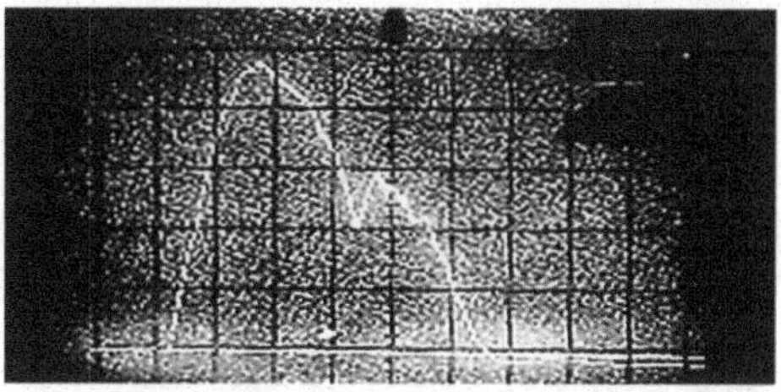

Fig. 8.6 Oscillogram of the current at $P_i = 19$ MPa [5]. Current is 300 kA/div, and time is 100 μs/div

The comparison of the oscillograms shows that although in both cases $L_{ie} = 1.5$ mm, the arcing mode at the experiment (Fig. 8.5) is characterized by the separate current pulses following through some time intervals, whereas at the experiment (Fig. 8.6) the arcing mode is characterized by the continuous current pulses.

Thus, the range of interelectrode distances L_{ie}, at which the optimal arcing mode is realized in the coaxial discharge chamber with refractory electrodes and high efficiency of energy transfer from the battery to the arc, is 1–3 mm.

It is necessary to mention the action of the initiation wire to the arcing mode. Although this aspect has not been studied in detail, one thing can be noted: the use of low-melting materials (aluminum) as a wire results in the "smoothing" of the voltage peak at the time of the wire explosion, the value of which can exceed the initial voltage at using the refractory materials (tungsten). Thus, under the conditions of the experiments carried out, the material and mass of the wire essentially affect only at the initial period of arc formation.

The overwhelming number of the experiments was carried out using copper wires of 0.01–0.05 g weight, and at only one experiment the weight of four tungsten wires was 4 g. For comparison, the average weight of eroded metal from the electrodes at experiment is ~ 10 g.

8.2 Arc at Fusible Electrodes

Except the interelectrode distance and initial gas pressure, the material of the electrodes is appreciable affected on the arcing [5]. The oscillograms presented in the previous section were obtained at the experiments with the electrodes of refractory highly erosion-resistant metals and alloys, such as Mo, MWC, WNC, WNF, and plasma-dusted tungsten.

The oscillograms obtained at the experiments with electrodes from the relatively fusible metals such as steel and copper are differed substantially from the above oscillograms (Fig. 8.7).

At these experiments, the interelectrode distance varied from 11.5 to 15 mm and initial gas pressure from 17 to 22 MPa. Attention is drawn to much larger

(a) (b)

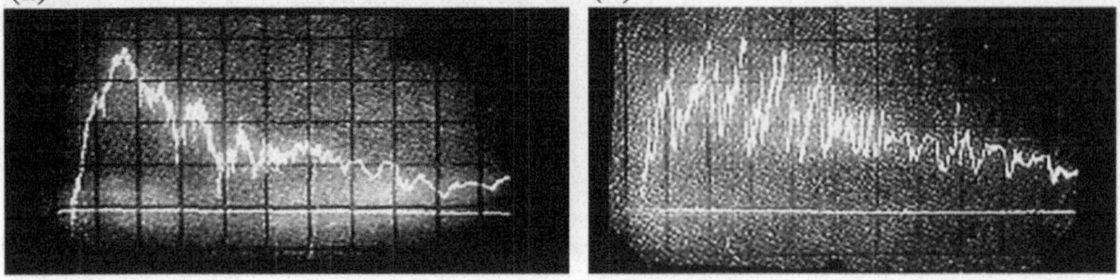

Fig. 8.7 Oscillograms of current and voltage at experiments with steel electrodes [5]. Current is 200 kA/div (**a**), voltage is 2 kV/div (**b**), time is 100 μs/div

interelectrode distance, than at the experiments with refractory electrodes. In this case, the arcing mode is similar the arcing mode with refractory electrodes at interelectrode distance of ~1 mm, which positive distinguishes this design of the discharge chamber from those considered in the previous section.

The characteristic features of these experiments: the high concentration of ions, electrons, atoms, metal particles in the gas, and the powerful electrode jets those contribute to the occurrence of the repeated breakdowns of the interelectrode gap at the early stage of the arc pulling. The arcing under these conditions is characterized by the continuous current pulses, when their amplitude is decreased monotonically as the voltage of the power supply is decreased.

Another feature of these experiments is the high and almost constant arc resistance, whose average value is ~10 mΩ. The coefficient of energy transfer from the power supply to the arc at these experiments is ~75%.

8.3 Multipulse Mode

The investigations aimed at increasing the efficiency of plasma generators were carried out mainly at the PPG-3 generator (Fig. 3.9). The design of the discharge chamber of this generator was developed taking into account all the experience accumulated over the long period of research. The main distinguishing feature of the camera is its reliability, for the achievement of which some additional elements were used, which not only increased the reliability of the installation, but also greatly simplified its maintenance.

The changes were connected not only to the elements of the discharge chamber, but also its internal configuration, and most important of them were increasing of the angle between the outer surface of the cathode and the inner surface of the anode from 45° to 90°. Such change of the configuration of the discharge chamber led to the appearance of the new arcing mode. The new mode, called the "multipulse", is characterized by multiple current pulses (up to 20) continuously following each by other [6]. Moreover, if the similar arcing mode was observed at a number of the experiments before, it was realized, most often, spontaneously, whereas in the

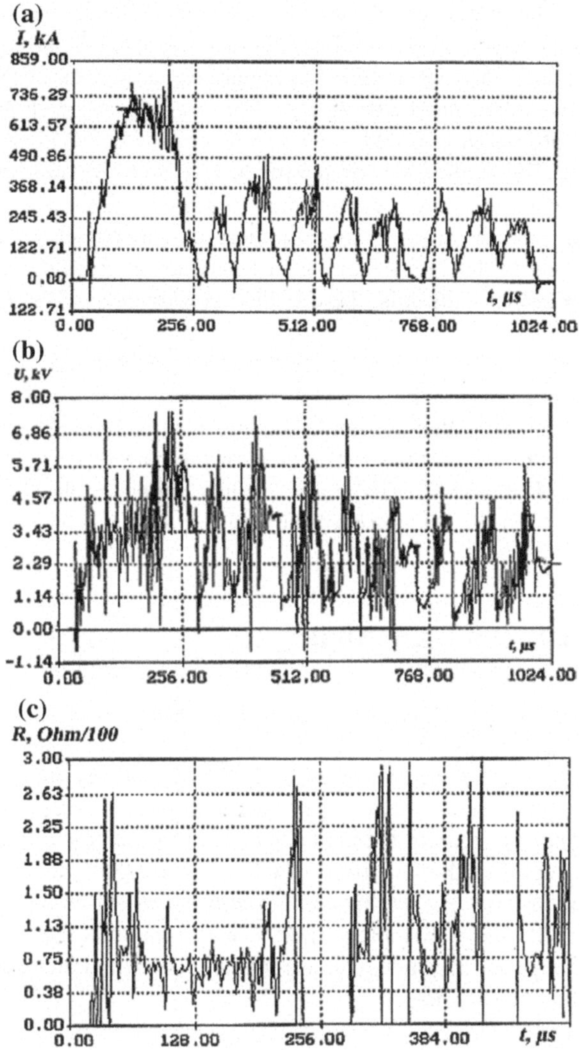

Fig. 8.8 Oscillograms of current, voltage drop across the arc, and arc resistance graph at the multipulse mode [5]

new discharge chamber the multipulse mode became stably reproducible under similar initial conditions of the experiments.

Figure 8.8 shows the oscillograms of current, voltage drop across the arc, and resistance graph at the experiment typical for the number of the similar experiments performed at the PPG-3 generator.

The initial conditions of the experiment: volume of the discharge chamber is 1.5 dm^3, initial gas pressure is 36 MPa, capacitance of the power supply is 0.076 F, initial voltage is 5 kV, interelectrode distance is 1.5 mm, anode is steel, and cathode is aluminum alloy.

The sequence of pulses is visible in the oscillogram of the current (Fig. 8.8a), the first pulse has maximal amplitude of ~680 kA and duration of ~200 μs, followed pulses 250–370 kA, what is approximately 50% of the amplitude of the first pulse and duration of 60–110 μs. The arc resistance (Fig. 8.8c) for all the process is stably high ~6 mΩ.

The multipulse arcing mode has both advantages and disadvantages. To the advantages can be attributed repeatedly appearing and extinguishing arcs, the motion of which in the discharge chamber leads to active mixing of the gas, and thus, to the intensification of heat transfer from the arc to gas.

Along with this, there is the mass transfer of the eroded metal of the electrodes caused by the motion of the gas (effect of the "ventilation" of the interelectrode gap Fig. 7.23). This helps to reduce the concentration of metal vapor (as well as other charge carriers), i.e., restoration the initial conditions preceding the next gap breakdown and arc ignition, and maintaining a stably high arc resistance, which is confirmed by the oscillograms of the current and voltage.

The only disadvantage of this mode is the long (≥1 ms) and practically uncontrolled duration of the arcing process, which in some cases is undesirable.

8.4 Two-Arc Mode

At currents above 1 MA, it was found that under the action of electrodynamic and gas-dynamic forces, the current lead is bended and, as a consequence, the inter-electrode distance is increased. To prevent this undesirable phenomenon, the series of the experiments was performed for ensuring the transverse stability of the current lead by creating the symmetrical current circuit formed by the second arc [5]. The scheme of the directions of the moments (M1, M2), currents (I1, I2), and magnetic fields (H1, H2) at this configuration of the discharge chamber is shown in Fig. 8.9.

By the autograph method, it was found that the two parallel arcings are occurred at approximately equal currents and outputted energies, what indicate that the goal had been achieved.

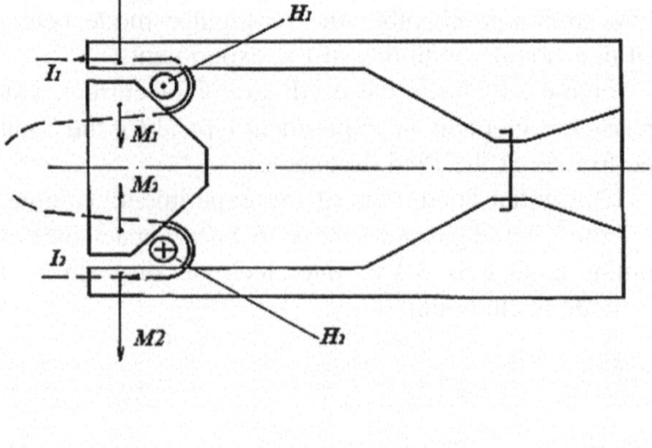

Fig. 8.9 Discharge chamber at two-arc discharge mode [5]

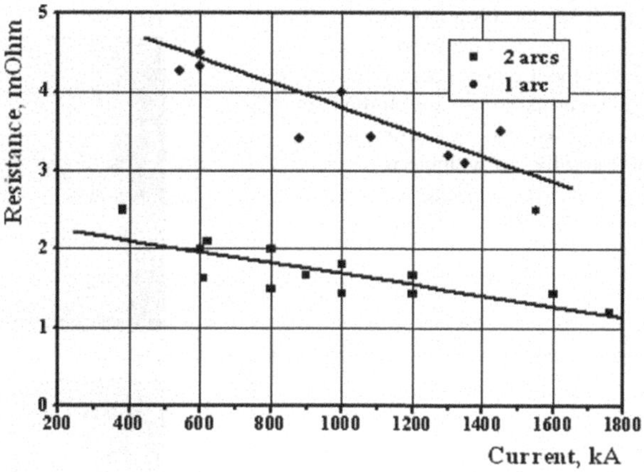

Fig. 8.10 Resistance of the load versus discharge current [5]

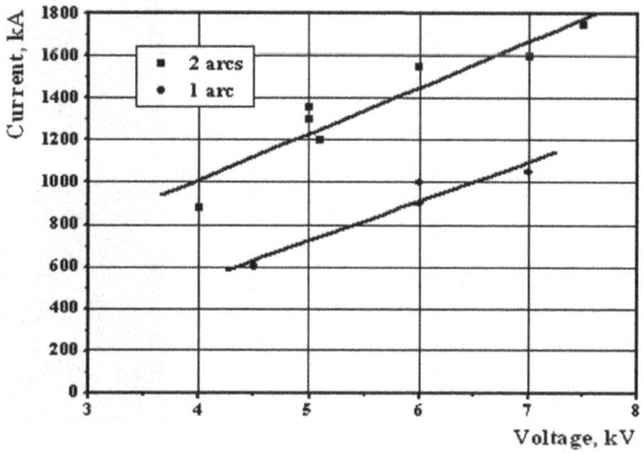

Fig. 8.11 Discharge current versus initial voltage [5]

However, the analysis of the data of these experiments has shown that this method of compensating of the mechanical load has significant drawbacks, expressed in approximately twofold reduction of the load resistance (Fig. 8.10) and in similar increase of the total discharge current (Fig. 8.11).

Last circumstances have caused the significant reduction of the coefficient the transfer of energy from the power supply into the load. Consequently, these experiments were stopped, and the problem of stability of current leads had to be solved by increasing of their mechanical strength.

8.5 Programmable Mode

At the PPG-3 generator were carried out the studies of the programmable energy input into the arc [7]. This section shows the characteristic features of the programmable capacitive bank discharge at forming the desired arcing.

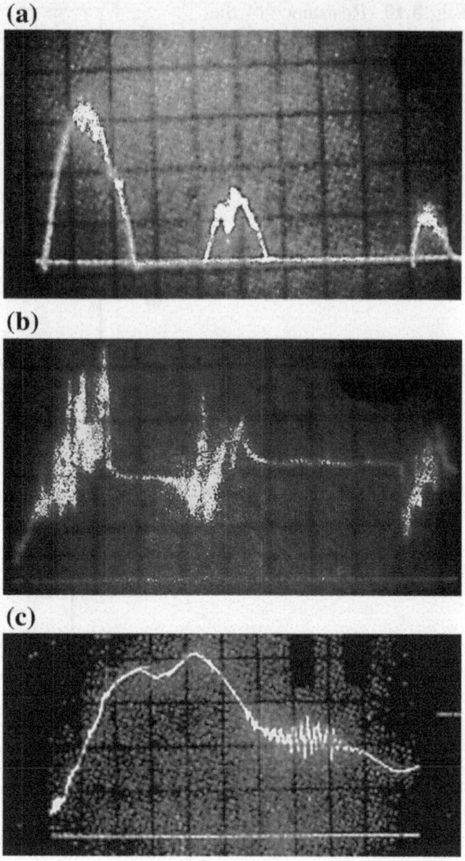

Fig. 8.12 Oscillograms of the current, voltage drop across the arc, and pressure at the experiment 3 + 2 with the delay of 0.4 ms [5]. Current is 300 kA/div (**a**), voltage is 2 kV/div (**b**), pressure is 100 MPa/div (**c**), and time is 100 μs/div

Figure 8.12 shows the oscillograms of the current, voltage drop across the arc, and pressure in the discharge chamber at the experiment with sequential (3 + 2) switching on of the battery modules at delay of 0.4 ms.

The experiment was carried out at initial voltage of 7 kV, hydrogen pressure of 40 MPa, discharge chamber volume of 0.6 dm^3, and interelectrode distance of 1.5 mm.

The oscillograms of the current and voltage (Fig. 8.12a, b) show that the switching on of two modules in the discharge circuit occurred at the required time at the residual voltage of the three modules of 4 kV. The third discharge pulse is the result of the spontaneous breakdown of the interelectrode gap at the voltage of all five modules of 4.2 kV.

The oscillogram of pressure (Fig. 8.12c) (the oscillogram has time delay of 0.4 ms) shows that at 0.7 ms, the noticeable growth of the pressure is observed, which corresponds to the input of additional energy of two modules into the arc.

However, the connection of the additional modules to the discharge circuit does not always lead to the repeated breakdown and the appearance of the arc, as evidenced by the increase of the voltage across the interelectrode gap at time of 0.4 ms.

Figure 8.13 shows the oscillograms of such experiment. At time of 0.4 ms, three sections are connecting to the discharge circuit, but there is no breakdown of the interelectrode gap, and since the current through the discharge gap no longer flows, but the dischargers are still initiated, these modules are recharged onto already partially discharged ones, what leads to the increase of their voltage from 3 to 5 kV.

Given this fact, at all subsequent programmable experiments, the switching on of the additional modules to the discharge circuit was performed at the time preceding the arc extinction.

Figure 8.14 shows the oscillograms of such experiment. The sequence of the modules switching on is 3 + 3 through the 0.1 ms.

The oscillogram of the current (Fig. 8.14a) shows that the connection of the additional modules occurred at the moment when the discharge current from the first three modules begins to decrease and the connection of these additional sections prolongs the discharge of the capacitive bank and maintains the discharge current at the level of the first maximum.

Figure 8.15 shows the oscillograms of the current and voltage drop across the arc at the experiment with synchronous switching on of all modules of the capacitive bank. The current at this experiment has one expressed maximum of 1.4 MA, after which it is monotonically decreased to zero.

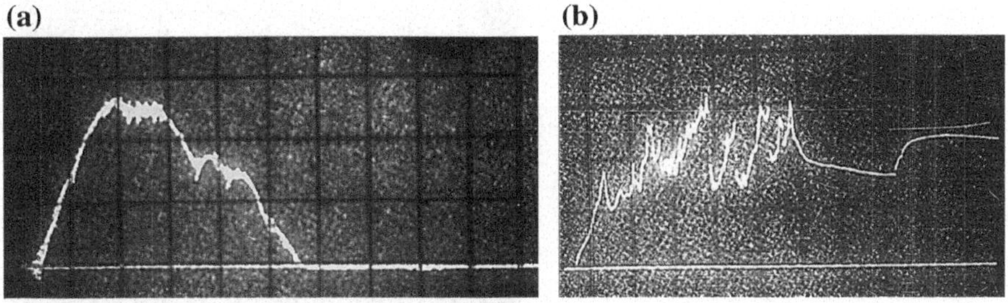

Fig. 8.13 Oscillograms of the current and voltage drop across the arc at recharging of the modules [5]. Current is 300 kA/div (**a**), voltage is 2 kV/div (**b**), and time is 50 μs/div

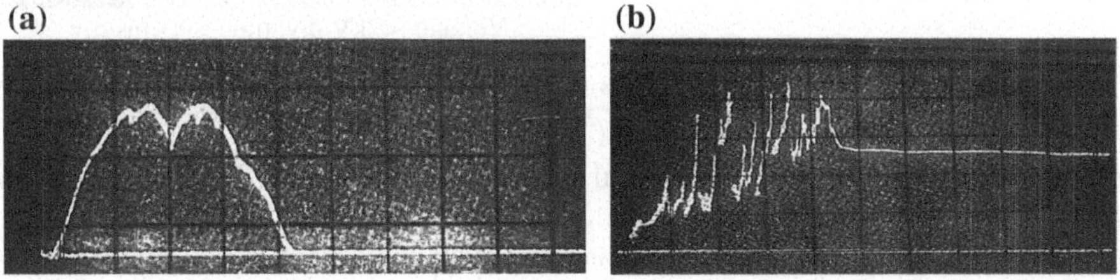

Fig. 8.14 Oscillograms of the current and voltage drop across the arc at the modules switching on of 3 + 3 through the 0.1 ms [5]. Current is 300 kA/div (**a**), voltage is 2 kV/div (**b**), and time is 50 μs/div

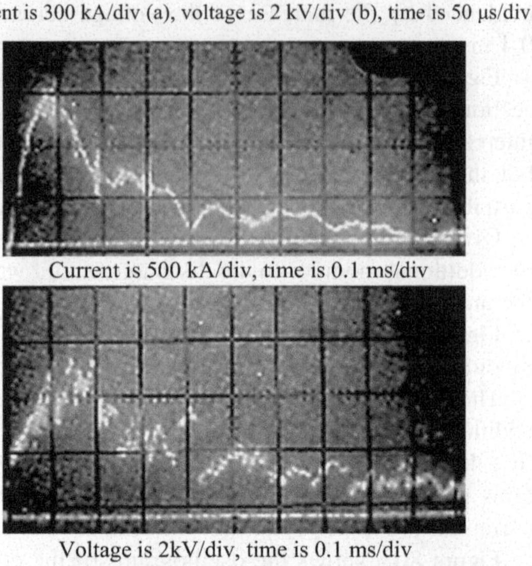

Fig. 8.15 Oscillograms of the current and voltage drop at the synchronous mode [5]

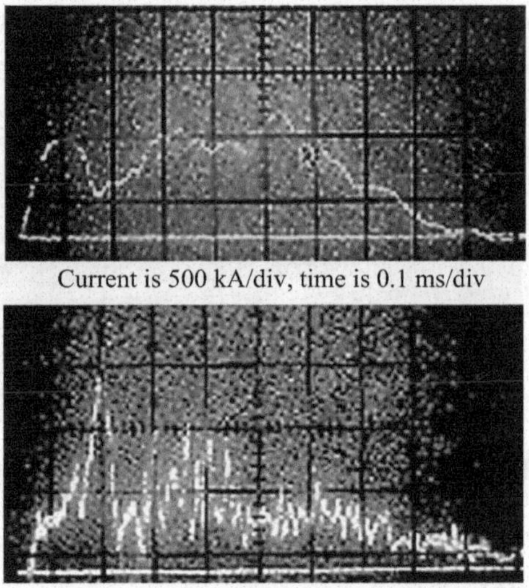

Fig. 8.16 Oscillograms of the current and voltage drop at the programmable mode [5]

Figure 8.16 shows oscillograms of the current and voltage drop at the programmable experiment with sequential connection of modules 2 + 2 + 2 through the 0.2 ms. Here the current has three maxima: 0.7, 0.8, and 1 MA.

The nature of the voltage drop change across the arc at the regime of synchronous switching on of all the modules is corresponded to the nature of the current change, i.e., the expressed maximum at the beginning of the process and the subsequent monotonic decrease.

At the programmable mode, the practically constant voltage drop across the arc throughout the arcing process is observed, due to the connection of the additional modules. Thus, the power in the programmable mode is actually constant during the arcing time, i.e., during of 1 ms.

The presented data show that it is possible to effectively influence the arcing process by programming of the input of the energy into the load.

References

1. A.D. Lebedev, in *I All-Union Seminar on Dynamics of the High-Current Arc Discharge in Magnetic field,* Novosibirsk, 1990 (in Russian)
2. V. Ts, G.A. Gurevich, V.L. Desjatov, B.A. Spektorov, B.S.Engelsht Urjukov, Rep. AS USSR **293**, 1102 (1987). (in Russian)
3. J.F. Bianchetta, K.R. Sivier, in *Second Symposium on Technic Research at Hyper Sound Speeds*, Denver (1962)
4. R. Rotert, K.R. Sivier, in *National Symposium on Hypervelocity Techniques* (1960)
5. V.A. Kolikov, Doctoral Thesis, St.-Petersburg, 2005 (in Russian)
6. A.V. Budin, A.A. Bogomaz, V.A. Kolikov, P.G. Rutberg, A.F. Savvateev, IEEE Trans. Mag. **35**, 189 (1999)
7. A.V. Budin, V.A. Kolikov, P.G. Rutberg, I.E.E.E. Trans, Plasma Sci. **34**, 1553 (2006)

Chapter 9
Arc Contraction: Modified Piza-Braginskii Critical Current

Abstract At study of the discharges in the hydrogen and helium at high and ultrahigh pressure, single and multiple contractions of the arc at increase of the current near its maximum were recorded. Contraction of the arc was accompanied by the decrease of its glow. This chapter presents the results of the experiments at which these phenomena were observed. Measurements of the arc parameters were carried out at the moment of the maximal discharge current, when the inductive component of the voltage drop across the arc is absent. In other experiments, this component was compensated by means of hardware. At the time of maximum contraction of the arc, coming before the maximum of the current, on the oscillograms of the current the "features" and on the oscillograms of the voltage drop the characteristic peaks are observed. Such "features" and peaks are characteristic for the Z-pinches, but under the significantly lower initial hydrogen pressure. Photometric sections of the arc brightness at various times show that the expansion and contraction of the arc are corresponded to increase and decrease of the brightness of the arc near its axis. Furthermore, it was found that the arc contraction is occurred with the subsonic speed, so the shock waves at these experiments did not appear.

9.1 Arc at Initial Hydrogen Pressure of 5 MPa

In this chapter are presented and analyzed the data of the experiments performed at the installation with preliminary adiabatic compression of the hydrogen at current rise rate of 10^{11} A/s. At one experiment with tungsten electrodes, the initial hydrogen pressure was 5 MPa, and current 500 kA.

Figure 9.1 shows the photo-scanning of the arc glow at the center of the inter-electrode gap (top left), photometric cross sections of the arc at the time of 9, 20, and 37 μs (lower left), graphs of the dependences of the arc diameter (*D*) on the time, brightness in the arc axis (*B*) (top right), and graphs of the current (*I*) and voltage drop across the arc (*U*) (bottom right).

V. Kolikov et al., *Powerful Pulsed Plasma Generators*,
Springer Series on Atomic, Optical, and Plasma Physics 101,
https://doi.org/10.1007/978-3-319-95249-9_9

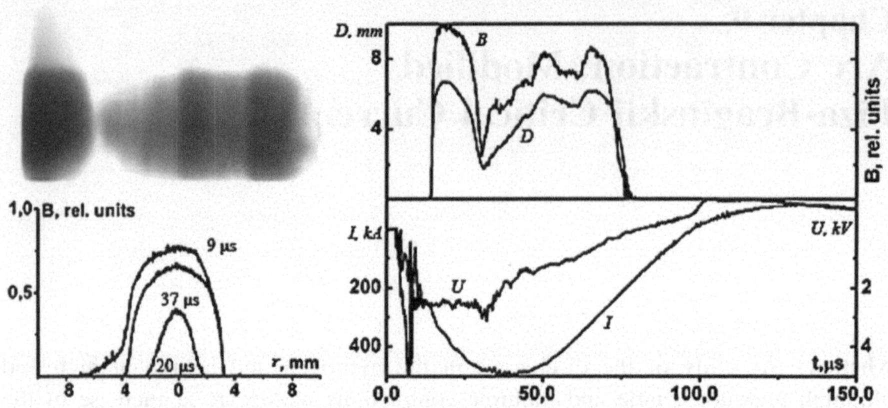

Fig. 9.1 Arc characteristics under initial hydrogen pressure of 5 MPa [1]

The photostreak of the arc glow shows the several contractions of the arc. The speed of the arc border motion at the time of maximal contraction is 6.4×10^2 m/s; thus, the shock wave at this experiment did not form. The electric field intensity in the arc E was determined taking into account the anode and cathode voltage drops U_{ac} from the relationship $U = El + U_{ac}$, where U is voltage drop across the arc, and l is interelectrode distance. The sum of the anode and cathode voltage drops U_{ac} was determined from the graph of the dependence of the voltage drop across the arc U versus interelectrode distance l at the extrapolation of the function to zero.

At the current of 10^5–10^6 A and its rise rate of $\sim 10^{10}$ A/s, the U_{ac} is ~ 2 kV. Unlike discharges under the low initial pressure, in our case, value of ΔU caused by the change of the inductivity of discharge gap at the arc contraction is not high

$$\Delta U = -I\frac{dL}{dt}; \quad L = \frac{\mu_0}{2\pi} \times \ln\frac{r_{ch}}{r(t)}, \tag{9.1}$$

where r_{ch} is radius of the discharge chamber, where the internal surface is the reverse current conductor, $\mu_0 = 4\pi \times 10^{-7}$ H/m.

$$\Delta U = -I\frac{\mu}{2\pi r(t)}\frac{dr}{dt}. \tag{9.2}$$

The value of the ΔU at the arc contraction speed of 6.4×10^2 m/s, current of 480 kA, arc radius $r_a = 0.2$ cm, and interelectrode distance $l = 1.2$ cm is 740 V. The experimentally registered $\Delta U = 680$ V confirms the contraction of the arc that is not related with the optical effects. For pinch discharges at low pressure, ΔU is much higher because of the high speed of arc contraction [2].

The experiments under the higher initial hydrogen pressure and higher discharge current were performed at the generator PPG-5.

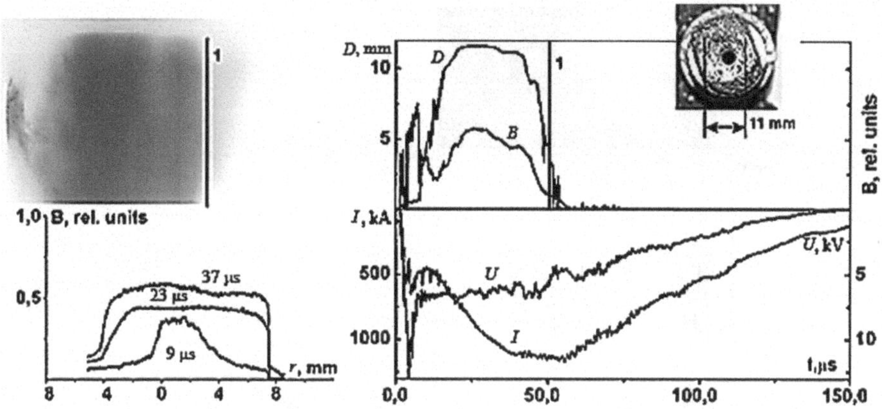

Fig. 9.2 Arc characteristics under initial hydrogen pressure of 32 MPa [1]

Figure 9.2 shows the photo-scanning of the arc glow in hydrogen at initial pressure of 32 MPa and current of 1.2 MA (top left), graphs of the photometric cross sections of the arc brightness at the time of 9, 23, and 37 µs (bottom left), graphs of the dependence of the arc diameter *D* and brightness *B* in the center of the arc versus time (upper right), and oscillograms of the current *I* and the voltage drop across the arc *U* (lower right). At the experiment, the steel electrodes of 20 mm diameter were used at interelectrode distance of 20 mm. There is erosion crater of 11 mm diameter on the end cathode surface, the central hole for installing the initiating wire.

The graph of the dependence of the arc diameter *D* on time shows the arc contraction at time of ~35 µs with corresponding decrease of the brightness *B*.

Figure 9.3 under the same initial conditions of the experiment shows the arc scanning (top) synchronized with the graphs of the brightness temperature of the arc in its center T_{br} at wavelength of 694 nm, arc diameter *D*, and with the oscillograms of the current *I* and voltage drop across the arc *U*.

The beginning of arc contraction corresponds to features on the oscillograms of the current *I* and to the voltage drop across the arc *U*.

It should be noted that the glow of the arc decreases when the arc is contracted. This phenomenon is absent at the discharges in the vacuum, where at contraction of various Z-pinches with the metal plasma, the intensity of radiation in the visible range of the spectrum increases. The decrease of the glow of the arc begins by 10–15 µs earlier than arc contraction.

Figure 9.4 shows X-ray intensity graphs from the sensors screened by the aluminum foil of 10 and 18 µm thick. The graphs were obtained at initial hydrogen pressure of 7 MPa, initial voltages of 12 kV (a), and 16 kV (b). Under these conditions quantums with energy of 0.8–1.5 keV were registered.

The increase of the signal from the X-ray sensor at the maximal arc contraction corresponds to the appearance of "features" on the oscillograms of the current *I* and the voltage drop across the arc *U*.

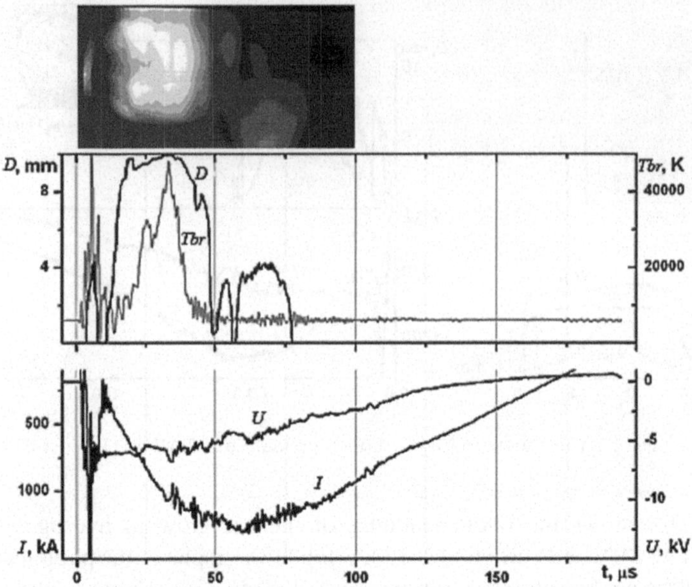

Fig. 9.3 Discharge in the hydrogen at initial pressure of 32 MPa [1]

Figure 9.5 shows the graphs of the arc brightness temperature changing versus time at various initial hydrogen pressures, obtained by the high-speed pyrometer at wavelength of 694 nm.

At this the beginning of the decrease of the arc brightness temperature ahead the instant in time of the "feature" on the current oscillogramm and the jump of the voltage by 4–8 µs, is registered at the wavelength of 694 nm and almost coincides at the wavelength of 550 nm. Comparison of the photostreak of the arc glow with pyrometry of the arc brightness temperature shows, that the beginning of the arc contraction is close to the time of the arc brightness temperature decay due to loss of transparency of the transition layer [3]. At the growth of initial hydrogen pressure the decrease of the arc brightness temperature and appearance of the "features" on the current and jump of the voltage shifts to the later time (Fig. 9.5).

The moment when the current reached the critical value I_{cr} correspond to the beginning of the arc contraction. This moment is between the times of the sharp decrease of the arc glow, which in Fig. 9.5 is noted by arrows for the experiments at various initial hydrogen pressures, and the time corresponding to the current feature and the voltage jump is shown in Fig. 9.4 of the vertical line.

The dependence of the maximum arc brightness temperature T_{br} versus initial hydrogen pressure P_i is presented in Fig. 9.6.

Figure 9.7 shows the dependences of the experimental current corresponding to the sharp decrease of the arc brightness (curve 1 and black squares), and the current

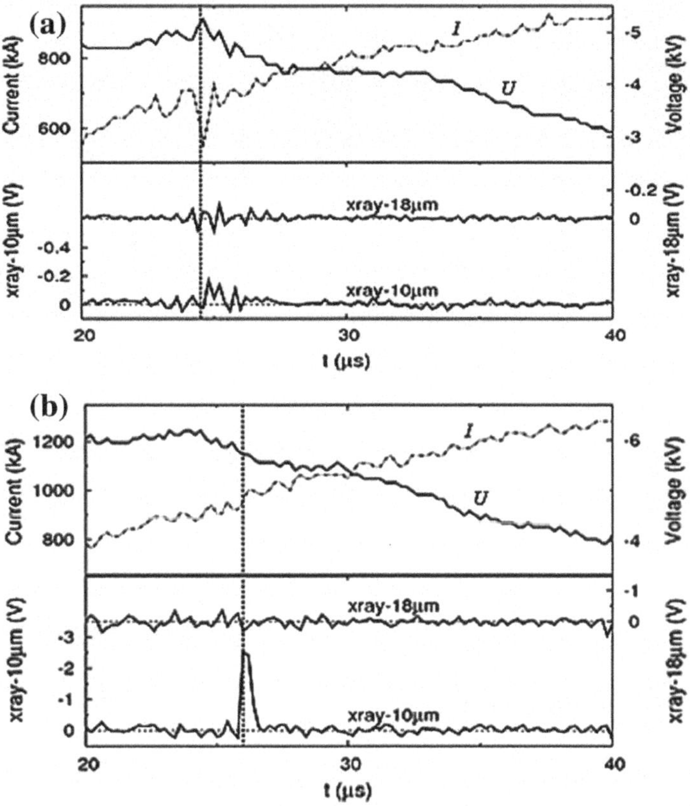

Fig. 9.4 Oscillograms of the current, voltage, and signal from the X-ray sensor near the moment of the maximal arc compression [1]

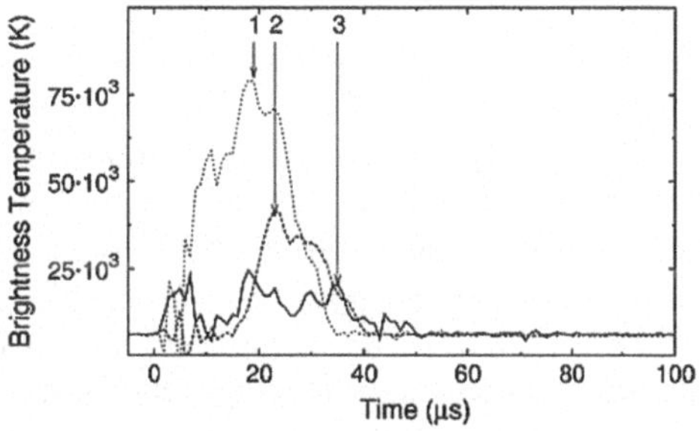

Fig. 9.5 Brightness temperature of the arc T_{br}, measured in the middle of the interelectrode gap at initial hydrogen pressure of 5 MPa (1), 15 MPa (2), and 30 MPa (3) [2]

corresponding to the moment of the feature in the current and jump in the voltage drop (curve 2 and empty squares) versus initial hydrogen pressure.

Thus, the experimental data of the critical current I_{cr} on the initial hydrogen pressure are between the curves 1 and 2 (Fig. 9.7).

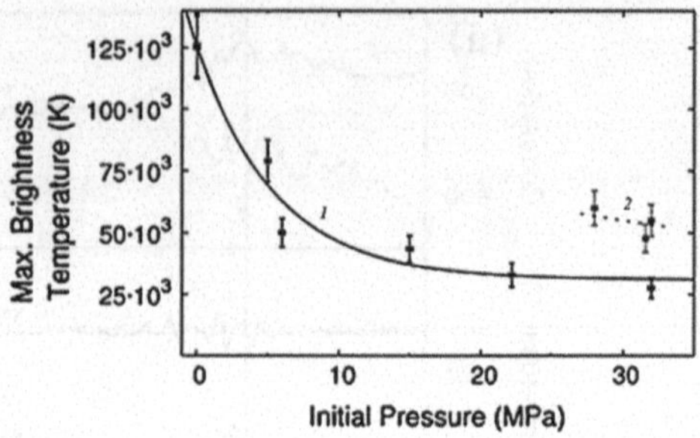

Fig. 9.6 Maximal arc brightness temperature T_{br} versus initial hydrogen pressure [1]. Interelectrode gap is 10 mm (1); interelectrode gap is 20 mm (2)

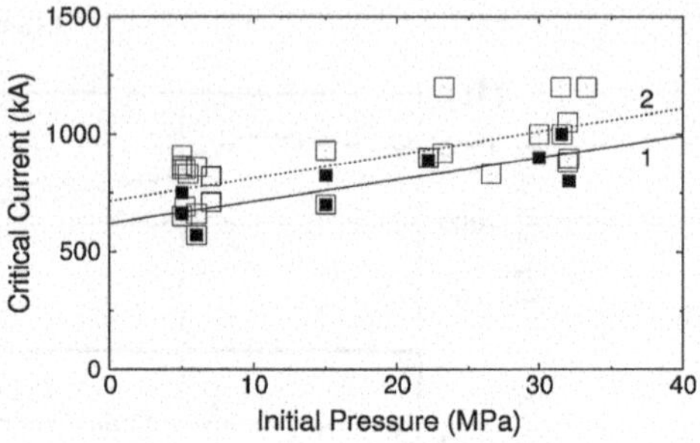

Fig. 9.7 Critical current I_{cr} versus initial hydrogen pressure [1]

9.2 Discharge in Helium

The discharge in the helium is characterized by higher brightness temperature. Figure 9.8 shows the results of the experiment performed at initial helium pressure of 25 MPa, steel electrodes of 20 mm diameter, and interelectrode distance of 20 mm.

The photo-scanning of the arc glow in the middle of the interelectrode gap is synchronized with the current (*I*) and the voltage drop (*U*). The dependence of the arc diameter on time was obtained at the various blackening density (*i*, *ii*, *iii*) of the photo-scanning.

The comparison of the oscillograms of the current, voltage, and photo-scanning of the arc glow shows that the moment of maximal arc contraction corresponds to the feature on the oscillogram of the current. Arc contraction occurs later than decrease of the arc glow.

To estimate the arc parameters and its contraction, the channel model is used, based on the assumption that all the current flowed in a region of radius r0 near the arc axis [4, 5].

For discharges in gases of high and ultrahigh pressure, such model is quite acceptable, since at increasing of the gas pressure, the arc is contracted and the thickness of the transition layer is decreased. The arc plasma is transparent for its own radiation. Outside the arc, the gas density is increased and its temperature is

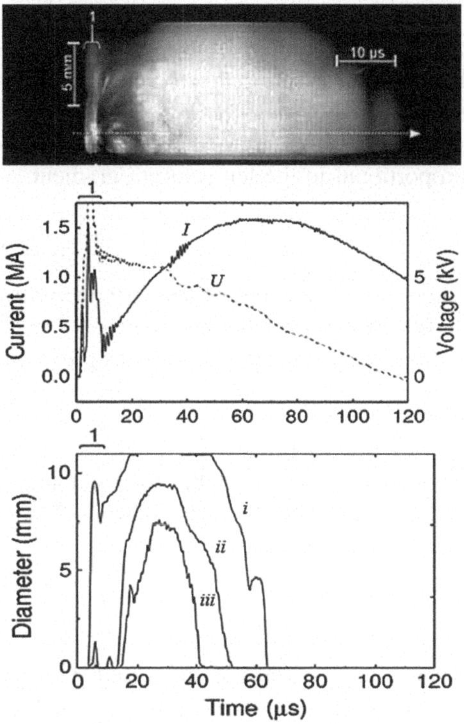

Fig. 9.8 Discharge in the helium [1]

decreased; therefore, part of the radiation is absorbed in the narrow transition layer at the boundary of the arc [6].

For example, at radius of the arc in the helium of few millimeters, the width of the transition layer at initial pressure of 15 MPa determined by the total absorption in the lines of the copper initiating wire is 3×10^{-3} cm [7, 8].

According to the data of the work [9], the thickness of the transition layer of the arc of 0.3 cm radius at initial hydrogen pressure of 10 MPa and current of 10 kA is 6×10^{-2} cm.

Absorption of radiation in the transition layer is increased at increasing of pressure; at this, the increase in radiation losses significantly affects the characteristics of the Z-pinch [10]. The increase of energy losses by the radiation from the arc leads to increase of the critical current I_{cr}, at which the contraction of the arc begins, and the brightness temperature of the arc is decreased.

The channel model assumes that the arc has few changing of the temperature and width of the transition layer, through which the heat transfer by the radiation and thermal conductivity takes place. Due to the high temperature of the arc, its central region is transparent to its own radiation, and because of its high radiant heat conductivity, the arc has the table-shaped temperature profile, to which is corresponded the equal conductivity over the cross section of the arc.

The estimations show that in the transition layer under these conditions the coefficient of radiative heat conductivity is more than by an order of magnitude higher than electron heat conductivity coefficient. As a consequence, all heat transfer in the transition layer is due to radiant heat conductivity.

The heat flux from the surface of the arc due to the heat conductivity S is proportional to the temperature gradient

$$S = -k \cdot \mathrm{grad} T, \tag{9.3}$$

where k is coefficient of heat conductivity.

Let us compare the coefficients of the radiant κ_l and electron κ_e heat conductivities, the formulas for estimating of which are taken from the work [3]

$$\kappa_l = 16\sigma T^3 l_r/3, \tag{9.4}$$

where l_r is mean free path of the radiation quantum carrying the main share of the energy in an optically dense plasma, T is temperature, and σ is Boltzmann constant

$$l_r = \frac{4.4 \times 10^{22} \cdot T^{\frac{7}{2}}}{n_i^2 Z(Z+1)}, \tag{9.5}$$

where n_i is ion concentration, and Z is ion charge.

The coefficient of electron heat conductivity κ_e is given by

$$\kappa_e = \frac{1.9 \times 10^{-5} \cdot T^{5/2}}{Z \cdot \ln \Lambda}, \tag{9.6}$$

where $\ln\Lambda$ is the Coulomb logarithm ~ 6 [3, 11].

For multiply charged metallic plasma at $T > 10^5$ K, $(Z + 1)$, we replace by Z, and using the approximation with the error of $\sim 10\%$, $Z = 1.6 \times 10^{-2} \sqrt{T}$ [12], we have $\kappa_l/\kappa_e = 1.8 \times 10^{28} T^3/n_i^2$. At $T > 10^5$ K, $n_i \leq 10^{20}$ cm^{-3}, what is typical for the considering range of temperature and concentration, we have $\kappa_l/\kappa_e \geq 10^3$. This indicates the predominant role of the radiation in the energy losses from the arc. Such estimation of the heat flux from the arc due to radiant heat conductivity is valid for the optically dense plasma.

At $l_f \gg r$, where l_f is photon mean free path averaged over the Planck, we can assume that the plasma is optically transparent, and the radiation power Q_r from the unit length of the arc of radius r is $Q_r = 4\pi\sigma T^4 r^2/l_f$

$$l_f = \frac{1.1 \times 10^{23} T^{7/2}}{n_i^2 \times Z(Z+1)^2 \times \bar{x}_1}. \tag{9.7}$$

where $\bar{x}_1$ is the average relative ionization potential $(\bar{I}/kT)$, and $\bar{I}$—average ionization potential, calculated according to statistical weights of ions with different degrees of ionization.

According to the data of the work [13] $\bar{x}_1 \sim 5$. Power of the electron heat conductivity is given by

$$Q_e = 2\pi r \kappa_e \frac{dT}{dr}, \frac{dT}{dr} \sim \frac{T}{0.1r}. \tag{9.8}$$

Setting $Z \sim (Z+1)$ and using the approximation $Z = 1.6 \times 10^{-2}\sqrt{T}$, after permutations at $n_i \sim 10^{19}\ \text{cm}^{-3}$, $T \sim 4 \times 10^5$ K, and $r = 0.3$ cm, we obtain $Q_r/Q_e \sim (10^{-28} n_i^2 r^2)/T$.

At the current of 1.2–1.6 MA, arc contraction is began at $n_i \sim 10^{19}\ \text{cm}^{-3}$, $T \sim 4 \times 10^5$ K, and $r \sim 0.4$ cm (Figs. 9.2 and 9.3), at this $Q_r/Q_e \sim 10^3$, which also indicates about the prevailing losses of the energy from the arc due to radiation. Therefore, for our conditions, taking into account the magnetic and external pressures at the absence of the shock waves, to determine the arc parameters at the equilibrium state, we had used the system of equations

$$\begin{cases} P_g = P_m + P_i \\ j = \sigma E \\ Q_{el} = Q_r \end{cases}. \tag{9.9}$$

where P_g is gas-kinetic pressure in the arc, P_m is magnetic pressure, P_i is pressure of the gas, j is current density, σ is conductivity, E is electric field intensity, Q_{el} is electric power, and Q_r is radiation power. It is universal system of the equations from which the Pisa-Braginskii relationship follows.

9.3 500 kA Arc at Tungsten Electrodes

For the hydrogen plasma, neglecting by the external pressure, which in our opinion corresponds to the conditions of the experiment (Fig. 9.1), in dependence on the parameters of the arc, the system of (9.9) should be as follow

$$\begin{cases} 2n_i kT = 1.6 \times 10^{-4}\frac{I^2}{r^2} \\ \frac{1.5\times10^{-4}T^{\frac{3}{2}}E}{\ln\Lambda} = \frac{I}{\pi r^2} \\ IE = \pi r^2\left(1.39 \times 10^{-34}T^{1/2} + 5.35 \times 10^{-29}T^{-1/2}\right)n_i^2 \end{cases}. \tag{9.10}$$

where I is current (A), E is electric field intensity in the arc (V/cm), T is temperature (K), r is arc radius (cm), n_i is ion concentration (cm^{-3}), and k is the Boltzmann constant.

In view of the high radiant heat conductivity, the temperature across the section of the arc of radius r is assumed to be the same.

In the first equation of the system, the plasma of the arc is assumed to be completely ionized, when the gas-kinetic pressure is balanced by the magnetic pressure.

The second equation is Ohm's law for the plasma with conductivity according to Spitzer [4]. At the times close to the maximal arc contraction when $n_i = n_e \sim 7 \times 10^{20}$ cm^{-3}, then $T \sim 4 \times 10^5$ K (Table 9.1), at this the Coulomb logarithm $\ln\Lambda \sim 6$.

In the third equation, the plasma is assumed to be transparent radiating the recombination Q_{rec} and the bremsstrahlung Q_{br} spectra, so the third equation can be represented as

$$IE = \pi r^2 (Q_{br} + Q_{rec}) = \pi r^2 K Q_{br}, \tag{9.11}$$

where according to the data of the work [14] $Q_{br} = 1.39 \times 10^{-34} T^{1/2} n_i^2$, $Q_{rec} = 5.35 \times 10^{-29} T^{-1/2} n_i^2$, $n_e = n_i$.

The coefficient K shows how many times the radiation from the unit of the arc volume is higher than the bremsstrahlung

$$K = \frac{(Q_{br} + Q_{rec})}{Q_{br}} = 1 + 3.8 \times \frac{10^5}{T}. \tag{9.12}$$

The system of (9.12) is used as a first approximation for estimating the possibility of the radiative contraction of the nonmagnetized dense hydrogen plasma with recombination-bremsstrahlung. The arc plasma is assumed to be transparent at its maximal compression, since the path length of the quantum of radiation by the Planck is much larger than the arc dimensions, and, as will be shown later, for initial pressure 5 MPa, at moment of maximal arc contraction, the absorption of radiation in the transition layer can be neglected.

As for the presence of metal in the arc, the estimations show that at I_{max} = 500 kA and P_i = 5 MPa, the conductivity and radiative characteristics will be determined mainly by the hydrogen. If the concentration of metal vapor of the ignition wire and electrodes $n < 10^{18}$ cm^{-3}, so the concentration of hydrogen ions is 1.6×10^{20} cm^{-3}. The average concentration of the metal wire vapor in the discharge chamber is $n \sim 10^{19}$ cm^{-3}. However, after wire explosion, the main mass of the metal vapor is concentrated in the front of the shock wave, so the metal concentration in the arc is lower. In addition, at the explosion of the wire, some of it remains in the form of nonevaporated fragments (Fig. 5.18) that also reduces the concentration of metal in the discharge chamber.

At the specific erosion of the electrodes metal is $\sim 10^{-4}$ g/C, the typical concentration of metal vapor, formed by electrode jets from rapidly moving spots is $n < 10^{18}$ cm^{-3}. The estimations show that at current of $\sim$500 kA and tungsten electrodes, the time of the electrode jet formation is $\sim 10^{-4}$ s, i.e., by two orders of magnitude higher than for discharges at current of $\sim$1.6 MA and steel electrodes. Thus, the electrode jet is appeared at the end of the first half-period of the discharge current and does not effect on the process of the arc contraction.

In addition, as will be shown below, for the metal plasma after reaching a current of 480 kA, the arc would continue to contract, which is not observed at the experiment.

Table 9.1 Critical currents and arc parameters near the current maximum

Time from maximal compression	r (cm)	I (kA)	E (V/cm)	σ $(\Omega \times \text{cm})^{-1}$	P (Pa)	T (K)	n_i (cm^{-3})	l_r (cm)	Q_J (W/cm)	Q_{rad} (W/cm)	I_{cr} (kA)	Arc action
−1.4 μs—t_0	0.15	470	1200	5.6×10^3	1.6×10^9	2.2×10^5	2.6×10^{20}	6	5.6×10^8	8.5×10^8	380	$I > I_{cr}$ compression
0—t_1	0.10	480	1600	9.5×10^3	3.7×10^9	4×10^5	7.7×10^{20}	26	7.7×10^8	7.4×10^8	490	$I \sim I_{cr}$ balance
+6 μs—t_2	0.15	500	300	2.4×10^4	1.7×10^9	9×10^5	7×10^{19}	2.2×10^4	1.5×10^8	6.5×10^7	760	$I < I_{cr}$ expansion

Doubling the equations of the system (9.10), we can show that the system has a stationary solution only at one current $I = I_{cr}$

$$I_{cr} = 3.9 \times 10^5 \sqrt{\ln\Lambda/K}. \tag{9.13}$$

This result is analogous to that obtained by Braginskii in the work [15] and also in the work [16]. Here it has been shown that for the completely magnetized hydrogen plasma with recombination-bremsstrahlung $I_{cr} \sim 1.6$ MA, i.e. when $K = 1$ and $\ln\Lambda \sim 15$.

Expression (9.13) was obtained for the nonmagnetized, dense, hydrogen plasma with recombination-bremsstrahlung. Under these conditions at $n_i \sim 10^{20}$ cm^{-3}, even for arc radius $r = 0.1$ cm and current ~ 0.5 MA, the frequency of Coulomb collisions ν_{ei} is higher than the cyclotron frequency ω.

Indeed, according to the data of the work [3],

$$\nu_{ei} = \frac{52 n_i \ln\Lambda}{T^{\frac{3}{2}}}, \quad \omega = 1.76 \times 10^6 \frac{I}{r}. \tag{9.14}$$

where n_i is ion concentration, T is temperature, H is magnetic field intensity, I is current, and r is arc radius.

At the maximum arc contraction, as will be shown below, at $r = 0.1$ cm, $E = 1{,}600$ V/cm, $I = 480$ kA, and $T = 4 \times 10^5$ K: $n_i = 7.7 \times 10^{20}$ cm^{-3}. In this case, $\nu_{ei} \sim 10^{15}$ s^{-1} and $\omega \sim 10^{13}$ s^{-1}.

The impact of the magnetic field on the conductivity of the arc plasma σ along the electric field vector is estimated from the relation [4]

$$\sigma = \frac{\sigma_0}{1 + \left(\frac{\omega}{\nu_{ei}}\right)^2}, \tag{9.15}$$

where σ_0 is conductivity of the plasma at absence of a magnetic field.

This relationship shows that the magnetic field does not effect on the conductivity.

From the system of (9.10), it also follows that

$$\frac{Q_{rad}}{Q_J} = \left(\frac{I}{I_{cr}}\right)^2, \tag{9.16}$$

therefore, when $I > I_{cr}$ the arc is contracted, and at $I < I_{cr}$, the arc is expanded, because at $I > I_{cr}$, the gas-kinetic pressure is decreased, and the magnetic pressure remains the same and vice versa. This expression is valid both for the cases considered by Braginskii and for the metal plasma with multiply charged ions [16]. Therefore, it is suitable for estimating the critical current.

At the purely bremsstrahlung, the system of (9.10) does not allow us to determine T and r; i.e., when the critical current is reached, the arc is contracted, and this is due to that $Q_J \sim (T^{3/2} r^2)^{-1}$ and $Q_{rad} \sim (T^{3/2} r^2)^{-1}$. The diameter of the arc is stabilized when Q_J grows due to the increase of the arc resistance, for example, due to the appearance of instabilities or when Q_{rad} decreases due to the opacity of the arc [17].

In our case, when the arc temperature is such that we must take into account the recombination radiation, the solution of the system at the critical current is due to the dependence of K on T (Formula 9.14). For maximal arc contraction at $r = 0.1$ cm, $E = 1600$ V/cm, and $I = 480$ kA, the temperature T can be determined from the system (9.10), and using T, by the Formulas 9.14 and 9.15, estimate the critical current I_{cr}. The critical current I_{cr} is higher than the effective current I by 30% at maximal contraction of the arc.

One possible reason for the error of such estimation may be the lack of correction for the nonideality of the plasma. Indeed, at $n_i \sim 4 \times 10^{20}$ cm^{-3} and $T \sim 4 \times 10^5$ K for hydrogen plasma, the corrections for nonideality on pressure are 10% and on conductivity is 25% [18]. Other reasons for the error of determining I_{cr} may be an inaccurate estimate of the radius of the conducting zone r and an error in the determination of the electric field intensity in the arc E. In addition, a small addition of metal vapor of concentration $n_i \leq 10^{18}$ cm^{-3} may affect the radiative characteristics of the arc.

More accurate values of I_{cr} had been determined for three times: before the maximal contraction $t_0 - 1.4$ μs, at maximal contraction t_1, and 6 μs after maximal contraction t_2 (Table 9.1).

The arc radius r was determined from the photo-scanning; the electric field intensity in the arc E was calculated based on the experimental data taking into account the near-electrode voltage drops and the inductance of the discharge circuit. Based on r and E, the conductivity σ and the magnetic pressure P_m were determined. The gas-kinetic pressure P_g is assumed equal to the magnetic pressure P_m, and the external pressure is neglected

$$P = 1.6 \times 10^{-4} I^2 / r^2, \quad \sigma = I / \pi r^2 E, \tag{9.17}$$

where I is current, P is pressure, and r is arc radius.

Based on the obtained data of the pressure P, conductivity σ, and according to the data of the work [18], the temperature T and the ion concentration n_i were determined. The critical current I_{cr} was calculated by the Formula (9.16).

For all three cases, the path length of the quantum of radiation according to Planck [3] is much larger than the arc diameter, and thus, the arc plasma can be considered as transparent. The power of recombination-bremsstrahlung of the hydrogen was calculated by the formula from the work [14]

$$Q_{rad} = (1.39 \times 10^{-34} T^{1/2} + 5.35 \times 10^{-29} T^{1/2}) \pi r^2 n_i. \tag{9.18}$$

The arc parameters and critical currents obtained from experimental data are presented in Table 9.1. At this, the characteristics of the arc were determined at the assumption that the plasma is hydrogen. If the metal predominates in the arc plasma, then at the same pressure and conductance corresponding to t_1, according to data of the work [19] $n_i = 6 \times 10^{18}$ cm^{-3}, $T = 220$ eV and mean free path of the radiation quantum averaged over the Planck $l_r = 20$ cm. In this case, for an optically thin body [3]

$$Q_{rad} = \frac{4\pi\sigma T^4 r^2}{l_r} \sim 1.5 \times 10^{12}\ \mathrm{W/cm}\ \text{и}\ Q_{rad} >> Q_J, \tag{9.19}$$

therefore the arc must continue to contract, which is not observed at the experiment. Thus, at initial hydrogen pressure of 5 MPa and current of 500 kA, the recorded contraction of the arc with tungsten electrodes occurs predominantly in the hydrogen at the mode of the radiative contraction. The decrease of the critical current in comparison with the Pisa-Braginskii current is connected with the decrease of the ln Λ by 2–3 times in comparison with high-temperature, low-pressure arcs, where this value is 10–15 and with recombination radiation, which leads to the increase of the K.

9.4 1.6 MA Arc at Steel Electrodes

At current of 1.6 MA and steel electrodes, due to the increase of the heat flux onto the electrode q and other thermophysical properties of the electrode metal, the time for the occurrence of the electrode jet is approximately by two orders of magnitude lower than at current of ~500 kA and tungsten electrodes and is several microseconds. Therefore, in this case, the arc in the metal vapor surrounded by the hydrogen of high density is considered. The advantage of metals with the high atomic weight used as electrodes is the possibility to achieve a high energy density because of higher degree of the arc contraction due to radiation and ionization energy losses [20, 21]. Because of the high (~100) value of the K due to the high emission in the lines of multiply charged ions [16], the critical current for the arcs in the vacuum is reduced to ~100 kA. The decrease of the critical current occurs even if only the recombination-bremsstrahlung of multiply charged ions is taken into account.

For the multiply charged semitransparent plasma, the system of (9.9) takes the form

$$\begin{cases} (1+Z)n_i kT = 1.6 \times 10^{-4}\frac{I^2}{r^2} \\ \frac{1.5\times 10^{-4} T^{\frac{3}{2}} E}{Z\ln\Lambda} = \frac{I}{\pi r^2} \\ IE = \frac{4\pi\sigma T^4 r^2}{l_f} \end{cases}. \tag{9.20}$$

where Z is average charge of the ion, which is approximated by the expression $Z = 1.6 \times 10^{-2}\sqrt{T}$ with an error of ~10% at temperature of 6×10^4–10^6 K and heavy particle concentration 10^{18}–10^{20} cm^{-3}, n_i is ion concentration, k is the

Boltzmann constant, I is current (A), r is arc radius, E is electric field intensity in the arc, l_f is the photon mean free path averaged over the Planck, σ is Stephan Boltzmann constant, and $\ln\Lambda$ is the Coulomb logarithm.

$$l_f = \frac{1.1 \times 10^{23} T^{7/2}}{n_i^2 Z(Z+1)^2 \bar{x}_1}, \tag{9.21}$$

where $\bar{x}_1 - I/kT$ is average relative ionization potential of the metal.

According to the data of the work [13] at the temperature of 6×10^4–10^6 K and concentration of the metal ions 10^{18}–10^{20} cm^{-3}: $\bar{x}_1 \sim 5$. At the temperature of 6×10^4–3×10^6 K and concentration of heavy particles of 10^{18}–10^{20} cm^{-3}: $n_i \sim 10^{19}$–10^{20} cm^{-3}, which gives the average error of $\bar{x}_1 \sim 20\%$ [19].

The third equation of the system (9.20) is applicable when $r \ll l_f$; further estimates show that this relation is satisfied.

Substituting these relations into the equations of the system (9.20) and assuming that $(1 + Z) \sim Z$, we obtain $I_{cr} = 0.7 \times 10^5 \sqrt{\ln\Lambda}$, since for the dense plasma $n_i \sim 10^{20}$–10^{21} cm^{-3}, $T \sim 10^6$ K, $\ln\Lambda \sim 6$, then $I_{cr} \sim 170$ kA. This value must be somewhat lower if we take into account the radiation in the lines of multiply charged ions. For example, in the works [22, 23], the I_{cr} for the iron vapor plasma, at which the radiative contraction was observed, is less than 150 kA. This estimation shows that the critical current for the multicharge ions plasma decreases at least by an order of magnitude in comparison with the classical value I_{cr} obtained first by Braginskii.

We assumed that the contraction of the arc observed in the photo-scanning (Figs. 9.2, 9.3, and 9.8) is due to the achievement of I_{cr}. However, the experimental current $I_{exp} = 0.5$–1.2 MA, at which the contraction begins, is to be almost by an order of magnitude higher than that generally accepted for vacuum low-inductance arcs with the metal plasma of multiply charged ions.

In our opinion, this increase of I_{cr} is associated with decrease of the emissivity of the arc, which is caused by the absorption of the radiation by the transition layer between the arc and surrounding gas. To confirm this, the critical current under our experimental conditions was estimated.

The dependence of K on the absorption of radiation in the transition layer of the arc in formula (9.22) was taken into account for the arc with the metal plasma at initial hydrogen pressure of 32 MPa [16]

$$I_{cr} = 0.27\sqrt{\ln\Lambda / K}, \tag{9.22}$$

where $\ln\Lambda$ is the Coulomb logarithm, and K is the ratio of the total power of the volume radiation to the power of the volume bremsstrahlung.

The arc radius r was determined from the photostreak (Fig. 9.3). In this case, it was necessary to determine which brightness is corresponded to the current radius. The arc diameters, determined at the different density of the filters, were compared with the oscillograms of the current and voltage. The brightness corresponding to the current radius was chosen so that the relative change of the radius $\Delta r/r$ correspond to the registered $\Delta E \sim 500$ V/cm. At this, we have used the relations

$$E = I/\pi r^2\sigma; \quad \Delta E = \frac{\partial E}{\partial r}\Delta r + \frac{\partial E}{\partial r}\Delta\sigma = 2E\frac{\Delta r}{r} + E\frac{\Delta\sigma}{\sigma}, \tag{9.23}$$

where E is electric field intensity in the arc, determined from the voltage drop taking into account the near-electrode drops and the inductance of the discharge circuit, σ is conductivity, r is arc radius.

The change of the arc inductance corresponds to an order of magnitude less change of ΔE. To the current radius is corresponded the relative brightness of 0.7 on the photo-scanning of the arc glow. The arc radius oscillation, in our opinion, is related to the interaction of the magnetic and gas-kinetic pressures [24–26].

Based on the experimental values of r, E, and I, the conductivity σ and the magnetic pressure P_m were determined assuming that P_m is equal to the gas-kinetic pressure P_g, while the external pressure was neglected. According to the data of the work [19], the temperature T and the ion concentration n_i were determined from pressure and conductivity.

Before the decrease of the arc glow, its temperature determined from the conductivity and pressure was ~9 eV. If we assume that the arc radiates as the absolutely black body, then the equality between the power of the arc radiation Q_{rad} and the electric power Q_J is satisfied. Then, the sharp decrease of the arc glow is observed associated with increase of the transition layer temperature and decrease of its transparency; at this, the arc temperature rises.

After ~6 μs from the decrease of the brightness temperature of the arc, its contraction occurs. The brightness temperature of the arc $T_{br} \sim 1$ eV, $E = 1100$ V/cm, $r = 0.45$ cm, and $I = 1.2$ MA. These values correspond to $n \sim 1.7 \times 10^{19}$ cm^{-3}, $T \sim 40$ eV, and the mean free path of the Planck radiation quantum $l_f \sim 13$ cm. At such length of the path of the radiation quantum, the arc plasma is transparent, and the radiation power per unit length without taking into account absorption in the transition layer

$$Q_{rad} = 4\pi\sigma T^4 r^2 l_f^{-1} = 1.3 \times 10^9 \text{W/cm}. \tag{9.24}$$

At this $Q_J = \text{IE} = 1.3 \times 10^9$ W/cm.

Let us determine what the parameters of the transition layer should be, so that the current 1.2 MA, at which contraction begins, could be considered critical, i.e., the condition

$$Q_{rad} \times \exp(-n_n\sigma_i l) = Q_J. \tag{9.25}$$

where $\exp(-n_n\sigma_i l)$ takes into account the absorption of the arc radiation in transition layer of thickness l.

The concentration of neutral hydrogen atoms in the transition layer $n_{na} = 2.1 \times 10^{20}$ cm^{-3} was determined from the experimental brightness temperature $T \sim 1$ eV under the condition of thermodynamic equilibrium in the transition layer and at pressure of 32 MPa. It is assumed here that the main share of the energy, in the case of metallic plasma, is emitted from the arc by quanta with energy $hv \sim 1.5\ kT$ [20] with photoionization absorption cross section

$$\sigma_{\rm i} = 5.42 \times 10^{-17} (hv_0/hv)^{3.5}, \tag{9.26}$$

where hv_0 is hydrogen ionization potential of 13.6 eV [3].

Then, $\sigma_{\rm i} = 3 \times 10^{-19}$ cm^2. If we assume that the relation (9.26) holds, then

$$\exp(-n_{\rm na}\sigma_i l) = Q_J/Q_{\rm rad} = 2.5 \times 10^{-2}, \tag{9.27}$$

hence we have $n_{\rm na}\sigma_{\rm i}l = 3.7$ and $l = 5.8 \times 10^{-2}$ cm.

From comparison of the sizes of the current and glow zones determined from the photo-scanning, we obtain a close value $l \sim 0.1$ cm (Fig. 9.3), which coincides with the data of the work [27], where $l = 0.06$ cm.

The estimation of the parameters of the transition layer shows that the relation (9.26) can be satisfied, and the current of 1.2 MA can be considered as critical for the beginning of the radiative contraction of the arc. At the process of the arc contraction, it is heated further. So, when the arc is contracted to $r = 0.16$ cm, from the values of conductivity and pressure, it follows that $T = 260$ eV, $E = 1{,}200$ V/cm, $n_{\rm i} = 10^{19}$ cm^{-3}, $l_{\rm f} = 17$ cm. The radiation power in this case is higher than the Joule heating power

$$Q_{\rm rad} = 8.5 \times 10^{12} \cdot \exp(-n_{\rm na}\sigma_{\rm i}l) \gg Q_J = 1.4 \times 10^9 \text{W/cm} \tag{9.28}$$

where $\sigma_{\rm i}l$ is absorption cross section in the transition layer corresponding to the quantum energy of 1.5 kT at $T = 260$ eV, since $\sigma_{\rm i}l \ll \sigma_{\rm i}$ and $exp\ (-n_{\rm na}\sigma_{\rm i}l) \approx 1$.

At further reduced of the arc radius, the condition $Q_{\rm rad} > Q_J$, which is necessary for radiative compression, is fulfilled.

Thus, before the arc contraction, the critical current is ~1.2 MA, and at further reduction of the arc radius, the conditions for its further radiative contraction are fulfilled.

According to the work [28], the decrease of the emissivity of the arc, which may be due to various causes, leads to increase of $I_{\rm cr}$. For example, in the work [29], the mode of radiative contraction of the arc in hydrogen and helium was considered, where it was shown that self-absorption of radiation in the arc leads to increase of $I_{\rm cr}$.

In our case, one of the reasons for the decrease of the emissivity of the arc can be decrease of K (Formula 9.24) due to the absorption of radiation in the transition layer.

In this case, we write the formula (9.24) as

$$I_{cr} = 0.27\sqrt{\frac{\ln\Lambda}{K\exp(-n_{na}\sigma_i l)}}. \quad (9.29)$$

The presence of the transition layer according to formula (9.29) leads to increase of I_{cr} by $[\exp(-n_n\sigma_i l)]^{-1/2}$ times. In our case, $\exp(-n_n\sigma_i l) \sim 2.5 \times 10^{-2}$ and critical current increases approximately by an order of magnitude in comparison with the current of the arcs in vacuum.

Summarizing the results of the estimations, we can say that the observed contraction of the arc can be of radiative nature, and the growth of I_{cr} is due to the high density of the surrounding hydrogen.

9.5 Soft X-rays Irradiation

The above estimations show that the plasma temperature in the arc at times close to the maximal arc contraction is close to one hundred of electron volts, which must be accompanied by intense soft X-ray irradiation (SRI).

To register the SRI, the special diagnostic system was developed [30], by means of which SRI was registered at experiments under the current of 1.0–1.6 MA and initial hydrogen pressure of 5–7 MPa. Registration of SRI at higher hydrogen pressures could not be performed because of the large thickness of the walls of the discharge chamber of the plasma generator.

As a result, it was found that at the moment of maximal contraction of the arc, the energy of SRI quanta is ~1 keV, at this the plasma temperature reaching of hundreds of electron volts. Measurements have shown that SRI accompanies the whole process of current flow, and features on the oscillograms of the current correspond to bursts of SRI intensity.

The results of the estimates of the ratio of SRI intensities passing through aluminum filters of 10 and 18 μm thicknesses and the hydrogen layer of 25 cm thickness at pressure of 7 MPa are presented in Table 9.2.

Table 9.2 Ratio of SRI intensities

$h\nu$ (eV)	Al 10 μm + H_2	Al 18 μm + H_2	I_{10}/I_{18}
400	$<10^{-18}$	10^{-34}	10^{16}
600	1.9×10^{-8}	6.2×10^{-13}	3×10^4
700	5×10^{-6}	6×10^{-9}	800
800	1.5×10^{-4}	1.2×10^{-6}	130
900	1.1×10^{-3}	3.4×10^{-5}	33
1000	4×10^{-3}	3.1×10^{-4}	13
1200	5×10^{-2}	6×10^{-3}	8

For the calculation, the data of the works [31–33] were used. The registered ratio of SRI signals corresponds to quantum energy of 1.0–1.2 keV (Table 9.2).

Figure 9.4 shows the oscillograms of the signals from X-ray sensors together with oscillograms of the current and voltage. Filters for sensors are made of aluminum foils of thicknesses: 1 foil of 10 μm and 2 foil of 18 μm. The vertical line corresponds to the moment of maximal contraction of the arc. The peak observed at this moment on the voltage oscillogram is much higher than the value that could be caused by the change of the arc inductance due to reduction of its diameter.

Figure 9.4a shows that the contraction of the arc clearly corresponds to features on the oscillograms of the current and voltage. In this case, the registered ratio of the SRI intensities $I_{10}/I_{18} < 6$ that corresponds to $h\nu > 1.2$ keV.

At the experiment (Fig. 9.4b), contraction of the arc is less pronounced, which is confirmed by the signal from the SRI sensor. In this case, $I_{10}/I_{18} \geq 20$ that corresponds to $h\nu \leq 1$ keV. It is necessary to note, that this calculation is executed under assumption, that the arc plasma consist from metal vapour, and the main part of energy losses is radiated by quantums with energy of ~1.5 kT [19].

The output of radiation quanta with this energy was recorded from the low-inductance spark in the vacuum [34].

9.6 Heating of Hydrogen by X-ray Irradiation

Figure 9.9 shows the dependence of the efficiency η of the energy transfer from the arc to the hydrogen on the initial pressure P_i, at various electric energies inputted into the arc.

Curve 1 corresponds to the dependence of η at inputted energy in the arc of ≥300 kJ, curve 2 at energy ≤300 kJ, and curve 3 at constant energy of SRI quanta.

Dependencies have the growing character. The value of η was determined as the ratio of the energy transferred to the hydrogen to electric energy inputted into the arc. The energy transferred to hydrogen was determined on a difference between initial and final pressures of gas, the electrical energy as $\int IU dt$. At estimating the energy of X-ray quanta, it was assumed that the hydrogen heating takes place only by means of them. In this case, the radiation energy onto the chamber wall

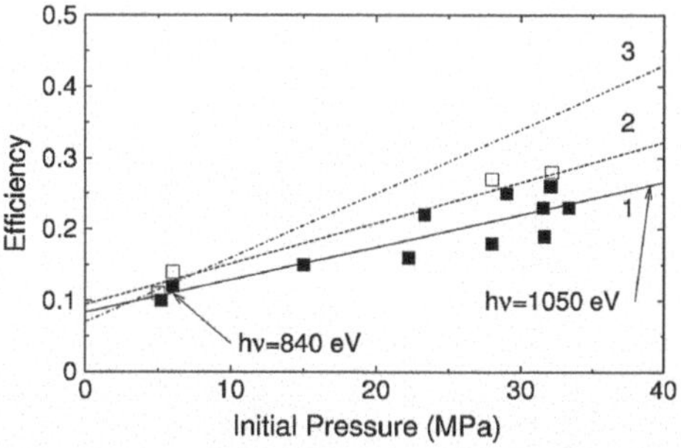

Fig. 9.9 Efficiency of the energy transfer from the arc to hydrogen versus initial pressure [1]

$$I = I_0 \exp(-n\sigma l), \tag{9.30}$$

where n is concentration of hydrogen molecules, l is distance from the arc to the chamber wall, and σ is absorption cross section of quantum [3].

$$\sigma = 5.4 \times 10^{-17} \left(\frac{h\nu_0}{h\nu} \right)^{3.5}. \tag{9.31}$$

The average energy of SRI quantum estimated from the ratio $1 - \eta = \exp(-n\sigma l)$ at initial pressure of 5 MPa is ~840 eV, and at 35 MPa is ~1.050 eV.

Assuming $h\nu \sim 1.5$ kT for metal plasma, then at increasing of initial hydrogen pressure, the temperature of the central zone of the arc is increased too. If the energy of the SRI quanta were constant, then due to the growth of the initial hydrogen pressure, the dependence of η on P_i would correspond to curve 3 (Fig. 9.9). Thus, at the increasing of the radiation "imprisonment," the temperature in the axial zone of the arc is increased.

The dependence of η on the electric energy inputted in the arc (Fig. 9.10) shows that at decrease of the energy input to the arc, the efficiency of energy transfer from the arc to the hydrogen is increased due to decrease of the arc plasma temperature and, as a consequence, increase of the cross section of the absorption of the quantum.

9.7 Arc Glow Change at Contraction

The glow of the arc after its maximal contraction is decreased and then remains at a level corresponding to the brightness temperature $T_{br} = (10\text{–}15) \times 10^3$ K (Fig. 9.5). This means that the transition layer glows itself heated by SRI, at this the

Fig. 9.10 Efficiency of the energy transfer from the arc to the hydrogen versus energy inputted into the arc [1]

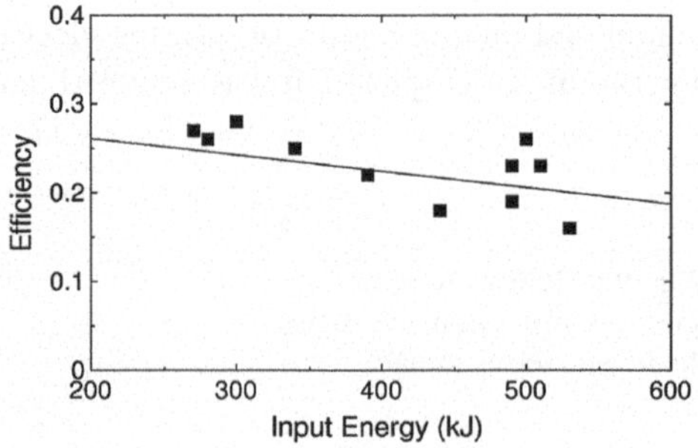

Table 9.3 Arc temperature

T_{br} (K)	T_a (eV)
20,000	29
15,000	87

radiation from the arc in the visible range of the spectrum is completely absorbed. In this case, the mean free path of the radiation quantum in the visible range l_λ becomes equal to, and then less than, the width of the transition layer [3]

$$l_\lambda = \frac{1}{x} = \frac{T_{br}^2 x^3}{0.96 \times 10^{-7} n e^{-(x_\lambda - x)}}, \tag{9.32}$$

where $x = h\nu/kT_{br}$, $x_1 = h\nu_1/kT_{br}$, $h\nu = 1.79$ eV, $h\nu_1 = 13.6$ eV, and n is concentration of hydrogen atoms.

In its turn, the mean free path of the SRI quantum is equal to the thickness of the transition layer Δx. According to the data of the work [3], $\sigma = 5.42 \times 10^{-17} (h\nu_1/h\nu)^{3.5}$ for $h\nu \gg h\nu_1$. For hydrogen, $h\nu_1 = 13.6$ eV. Then,

$$\Delta x = 1/n\sigma = \left[5.42 \times 10^{-17} n (13.6/1.5T)^{3.5}\right]^{-1}. \tag{9.33}$$

It is assumed that the main share of SRI energy is transferred by quanta with energy $h\nu = 1.5\ kT$, where T is average arc temperature.

From the relation $l_\lambda = \Delta x$ for $\lambda = 694$ nm ($h\nu = 1.79$ eV) at the arc brightness temperature T_{br} changing (from 1.5 to 2) $\times 10^4$ K, the arc temperature T_a is changed from 87 to 29 eV (Table 9.3).

For $\lambda = 550$ nm, the temperature is slightly higher, as radiation is emitted from the deeper layers of the arc.

The similar phenomenon is observed in the powerful shock waves. The brightness temperature of the powerful shock wave becomes smaller than the true temperature behind the front. So, for the shock wave in the air at true temperature behind the front of $\sim 9 \times 10^4$ K, the brightness temperature is $\sim 2 \times 10^4$ K [3].

References

1. A.A. Bogomaz, Doctoral Thesis, St.-Petersburg, 2012 (in Russian)
2. S.J. Luk'janoiv, *Hot Plasma and Controlled Nuclear Synthesis* (Moscow, 1975) (in Russian)
3. Y.B. Zel'dovich, Y.P. Raizer, *Physics of Shock Waves and High-Temperature Hydrodynamic Phenomena,* (New York, 1966)
4. Y.P. Raizer, *Physics of the Gas Discharge* (Moscow, 1987) (in Russian)
5. V.L. Granovskiy, *Electric Current in Gas. Established current* (Moscow, 1971) (in Russian)
6. J.K. Bobro, V.V. Vikhrev, I.P. Fedotov, Rus. J. Plasma Phys. Rep. **14**, 1222 (1988). (in Russian)
7. F.G. Baksht, A.M. Voronov, V.N. Zhuravlev, Sov. J. Tech. Phys. **61**, 53 (1991). (in Russian)
8. A.M. Voronov, V.L. Gorjachev, V.N. Zhuravlev, Tech. Phys. Lett. **19**, 35 (1993)

9. A.T. Onufriev, V.G. Sevastjanenko, Sov. J. Appl. Mech. Tech. Phys. **2**, 17 (1968). (in Russian)
10. V.V. Vikhrev, S.I. Braginskiy, *Questions of Plasma Theory* (Moscow, 1980) (in Russian)
11. N. Kroll, A. Trajvelpils, *Bases of Plasma Physics* (Moscow, 1975) (in Russian)
12. Y.G. Kalinin, A.S. Kingsep, V.P. Smirnov, YuL Bakshaev, A.V. Bartov, P.I. Blinov, S.A. Dan'ko, L.G. Dubas, A.V. Korel'ski, V.D. Korolev, V.I. Mizhiritski, G.I. Ustroev, A.S. Chernenko, R.V. Chikin, Ay Shashkov, Z. Li, X. Hua, X. Peng, S. Feng, C. Guo, Plasma Phys. Rep. **32**, 656 (2006)
13. S.G. Alikhanov, I.K. Konkambarv, B.S. Estrin, Sov. J. High Temp. **5**, 410 (1967). (in Russian)
14. I.M. Podgornuiy, *Lectures on Diagnostics of Plasma* (Moscow, 1968) (in Russian)
15. S.I. Braginskiy, Sov. Phys. JETP **6**, 494 (1958)
16. V.V. Vikhrev, V.V. Ivanov, K.N. Koshelev, Sov. J. Plasma Phys. **8**, 688 (1982)
17. K.N. Koshelev, N.R. Pereira, J. Appl. Phys. **69**, 21 (1991)
18. O.V. Koryshev, D.O. Nogotkov, Yu.Yu. Protasov, V.D. Telekh, *The Thermodynamic, Optical, and Transport Properties of the Working Media of Plasma and Photon Power Plants* (Bauman Moscow State Tech. Univ., 1999) (in Russian)
19. B.V. Zamyshlyaev, E.L. Stupitskiy, A.G. Guz', V.G. Zhukov, *Composition and Thermodynamic Functions of Plasma, Handbook* (Moscow, Énergoatomizdat, 1984) (in Russian)
20. E.V. Aglickiy, V.V. Vikhrov, A.V. Gulov, *Multicharge Spectroscopy in High temperature Plasma* (Nauka, Moscow, 1991). (in Russian)
21. V.V. Jan'kov, Sov. J. Plasma Phys. Rep. **17**, 521 (1991) (in Russian)
22. A.N. Dolgov, V.V. Vikhrev, Plasma Phys. Rep. **31**, 259 (2005)
23. A.E. Gurei, A.A. Tikhomirov, A.N. Dolgov, D.E. Prokhorovich, A.S. Savjolov, Plasma Phys. Rep. **30**, 38 (2004)
24. B.E. Mejerovitch, *High Current Channel* (Moscow, 1999) (in Russian)
25. L.S. Solov'ev, Sov. J. Plasma Phys. Rep. **10**, 1045 (1984) (in Russian)
26. B.E. Mejerovitch, Sov. J. Plasma Phys. Rep. **11**, 1446 (1985). (in Russian)
27. A.T. Onufriev, V.G. Sevast'janov, Sov. J. Appl. Mech. Tech. Phys. **2**, 17 (1968). (in Russian)
28. J.W. Shearer, Phys. Fluids **19**, 1426 (1976)
29. A.E. Robson, Phys. Fluids **1**, 1834 (1989)
30. A.A. Bogomaz, A.V. Budin, V.V. Zabrodskiy, I.V. Kuznetsova, SYu. Losev, M.V. Petrenko, M.E. Pinchuk, P.G. Rutberg, Instr. Exp. Tech. **51**, 744 (2008)
31. B.L. Henke, E.M. Gullikson, J.C. Davis, Atomic Data Nucl. Data Tables **54**, 181 (1993) http://henke.lbl.gov/optical
32. http://physics.nist.gov/PhysRefData/contents-xray.html
33. M.A. Blokhin, I.G. Shveitser, *Directory of X-Ray Spectra* (Moscow, 1982) (in Russian)
34. O.A. Bashutin, A.S. Savjolov, E.D. Vovchenko, Plasma Phys. Rep. **35**, 813 (2009)

Chapter 10
Arc at Ultrahigh Pressure

Abstract Most publications presenting the results of studies of high-current pulsed arcs in gases describe the discharges in the vacuum or in gas at low pressure. At the same time, pulsed arcs in gases at high and ultrahigh initial pressures are the relatively poorly studied area of electrophysics. Our investigations of high-current pulsed arcs in super dense gases have revealed a number of characteristic phenomena inherent only for such arcs. Three installations have been developed to carry out the researches in this area: two two-stage electric discharge units with d*I*/d*t*—10^{10} A/s and one with d*I*/d*t*—10^{11} A/s. The two first installations were designed for operation at currents up to 500 kA and initial concentrations of normal molecules and atoms of gas to 3.3×10^{22} cm^{-3}. The PPG-5 plasma generator was designed to operate at current of up to 2 MA and initial concentration of molecules and atoms up to 10^{22} cm^{-3}. The aim of these studies was generation and diagnostic of the hydrogen and helium plasmas with the charged particles concentration up to 10^{21} cm^{-3} and temperature up to 50 eV. The results of the experiments show that the increase of the initial concentration of hydrogen molecules makes it possible to reach the temperature of the central arc zone above 30 eV at concentration of the electrons is 10^{19} to 10^{21} cm^{-3}. Since the plasma with such parameters at duration of steady state of $\sim$10 μs is the source of UV and soft X-ray radiation, the diagnostic of these radiations is added to the number of the basic problems. It is determined that increase of the initial gas pressure causes the increase of the electric field intensity in the arc and its contraction. Other important conclusion is that at such initial gas pressures, the increase of the discharge current from 100 to 500 kA and its rise rate from 10^9 to 10^{10} A/s leads to the increase of the near-electrode voltage drops. The coefficient of the energy transfer from the arc to gas at initial pressure $\geq$20 MPa reaches 90%.

10.1 Arc in Pre-compressed Gas

Study of the arc at ultrahigh initial gas pressure at two-stage electric discharge units (Chap. 3.2) was performed. At the installation with rise rate of current of 10^{10} A/s, the arc in hydrogen was investigated at adiabatic compression up to 350 MPa. It is

V. Kolikov et al., *Powerful Pulsed Plasma Generators*,
Springer Series on Atomic, Optical, and Plasma Physics 101,
https://doi.org/10.1007/978-3-319-95249-9_10

established that the diameter of the arc does not exceed 3–4 mm, and its temperature reaches of 1.8×10^5 K [1]. At increase of the initial concentration of hydrogen molecules from 10^{21} to 3×10^{22} cm^{-3}, the electric field intensity in the arc is increased from 300 to 1300 V/cm.

Similar studies were performed at the installation with the rise rate of the current 10^{11} A/s and current up to 500 kA. As the result of adiabatic compression, the initial concentration of hydrogen molecules was 2.3×10^{22} cm^{-3}.

Figure 10.1 shows the photo-scanning of the arc glow (top left), the pressure in the discharge chamber (bottom left), the current I and voltage U oscillograms (bottom right), and the graphs of the variation of arc diameter (D), brightness of the arc (B), and pressure in the discharge chamber (P) (top right). The experiment was carried out with tungsten electrodes of 6 mm diameter with interelectrode distance of 12 mm.

The process of adiabatic compression of hydrogen from 9.5 to 110 MPa had been recorded on the oscillogram of the pressure in the discharge chamber (bottom left). At the moment of maximal compression of hydrogen, the arc is initiated, as a result of that the pressure increases from 110 to 300 MPa. The change of the pressure during adiabatic compression of hydrogen is expressed by the equation

$$P_1 = P_i(V_i/V_f)^{\gamma}, \tag{10.1}$$

where γ is adiabatic index (1.4), P_1 is pressure after compression (110 MPa), P_i is initial pressure (9.5 MPa), V_i is initial volume of hydrogen, and V_f is final volume of hydrogen.

The distinctive feature of the pulsed arcing in hydrogen at initial pressure of several hundred megapascals is the oscillations of large amplitude of the voltage drop across the arc and pressure with frequency of 150–200 kHz. The amplitude of

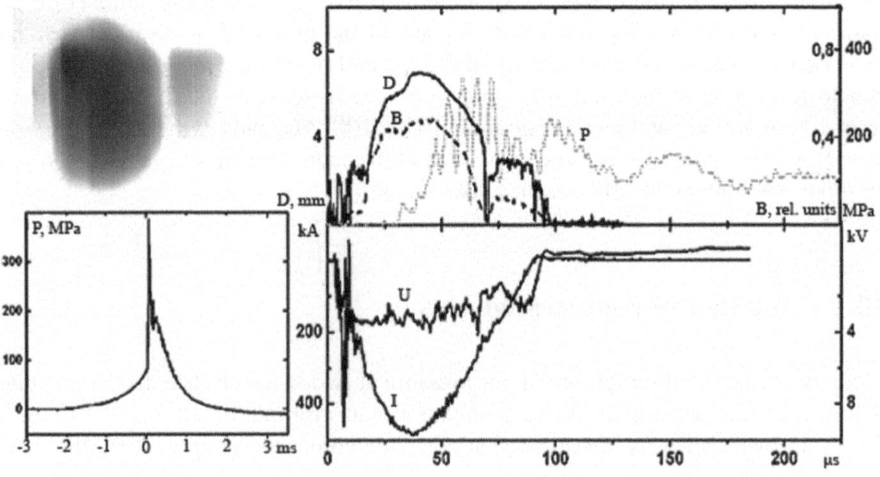

Fig. 10.1 Characteristics of the experiment at initial hydrogen pressure of 110 MPa [2]

the oscillations in some cases reaches and even exceeds the average value of the voltage drop and pressure. At lower initial pressures of 5–30 MPa, the amplitude of voltage drop and pressure oscillations does not exceed a few percent of their average value.

The photo-scanning of the arc glow and the oscillograms of the current and voltage drop (Fig. 10.1) show that the arc contraction begins after the maximum of the current. The pressure oscillogram shows that the beginning of arc contraction is sinphase with the appearance of intense shock waves with frequency of ~150 kHz. The end of the arc contraction corresponds to sharp decrease of the voltage and pressure oscillations. Indirect confirmation of the fact that the voltage and pressure oscillations are due to shock waves is that with increasing of initial hydrogen pressure their amplitude and number are increased.

Thus, for example, at initial hydrogen pressure of 12.6 MPa, there are no appreciable voltage oscillations, and the pressure oscillations are weak and have high frequency (Fig. 10.2).

At the experiment with adiabatic compression of hydrogen up to 84 MPa (Fig. 10.3), the amplitude of voltage drop and pressure oscillations is lower, in comparison with experiments where the pressure is much higher. So, the ratio of the hydrogen pressure at the experiments (Figs. 10.1 and 10.3) is ~1.3, and the ratio of the pressure oscillation amplitude is ~1.2. The oscillogram of pressure also shows

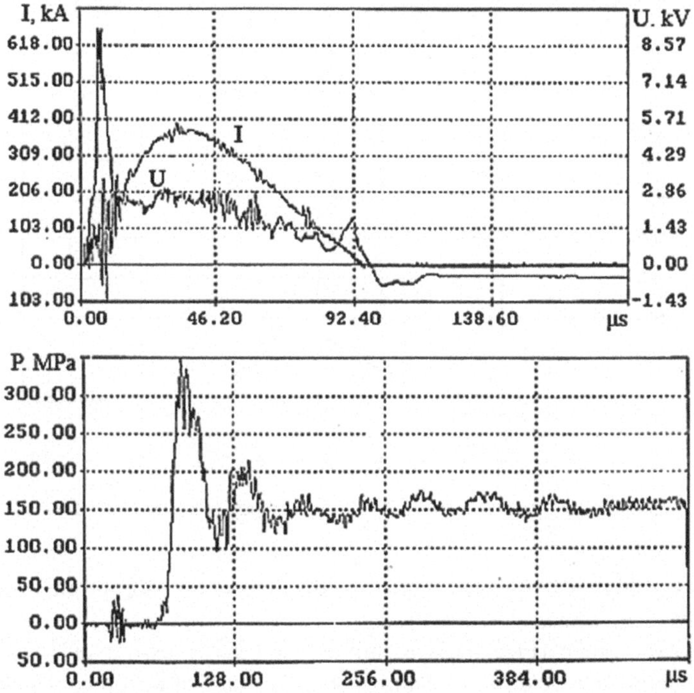

Fig. 10.2 Oscillograms of the experiment at initial hydrogen pressure of 12.6 MPa [2]

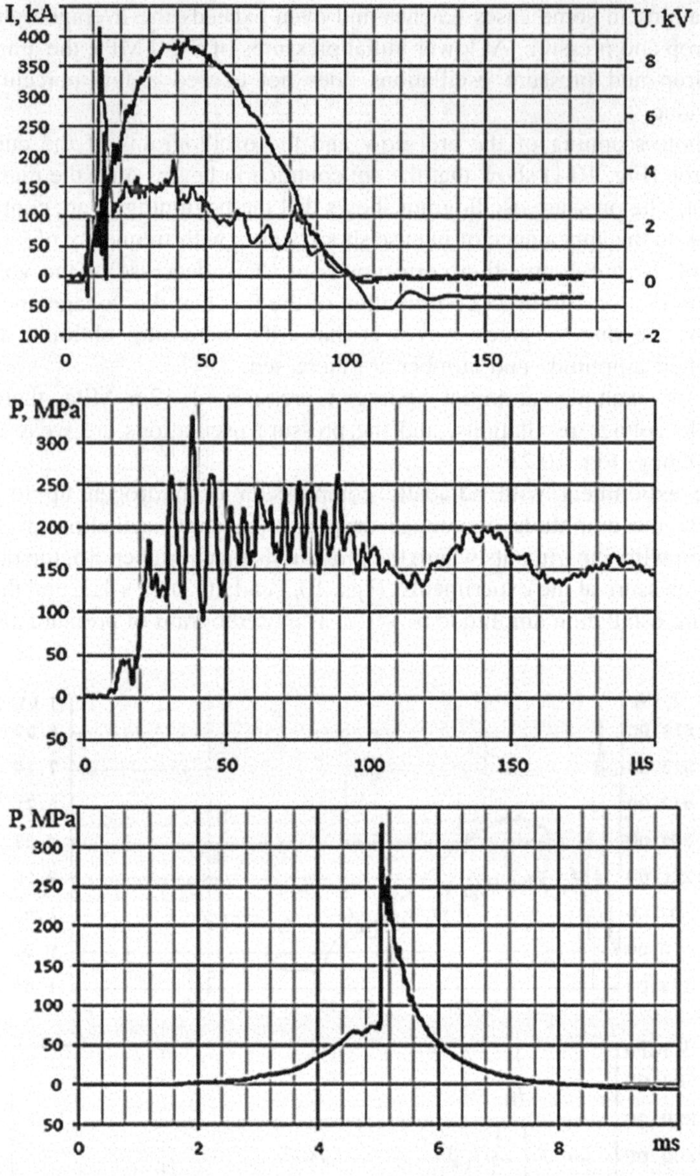

Fig. 10.3 Oscillograms of the experiment at initial hydrogen pressure of 84 MPa [2]

oscillations of higher frequency, which are disappeared at stop of the energy input into the arc (Fig. 10.3).

At the experiment (Fig. 10.5) at initial hydrogen pressure of 157 MPa, the amplitude of the voltage drop oscillations at initial stage of the process reaches 6 kV at average voltage of ~3 kV. To this time, when the amplitude of the voltage drop oscillations is highest, the feature on the current oscillogram is corresponded, which also occurs at number of other experiments.

To estimate the arc parameters, we use oscillograms of current and voltage, which in most cases are similar to the oscillograms of the experiment in Fig. 10.1. In addition, we use the results of estimates of the concentration and temperature of copper vapor in the arc from its conductivity and pressure, presented in the work [3].

The electric field intensity in the arc was determined from the slope of the curve of the dependence of the voltage drop across the discharge gap on its length.

It was established that the calculated and experimental parameters of the arc are close to each other in the case when it is assumed that the arc consists only of the vapor of the initiating copper wire.

Estimates of the time of occurrence of the electrode jet from tungsten electrodes at current of 490 kA (Chap. 5) give the value of ~ 200 μs, which on average is twice more than arcing time at series of these experiments.

At increasing of the initial hydrogen pressure up to 132 MPa at the experiments (Figs. 10.4 and 10.5), the voltage oscillations are more pronounced.

At the maximum of the discharge current, the arc radius determined from the photo-scanning is 0.35 cm, and the electric field intensity in the arc is 1.3×10^3 V/cm (Fig. 10.1). This corresponds to the specific conductivity of the arc plasma 960 $(\Omega \times \text{cm})^{-1}$.

The pressure in the arc is the sum of the magnetic pressure and hydrogen pressure in discharge chamber of 610 MPa. The hydrogen pressure in the discharge chamber was averaged by the data of two pressure transducers, by one of which recorded the pressure change at the adiabatic compression of hydrogen and subsequent heating of hydrogen by the arc, and by the second recorded the pressure in synchronism with the current and voltage.

The oscillograms in Fig. 10.1 show that the hydrogen pressure in the discharge chamber is 300 MPa. When the plasma consists of copper vapor, then with an error of $\sim 10\%$ for pressure in the arc of 610 MPa and conductivity of the plasma 960 $(\Omega \times \text{cm})^{-1}$ are corresponded the concentration of ions n_i $\sim 2 \times 10^{20}$ cm^{-3}, temperature $T \sim 9 \times 10^4$ K, and average ion charge $\bar{z} = 2.6$.

According to the estimations made earlier, at such high gas density the critical current, at which contraction of the arc begins, should be few megaamperes. In this case, the radiative contraction of the arc does not occur, and the change of the arc sizes must occur when inputted energy is equal to radiated energy.

Let us estimate the radiation power from the arc. According to the data of the work [4], the mean free path of the Rosseland l_r at multiple ionization is defined as

$$l_r = \frac{4.4 \times 10^{22} T^{7/2}}{n_i^2 \bar{Z} (\bar{Z} + 1)^2}. \qquad (10.2)$$

where T is arc temperature, n_i is ion concentration, and $\bar{Z}$ is average ion charge.

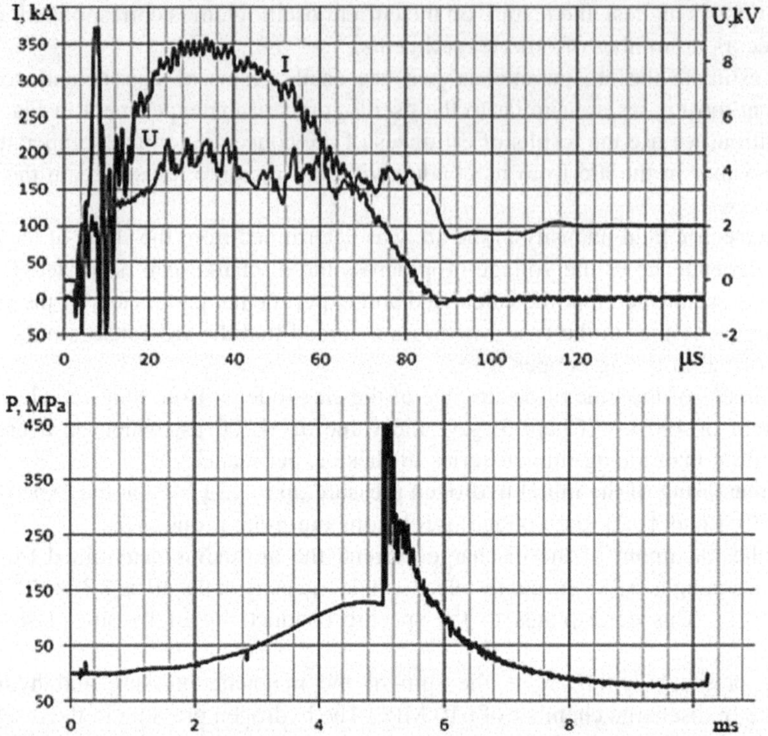

Fig. 10.4 Oscillograms of the experiment at initial hydrogen pressure of 132 MPa [2]

For the copper plasma $l_r = 10^{-2}$ cm, what is much less than the diameter of the arc (0.7 cm). Therefore, the arc radiates as the absolutely black body. In this case, the radiation power N_r is

$$N_r = 2\pi r l \sigma T^4 \frac{\int_0^x \frac{x^3 dx}{e^x - 1}}{\int_0^\infty \frac{x^3 dx}{e^x - 1}}; \quad x = \frac{h\nu}{kT}. \tag{10.3}$$

where r is radius of the arc, l is arc length, T is arc temperature, and $h\nu = 13.6$ eV is the ionization potential of the hydrogen.

Substituting $r = 0.35$ cm, $l = 1.2$ cm, and $T = 9 \times 10^5$ K, we have

$$\int_0^x \frac{x^3 dx}{e^x - 1} = 0.7; \quad \int_0^\infty \frac{x^3 dx}{e^x - 1} = 6.49. \tag{10.4}$$

At $\sigma = 5.67 \times 10^{-12}$ W/cm^2, the radiation power $N_i = 1.61 \times 10^8$ W at electric power of the arc $N_e = IUl = 7.6 \times 10^8$ W.

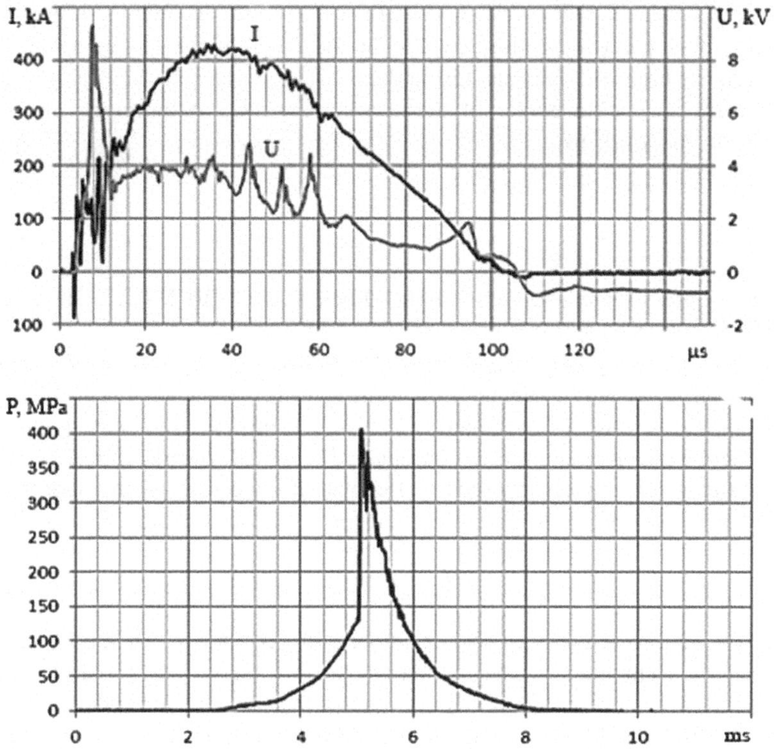

Fig. 10.5 Oscillograms of the experiment at initial hydrogen pressure of 157 MPa [2]

Let us show that short-wave quanta of radiation with energy 4 kT, by which in the case of radiant heat conductivity is transferred the main share of the energy, cannot go beyond the boundaries of the arc. According to the data of the work [5] to the hydrogen density of 4.16×10^{-2} g/cm^3 and pressure of 300 MPa, are corresponded the average hydrogen temperature in the discharge chamber of 1160 K and isothermal speed of sound 2.68×10^3 m/s.

The mean free path l_r of radiation quanta with energy of 4 kT*l*

$$l_r = 1/n\sigma. \tag{10.5}$$

where n is concentration of hydrogen, $\sigma = 8.3 \times 10^{-18}\,(h\nu_0/4\text{ kT})^3$ [4].

At $h\nu_0 = 13.6$ eV and 4 kT = 34.5 eV: $\sigma = 5.1 \times 10^{-19}$ cm^2. Then at $n = 1.2 \times 10^{22}$ cm^{-3}, the mean free path of quanta $l_r = 1.6 \times 10^{-4}$ cm, and, consequently, the radiation is actually "locked" in the arc. Therefore, at the temperature of 10^5 K almost all radiation from the arc by quanta with energy above the ionization potential of hydrogen will be absorbed in the transition layer, whose thickness by several orders of magnitude is less than the diameter of the arc.

The power balance will be satisfied when we take into account the power for the formation of shock waves, the pressure in the front of which according to the oscillogram in Fig. 10.1 reaches 150 MPa.

The average specific power of the shock wave N_{sw} at harmonic oscillations according to the work [6]

$$N_{sw} = \frac{1}{2}\frac{P_f^2}{\rho c}. \tag{10.6}$$

where P_f is pressure in the front of the shock wave (150 MPa), ρ is hydrogen density (10^{-2} g/cm^3), and c is speed of sound in hydrogen.

The speed of sound determined on the time of passage by the shock wave from the explosion of the wire to the pressure transducer place is 2.72×10^3 m/s, and by the density and pressure of hydrogen is 2.68×10^3 m/s [7]. Substituting the values of P_f, ρ, and c into formula (10.6), we obtain $N_{sw} = 10^7$ W/cm^2. Since the specific power of the shock wave is recorded on the wall of the discharge chamber, the total power of the shock wave N_t assuming that the wave has cylindrical shape

$$N_t = I2\pi r_{ch} l, \tag{10.7}$$

where r_{ch} is radius of the discharge chamber and l is interelectrode distance.

For $r_{ch} = 2.1$ cm and $l = 1.2$ cm, the total shock wave power $N_t = 1.6 \times 10^8$ W. Since the arc periodically contracts and expands, the shock waves reflected from the wall of the discharge chamber and cumulating in the center of the arc must appear. This is confirmed by recorded oscillations of the SRI intensity, whose frequency coincides with the frequency of the voltage oscillations.

The power of the shock waves cumulating in the center of the arc is also estimated from formula (10.6). Since for the shock waves of cylindrical shape $P_f \sim r^{-1/2}$, and at $r = 2.1$ cm, the pressure on the chamber wall is $P_f = 150$ MPa, then on the arc surface at $r = 0.35$ cm, the pressure $P_f = 370$ MPa. The density of copper vapor in the arc ρ at their concentration of 2×10^{20} cm^{-3} is 2.1×10^{-2} g/cm^3.

The speed of sound c in copper vapor is determined by the relation

$$c = \sqrt{\gamma P/\rho}, \tag{10.8}$$

where P is pressure in the arc (610 MPa) and γ is adiabatic index (1.1) [8].

Then $c = 5.6 \times 10^3$ m/s, and total shock wave power $N_t = I(2\pi r)/2 \sim 1.51 \times 10^8$ W.

The shock wave power $N_{sw} = 1.6 \times 10^8$ W was determined taking into account that the radiation is "locked" by the size of the discharge chamber, according to more accurate calculations [6], the total shock wave power N_t due to oscillations of the arc of length $h \equiv l$ is expressed as

$$N_t = \frac{2I\pi r^2 \lambda}{h}, \tag{10.9}$$

where I is radiation intensity at distance r from the arc axis and λ is wavelength of the radiation.

Substituting in the formula (10.9) values $I = 10^7$ W/cm^2, $r = 2.1$ cm, $\lambda = cT = 1.8$ cm, and $h = 1.2$ cm, we have $N_t = 4.2 \times 10^8$ W.

The power balance for the arc in the presence of shock waves $IEl = 7.6 \times 10^8$ W; $N_{tl} + P_w + N_t \sim 7.3 \times 10^8$ W, where N_{tl} is power of radiation passed through the transition layer, p_w is shock wave power on the chamber wall, and N_t is total shock wave power.

As shown earlier, the energy of the shock wave is transformed into energy of the SRI. Since the total radiation power of the arc and shock wave is 7.3×10^8 W, and the electric power of the arc is 7.6×10^8 W, the power balance is satisfied with error of $\sim$5%, with significant share of the electric energy inputted in the arc is transformed into the shock wave energy.

10.2 Arc at Current of 220 KA

Study of the pulsed electric arc in the hydrogen at high initial pressure was performed at the two-stage electric discharge installation with preliminary gas compression. Initial conditions of experiments are next:

- Energy in the capacitive storage up to 300 kJ
- Discharge current up to 220 kA
- Rise rate of the current $\sim 10^8$ A/s
- Initial voltage up to 6.0 kV
- Concentration of hydrogen molecules after compression up to 3.3×10^{22} cm^{-3}.

During the experiments, the electrical parameters of the arc and hydrogen pressure were measured. Estimated parameters under these initial conditions are as follows: The arc temperature is $(1.3–1.8) \times 10^5$ K, concentration of charged particles $(0.7–1.2) \times 10^{20}$ cm^{-3}, and arc radius is 0.14–0.2 cm.

The two-stage electric discharge installation (Fig. 3.12) with dI/dt of 10^{10} A/s was used to perform the researches. At its development, the combination of the principles piston adiabatic compression of hydrogen and pulsed electric discharge was used. The power source was the capacitive energy storage CPS-10 with capacitance of 0.11 F and maximal voltage of 10 kV [9].

The installation operates as follows: Before the experiment, the discharge chamber and the compression channel are filled with hydrogen up to pressure of 10–25 MPa. Adiabatic compression of hydrogen was performed by the piston moving along the compression channel under the action of gases formed at combustion of the powder charge. As a result of the compression, the hydrogen pressure in the discharge chamber increases up to 250–350 MPa, what corresponds to the concentration of hydrogen molecules (2.0–3.3) × 10^{22} cm^{-3}. The maximal hydrogen pressure after heating by the electric arc is 600 MPa, and the average hydrogen temperature is 3000 K.

The igniting copper wire of 0.3–0.6 mm diameter was installed between the cathode and the housing of the discharge chamber, which is anode, interelectrode distance of 2–20 mm. It is envisaged the change of the electrode system to ensure the stability of arcing and high efficiency of energy transfer from the arc to the gas [10].

The aim of the study was to determine the effect of the initial concentration of hydrogen molecules on the parameters of the electric arc. For this, based on the experimental data, the arc diameter, its temperature, and the conductivity of the plasma were estimated.

Typical oscillograms of the discharge current and voltage drop across the arc at initial concentration of hydrogen molecules of 2 × 10^{22} cm^{-3} are shown in Fig. 10.6.

The absence of pronounced voltage jumps testifies to the stability of the arcing process, what is the characteristic feature of the discharge under these initial conditions. As was shown earlier and confirmed at this series of the experiments, the voltage drop across the arc depends mainly on the initial concentration of hydrogen molecules or, equivalently, on the initial hydrogen pressure.

The oscillogram of the pressure characterizing the process of the adiabatic compression of hydrogen and its subsequent heating by the arc in the discharge chamber is shown in Fig. 10.7.

The section of the pressure curve 1 → 2 corresponds to the adiabatic compression of hydrogen. At the time of 5.1 ms the diaphragm opens, which is the signal to connect the power supply to the discharge circuit that causes the wire explosion and initiation the electric arc. Section 2 → 3 corresponds to the increase of the hydrogen pressure due to its heating by the arc and its expiration from the discharge chamber.

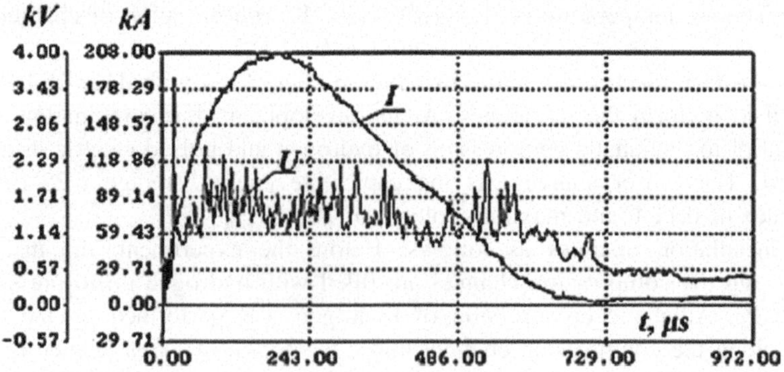

Fig. 10.6 Oscillograms of the discharge current (*I*) and voltage drop across the arc (*U*) [2]

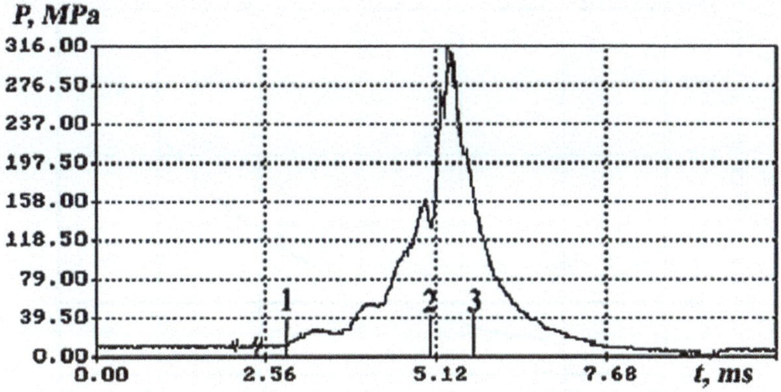

Fig. 10.7 Pressure of hydrogen in the discharge chamber [2]

The graph of voltage drop across the arc versus interelectrode distance at concentration of hydrogen molecules (1.5–3.3) × 10^{22} cm^{-3}, steel (Fe) and tungsten (W) electrodes is presented in Fig. 10.8.

Figure 10.8 shows that the dependence of the voltage drop across the arc U on the interelectrode distance is close to linear and can be described with acceptable accuracy by the expression

$$U = U_e + El_{ie}, \tag{10.10}$$

where U_e are total near-electrode voltage drops, E is electric field intensity in the arc, and l_{ie} is interelectrode distance.

Analysis of the experimental data and data obtained earlier [11–14] shows that the initial concentration of hydrogen molecules strongly affects the electric field

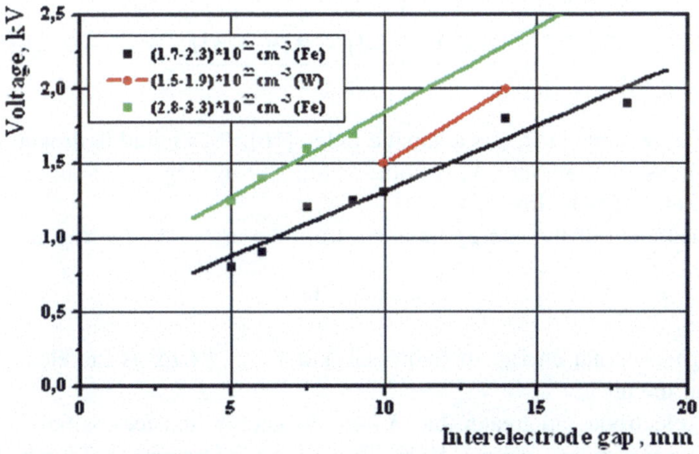

Fig. 10.8 Voltage drop across the arc versus interelectrode distance [2]

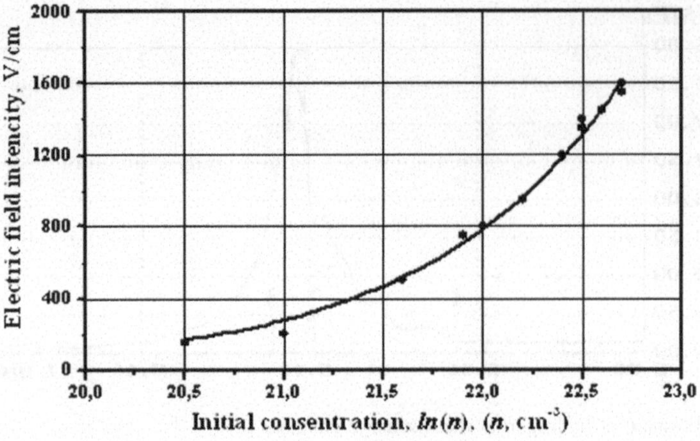

Fig. 10.9 Electric field intensity in the arc versus the initial concentration of hydrogen molecules [2]

intensity in the arc. At the initial concentration (1.7–2.3) × 10^{22} cm^{-3}, the average electric field intensity is 700 V/cm, and the total near-electrode voltage drops are 450–500 V. At increase of the initial concentration up to (2.8–3.3) × 10^{22} cm^{-3}, the average electric field intensity in the arc increases up to 1300 V/cm, at this the total near-electrode voltage drops remain the same 450–500 V (Fig. 10.9).

Determination of the diameter of the arc used in further calculations was performed by the "autograph" method, i.e., by measuring the spot of binding the arc to the surface of the cathode. The spot size did not exceed 4 mm at cathode of 8 mm diameter, i.e., it was assumed that the radius of the arc is ~2 mm.

The efficiency of energy transfer from the arc to the hydrogen is defined as the ratio of the internal energy of hydrogen to the energy inputted in the arc. The internal energy of the hydrogen was calculated by formula

$$W_g = \frac{P(V - \alpha M) - P_i(V_i - \alpha M)}{\gamma - 1}. \quad (10.11)$$

where α is covolume [15], γ is adiabatic index [16], P_i is initial hydrogen pressure, P is final hydrogen pressure, V_i is initial volume of hydrogen, V is final volume of hydrogen, and M is hydrogen weight.

The coefficient of the energy transfers from the arc to the hydrogen

$$\eta = W_g/W_a. \quad (10.12)$$

where W_g is internal energy of hydrogen and $W_a = \int UI\mathrm{d}t$ is the electric energy inputted in the arc.

At interelectrode gap greater than 5 mm, the energy transfer coefficient from the arc to the hydrogen was close to 100%. The average temperature of the hydrogen at these experiments did not exceed 1300 K, and maximal energy inputted in the arc is 140 kJ:

Estimation of the parameters of the arc in the hydrogen at ultrahigh pressure was performed based on the data, at discharge current I, electric field intensity in the arc E, and hydrogen pressure P at the current maximum.

Previously, it was assumed that the radius of the arc should be close to the radius of the spot of binding the arc to the cathode, i.e., ≤ 2 mm. In this case, estimations made based on conductivity and hydrogen pressure give the arc temperature of $\sim 10^5$ K, and the ion concentration, taking into account the total ionization of hydrogen $\sim 10^{20}$ cm^{-3}. The results of calculations of the thermodynamic and transport properties of the ultrahigh-pressure hydrogen plasma obtained in the work [17] were used. These results were used also for more accurately determining the parameters of the arc from the solution of the system of equations, received from (9.11) and (9.12):

$$2n_0kT_0 = P + 1.6 \times 10^{-4} \cdot \frac{I^2}{r_0^2}$$

$$IE = 5.35 \times 10^{-29} \pi n_0^2 T_0^{-0,5} r_0^2. \qquad (10.13)$$

$$\frac{E}{I} = \frac{1}{\pi r_0^2 \sigma(T_0, n_0)},$$

where P is hydrogen pressure, I is current, E is electric field intensity in the arc, r_0 is arc radius, T_0 is arc temperature, σ is arc plasma conductivity, n_i and n_e are ion and electron concentrations, and k is the Boltzmann constant.

According to the data of the work [17] at $T \sim 10^5$ K and $n_0 \sim 10^{20}$ cm^{-3}: $n_\mathrm{i} = n_\mathrm{e} \sim n_0$ (owing to full ionization of hydrogen plasma), and the correction for the nonideality in the value of the pressure inside the arc is negligible, what was used in the first equation of the system (10.13).

200–400 MPa, необходимо в отличие от разрядов при атмосферном давлении.

The first equation of the system (10.13) assumes the equality of pressure inside the arc and external pressure, which is the sum of the pressure of the hydrogen surrounding the arc and the pressure of the own magnetic field of the current. It is necessary to consider the pressure of the hydrogen P, which in our case is 200–400 MPa, in contrast to discharges at atmospheric pressure.

The second equation of the system (10.13) assumes that the electric power per unit arc length is equal to the power of losses from the arc with recombination radiation. In view of the fact that the path length of radiation quantum averaged over Rosseland for $T_0 \sim 10^5$ K and $n_0 \sim 10^{20}$ cm^{-3} is ~ 30 cm [18], the arc plasma is transparent and its radiation is volumetric. At the boundary of the arc, i.e., in the transition layer, the temperature decreases, and the density of the hydrogen is increased, so the path length of the radiation quanta decreases sharply. So, for $T \sim 4 \times 10^4$ K and $n_0 \sim 10^{21}$ cm^{-3}, the mean free path of radiation quanta is

$\sim 10^{-3}$ cm [4]. At this, energy from the arc into the hydrogen is transferred by radiant heat conductivity.

It was assumed that the Joule heating of hydrogen in the transition layer is negligible because of its low conductivity. The reason for this conclusion is the presence of the overheating zone in the arcs at ultrahigh gas pressure [19, 20]. The results of estimating the arc parameters at experiments at various initial concentrations of hydrogen molecules and the corresponding them electric field intensity in the arc are presented in Table 10.1.

At increasing the initial concentration of hydrogen molecules, the electric field intensity in the arc increases, and the radius of the arc decreases. In the made estimations, the equality between the electric power and the power of the radiation from the arc is assumed. However, as noted in the work [21], the time of diffusion of radiation through the transition layer at such plasma parameters may be longer than the time of the input of energy into the arc. In this case, the radiation "imprisonment" takes place, and the arc temperature can be much higher.

These estimations are based on the assumption that the wire metal, as in the work [19], does not affect the characteristics of the arc. However, at some experiments a wire with a diameter of 0.4 mm was used. The radius of the expansion of the vapors and fragments of the wire r_0 can be estimated at the equality between the internal energy of the vapor of the wire metal and the energy expended for expanding the hydrogen surrounding the wire after the hydrogen compression

$$\pi r_w^2 \frac{nkT}{\gamma - 1} = \pi r_0^2 P. \tag{10.14}$$

where k is the Boltzmann constant, n is vapor concentration of the wire metal before the arc expansion (6×10^{22} cm^{-3}), T is arc temperature (1 eV) [22], P is pressure in the discharge chamber at time of compression end and arc ignition (160 MPa), γ is adiabatic index (1.1), and r_w is radius of the igniting wire.

In this case r_0 is ~ 0.2 cm, and concentration of metal vapor $\sim 10^{21}$ cm^{-3}. The conductivity of the plasma at the current maximum and arc radius of 0.2 cm, determined from the experimental data, is ~ 1200 $(\Omega \times \text{cm})^{-1}$. The pressure in the arc is equal to the sum of the external gas-kinetic pressure of hydrogen of 300 MPa and the pressure of the own magnetic field of the current of 90 MPa, i.e., 390 MPa.

To aforementioned values of pressure and concentration of metal vapors, according to the results of calculations, presented in the work [8], corresponds the conductivity of ~ 1200 $(\Omega \times \text{cm})^{-1}$, and the temperature of 8.4×10^4 K, which coincides with our experimental data.

Table 10.1 Parameters of the arc

Test number	E (V/cm)	n_0 (cm^{-3})	T_0 (K)	r_0 (cm)
1	600	7.0×10^{19}	1.5×10^5	0.20
2	720	1.0×10^{20}	1.8×10^5	0.15
3	1170	1.2×10^{20}	1.3×10^5	0.14

The mean free path of radiation quantum over the Rosseland at 8.4×10^4 K is 10^{-2} cm, i.e., the arc plasma is opaque. If we assume that the outer surface of the arc radiates as the absolutely black body, then the balance between the radiation power and the arc power at radius of $r_0 = 0.2$ cm comes at the external arc layer temperature of 6.7×10^4 K.

Thus, it is possible that all current flows through the vapor of the wire metal in the arc of 0.2 cm radius at vapor concentration of $\sim 10^{21}$ cm^{-3}. At this radius of the arc, the experimental conductivity value coincides with the calculated at temperature of 8.4×10^4 K. The arc is screened by the transition layer with temperature of 6.7×10^4 K. In this case, the increase of the electric field intensity in the arc at increasing of initial hydrogen pressure can also be associated with the decrease of the arc diameter, which (decrease) leads to the increase of the current density in the arc. The decrease of the arc diameter is associated with the decrease of the radius of expansion of the wire metal vapor at increase of the initial hydrogen pressure.

However, the option assuming that the all current flows across the metal vapors is unlikely. At low ($\sim 10^8$ A/s) current rise rates characteristic for the performed experiments, in contrast to experiments at high current rise rates [23–25], the process of evaporation of the wire metal due to thermal and magneto-hydrodynamic instabilities can be accompanied by its partial destruction (fragmentation). Thus, because of this phenomenon, the concentration of metal vapors will be much less than 10^{21} cm^{-3}.

The estimations show that at metal vapor concentration of 10^{18} to 10^{19} cm^{-3} and pressure of 300–400 MPa, the equilibrium vapor temperature of the metal should be of $(2\text{–}10) \times 10^5$ K. In this case, the power of the losses by radiation at the arc diameter determined from its conductivity must exceed electric power by several orders of magnitude.

Thus, in any of the considered cases, the calculated radius of the arc does not exceed 0.2 cm, and the increase of the electric field intensity in the arc at increasing of initial hydrogen pressure is associated with decrease of the arc radius and increase of the current density.

The high efficiency of energy transfer from the arc into the hydrogen is associated with the absorption of radiation by metal vapors at the subsequent transfer of energy into the hydrogen, and also with the absorption of radiation by the hydrogen molecules.

10.3 Arc at Current of 500 KA

This chapter presents the results of the investigation of the high-current arc in high-density hydrogen at the modernized installation with preliminary adiabatic compression of hydrogen. The experiments were performed under the next initial conditions:

- Energy of the capacitor storage up to 450 kJ
- Voltage of the power supply up to 14 kV
- Discharge current up to 500 kA
- Current rise rate 10^9 to 10^{10} A/s
- Initial concentration of hydrogen molecules up to 2×10^{22} cm^{-3}.

The main result of the study is that the increase of the initial concentration of hydrogen molecules and the current rise rate cause the decrease of the arc diameter and increase of its temperature.

The two-stage electric discharge installation with dI/dt of 10^{11} A/s and discharge current of up to 5×10^5 A (Fig. 3.13) is the upgrade of the installation (Fig. 3.12). In this installation, there is window of 3 mm diameter for recording the optical parameters of the arc, in particular, the photo-scanning of the arc evolution.

The oscillogram of the preliminary compression of hydrogen and the subsequent increase of its pressure due to heating by the electric arc in the discharge chamber is similar to that shown in Fig. 10.7. In a number of experiments at this series, both single and multiple contractions of the arc were recorded. To the moments of the arc contraction are corresponded the abrupt increase of the voltage drop across the arc and feature on the oscillogram of the current.

Figure 10.10 shows oscillograms of the current and voltage drop across the arc of experiments with various current rise rates, but with similar other initial conditions.

The oscillograms show that both the discharge current and the voltage drop across the arc multiply (more than twice) increase at increasing the current rise rate from 10^9 to 10^{10} A/s. At this, on both oscillograms there are characteristic voltage peaks and current features, which are more pronounced on oscillograms with higher current rise rate (Fig. 10.10b).

Based on the data of the oscillograms, the dependences of the voltage drop across the arc on the interelectrode distance were plotted (Fig. 10.11).

To determine the near-electrode voltage drops at various current rise rates, the extrapolation of the voltage drop curves to zero value of the interelectrode distance was performed.

Figure 10.12 shows that the electric field intensity in the arc increases at the growth of the initial concentration of hydrogen molecules.

Figure 10.13a shows the oscillograms of the current and voltage drop and the photo-scanning of the arc glow in the visible range of the spectrum at initial hydrogen pressure of 5 MPa, and the same oscillograms in Fig. 10.13b at pressure of 110 MPa. The aperture of the photoregistrator was located perpendicular to the axis of the arc at the middle between the electrodes.

The short-term ($\sim$6 μs) contraction of the arc on the image of photo-scanning is observed (Fig. 10.13a, b). At pressure of 5 MPa, the diameter of the arc decreases from 7 to 1.6 mm, and at pressure of 110 MPa from 7 to 3.8 mm. The more accurate estimation of the arc diameter at the time of maximum contraction is impossible that at this time the intensity of the arc radiation in the visible spectral range decreases to zero.

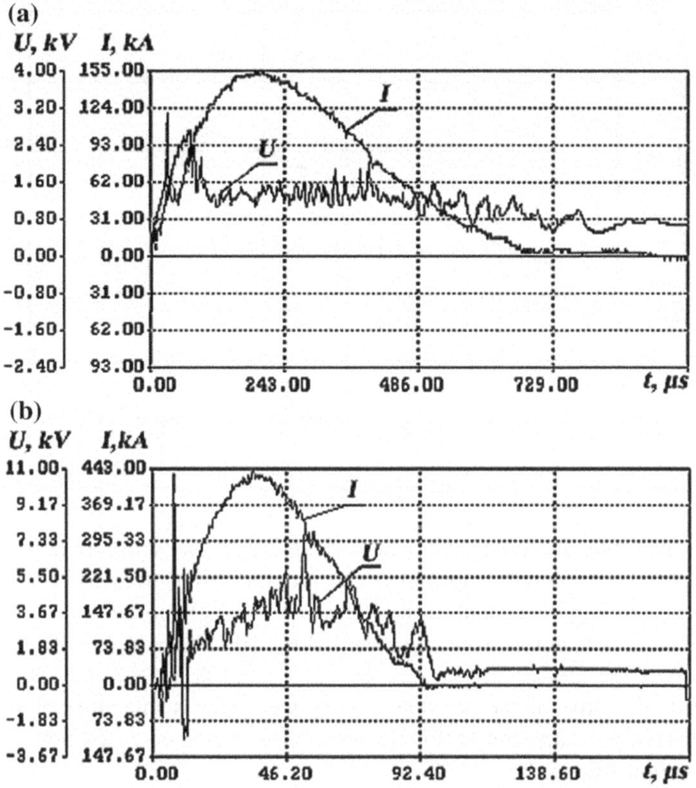

Fig. 10.10 Oscillograms of the current and voltage drop [2]. dI/dt is 10^9 A/s (**a**), dI/dt is 10^{10} A/s (**b**)

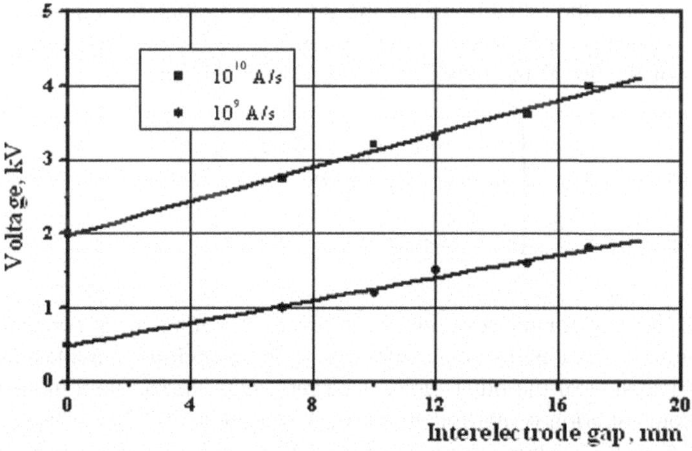

Fig. 10.11 Voltage drop across the arc versus interelectrode distance [2]

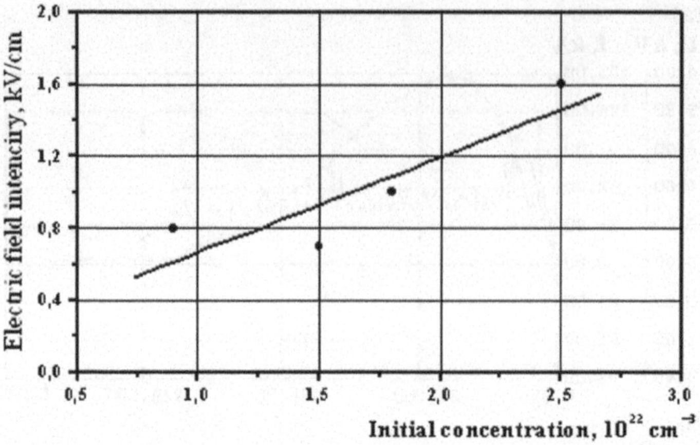

Fig. 10.12 Electric field intensity in the arc versus initial hydrogen concentration [2]

To the maximal arc contraction is corresponded the short-term peak on the oscillogram of the voltage drop across the arc and slight decrease of the current (feature), and the maximal yield of the X-ray. At some experiments, several voltage peaks and current features were registered (Fig. 10.13a), which indicate multiple contractions of the arc.

As a rule, the time of the arc contraction beginning is increased at increase of initial hydrogen pressure, but at this the arc contraction is occurred both before and after the current maximum.

We estimate the arc parameters for two cases, assuming that the arc:

- In pure hydrogen at initial concentration of hydrogen molecules $\sim 10^{22}$ cm^{-3}
- In the vapor of metal of the initiating wire and electrodes.

In the first case, the radiation inside the arc is recombination-bremsstrahlung, the plasma is optically transparent. The system of equations for estimating the parameters of the arc at the maximal contraction in this case

$$\begin{cases} IE = \left(An_i^2 T_a^{1/2} = Bn_i T_a^{1/2}\right)\pi r_a^2 \\ 2n_0 kT_a = P + C\frac{I^2}{r_a^2} \\ \frac{E}{I} = \frac{1}{\pi r_a^2 \sigma(T_a, n_a)} \\ n_i = n_e \sim n_a \end{cases} \tag{10.15}$$

where k is the Boltzmann constant, T_a is arc temperature, n_i is concentration of hydrogen ions, n_e is electron concentration, r_a is arc radius, σ is plasma conductivity, P is pressure of the outer gas, I is current, E is electric field intensity in the arc, n_a is concentration of hydrogen atoms, A is 1.39×10^{-34}, B is 5.35×10^{-29}, and C is 1.6×10^{-4}. This system is similar (9.11) with that difference, that in the second equation the pressure of surrounding gas was taken into account.

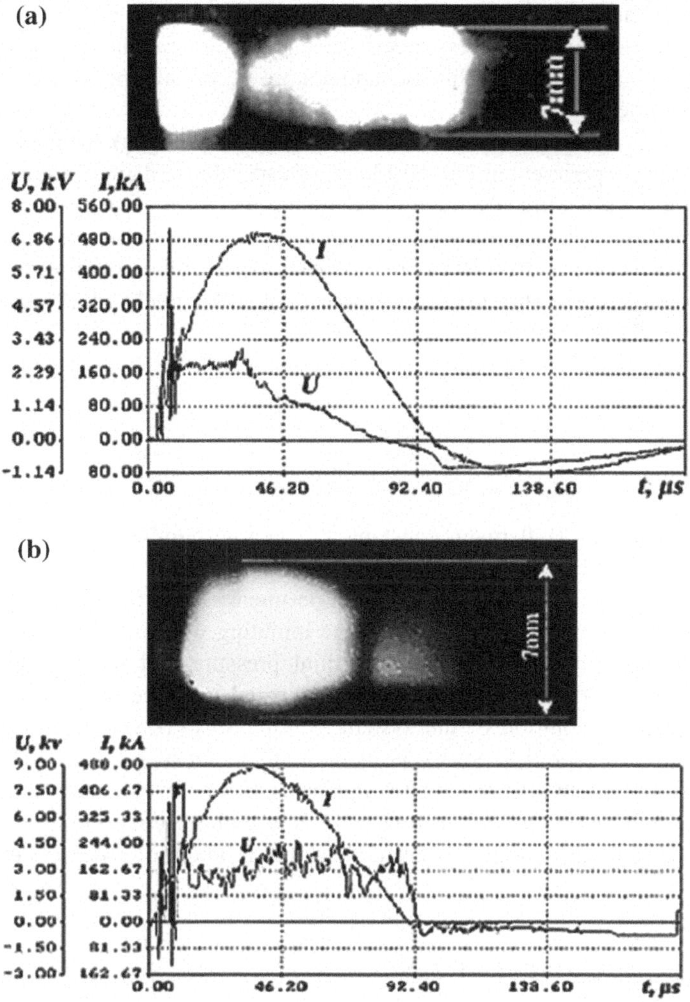

Fig. 10.13 Photo-scanning of the arc, oscillograms of current, and voltage drop across the arc [2]

The values of P, I, and E were determined from their oscillograms at the time corresponding to the contraction of the arc. The value of the conductivity σ of the hydrogen plasma was obtained from the work [17], of metallic plasma from the work [8].

The system of equations was solved at two values of the electric field intensity E: 1600 V/cm (Fig. 10.13a), and 3600 V/cm (Fig. 10.13b), in both cases $I \sim 500$ kA and $P = 300$ MPa.

Solutions of the system of equations for the first case:

- At $E = 1600$ V/cm: $T_a = 3 \times 10^5$ K, $n_a = 3.4 \times 10^{20}$ cm^{-3}, $r_a = 1$ mm
- At $E = 3600$ V/cm: $T_a = 3 \times 10^5$ K, $n_a = 10^{21}$ cm^{-3}, $r_a = 0.6$ mm.

The comparison of the obtained data shows that for the higher electric field intensity E in the arc is corresponded the smaller arc radius r_a, higher particle concentration n_a in the arc, at this the temperature T_a and number of particles in the arc are the same.

In addition, for this case (when the arc burns in pure hydrogen), the photo-scanning of the arc at the experiment in Fig. 10.13a is corresponded to a greater degree, i.e., the twice arc radius of 2 mm obtained by calculation is corresponded to the registered diameter of the arc of 1.6 mm at the time of arc maximal contraction.

In the second case, when the arc in the metal vapor of the initiating wire and the electrodes, we assume that the arc radiates as the absolutely black body. In this case, solving the system of equations

$$\begin{cases} IE = 2\pi r_a \sigma_b T_a^4 \\ (I+\bar{z})kT_a = P + C\frac{I^2}{r_a^2}, \\ \frac{E}{I} = \frac{1}{\pi r_a^2 \sigma(T_a, n_0)} \end{cases} \tag{10.16}$$

where σ_b is Stephan Boltzmann constant, Z is charge number, C is 1.6×10^{-4}.

we have $T_a = 1.25 \times 10^5$ K, $n_0 = 10^{20}$ cm^{-3}, and $r_a = 2$ mm.

The photo-scanning of the arc at the experiment (Fig. 10.13b), is more corresponded to the arc in the metal vapor of the initiating wire and the electrodes. The diameter of the initiating wire at this initial pressure and the current rise rate according to the estimations is 4 mm and is equal to twice the radius of 4 mm obtained from the solution of the system of equations (10.16), what agrees with experimentally measured at this experiment arc diameter of 3.8 mm. It is necessary to note, that at ultrahigh pressure (as against a mode of radiative contraction at pressure of 5–30 MPa) an arc all time is in near-equilibrium condition. Therefore the estimations of arc parameters are done at the moment of a current maximum.

10.4 Arc at Current of 2 MA

The purpose of this study was to develop the physical basis for generation and diagnosing of dense gas plasma with a charged particle concentration up to 10^{21} cm^{-3} and temperature of up to 50 eV generated at powerful pulsed arcs in hydrogen and helium at current up to 2 MA and initial pressure up to 30 MPa. The results of the experiments showed that an increase of the initial concentration of molecules and/or atoms makes it possible to reach the temperature of the arc central zone up to ≥ 30 eV at electron density of 10^{19} to 10^{21} cm^{-3}.

The registration of UV and SXR radiations from the arc showed their significant yield at steady-state plasma duration of ~ 100 μs.

The contraction of the arc near the maximum of the discharge current is occurred due to the achievement of the critical Pisa-Braginskii current. Earlier it was shown that at megaampere currents, the arc burns in metal vapor eroded from the electrodes. An increase of the critical Pisa-Braginskii current up to $\sim$1 MA in comparison with the theoretical value (100–200 kA) for arcs in the vacuum is due to the effective absorption of radiation from the arc by the high-density gas.

Methods of diagnostics. By the high-speed camera, the arc was photographed at the frame-by-frame and photo-scanning modes. Through the window with a light diameter of 8 mm, the integral spectrum of radiation from the arc was recorded. The resulting photographs were digitized using AserScan 450 and BearPaw 4800 TA II scanners. The scanners were calibrated with the step and linear optical attenuators of the LOT-Oriel Instruments with the optical density of 0.04–4.00, which provided an error of measurements of $\leq 5\%$. Using the reference brightness source, the photometric curve of the film was obtained through these same attenuators.

Based on the analysis of the photographs, it can be concluded that the arc is contracted, i.e., a pronounced arc channel is observed. In the all range of initial pressures, one or more arc contractions are observed both before and after the maximum of the current.

At processing of these experiments data, the distorting effect of the material of the optical window (polycarbonate) on the spectrum of the arc radiation was taken into account. In addition, polycarbonate strongly fluoresces upon irradiation in the visible and UV ranges, for example, under irradiation by light with wavelength of 550 nm.

In order to determine the brightness temperature of the arc, we compared the average intensity of its glow with the intensity of the reference capillary Podmoshensky source, which has brightness temperature of 4×10^4 K. At measuring of the brightness temperature was selected such range of the spectrum, which was free of absorption lines both for the reference source itself and for the arc under investigation. The brightness temperature of the transition layer measured at this manner was of $(2–4) \times 10^4$ K.

The discharge current was measured by Rogowski coil, voltage drop across the arc by resistive voltage divider. The error of measuring current and voltage did not exceed 5%. The maximal pulsed pressure in the discharge chamber was 450 MPa, which corresponds to the average gas temperature of 3000 K.

Figure 10.14 shows the oscillograms of the discharge current and the voltage drop across the arc at interelectrode distance of 10 mm, initial pressure of 40 MPa, and voltage of 12 kV.

A characteristic feature of this experiment is an insignificant change of the voltage drop across the arc during the discharge process. Peaks on the oscillogram of the voltage after the maximum current are due to contraction of the arc. The maximal electrical power of the arc is 8×10^9 W.

The oscillogram of pressure in the discharge chamber is shown in Fig. 10.15 at an initial hydrogen pressure of 20 MPa and voltage of 10 kV.

The time delay on the pressure oscillogram is due to the channel connecting the pressure sensor to the discharge chamber. The first peak of pressure corresponds to the front of the shock wave arising at the initial stage of arc formation.

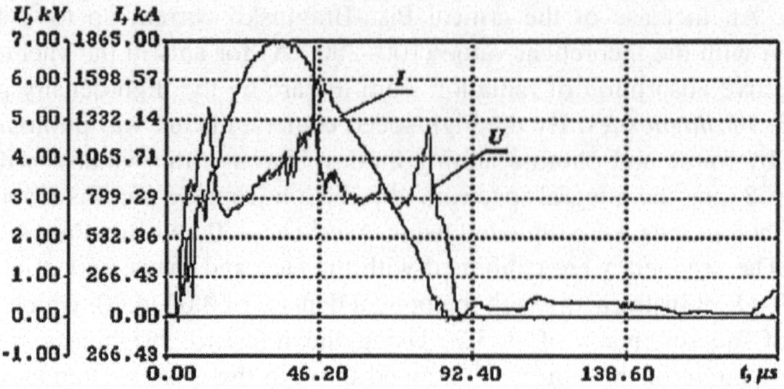

Fig. 10.14 Oscillograms of the discharge current I and voltage drop across the arc U [2]

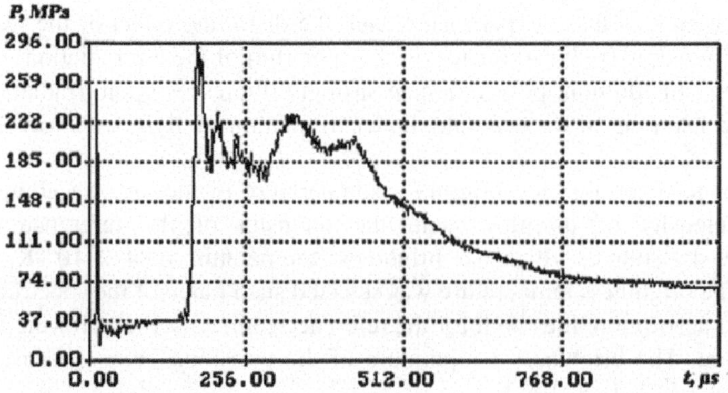

Fig. 10.15 Pressure in the discharge chamber [2]

10.5 Critical Pisa-Braginskii Current

The critical current of Pisa-Braginskii under these conditions is substantially higher than at experiments at low pressure (vacuum). To the study of the nature of this difference is devoted this chapter.

Arc in the hydrogen. The results of diagnosing the optical and electrical parameters of the arc in hydrogen at initial pressure of 32 MPa are shown in Figs. 10.16, 10.17, 10.18 and 10.19. At these experiments, we used steel electrodes of 20 mm diameter at interelectrode distance of 20 mm.

Figure 10.16 shows the photo-scanning of the arc glow by duration of 50 µs. The time of 37 µs corresponds to the arc contraction and the dimming of its glow.

Figure 10.17 shows the photometric cross sections of the arc glow brightness, obtained by processing of the photo-scanning in Fig. 10.16.

Fig. 10.16 Photo-scanning of the arc glow [2]

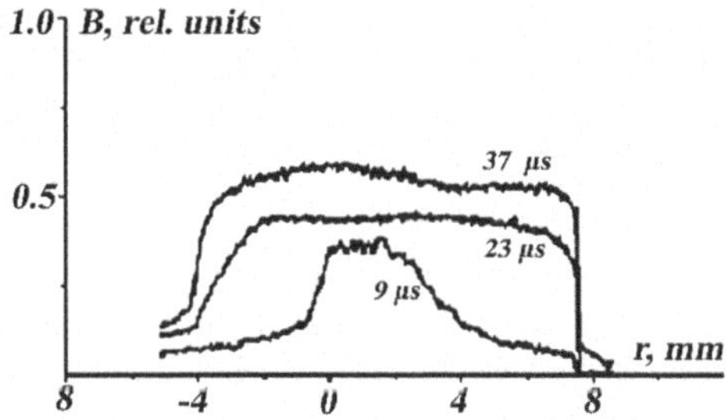

Fig. 10.17 Photometric cross sections of the arc glow brightness [2]

The arc according to the graph has the maximal diameter of 12 mm, the arc glow brightness is uneven in the time, and the arc is somewhat asymmetric.

On the photometric curves of the photo-scanning (Fig. 10.18), the small contraction of the arc diameter (D) is observed with decreasing brightness of its glow (B), at this decrease of the brightness of the glow begins by 10–15 μs earlier than the contraction of the arc.

On the end of the cathode, an erosion crater of 11 mm diameter is formed during arcing (the central hole is intended for installation of the initiating wire) (Fig. 10.18).

Since the diameter of the arc remains practically unchanged for the time of $\sim$30 μs (Fig. 10.18), i.e., the main part of the arcing process, we therefore assume that the arc diameter is equal to the diameter of the crater. At estimating the parameters of the arc, we will use just this value of the arc diameter.

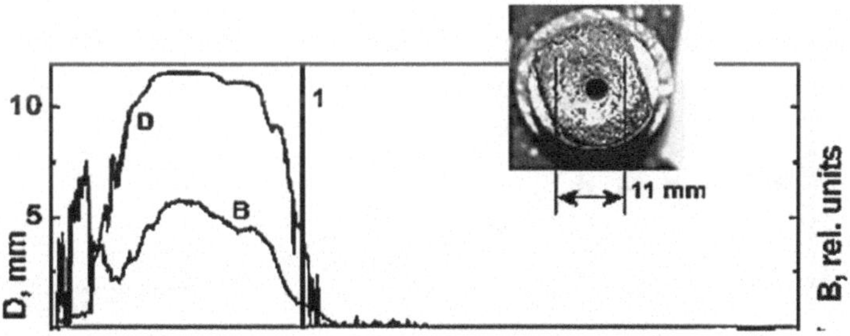

Fig. 10.18 Diameter of the arc (D), the arc glow brightness (B), and the photo of the end surface of the cathode [2]

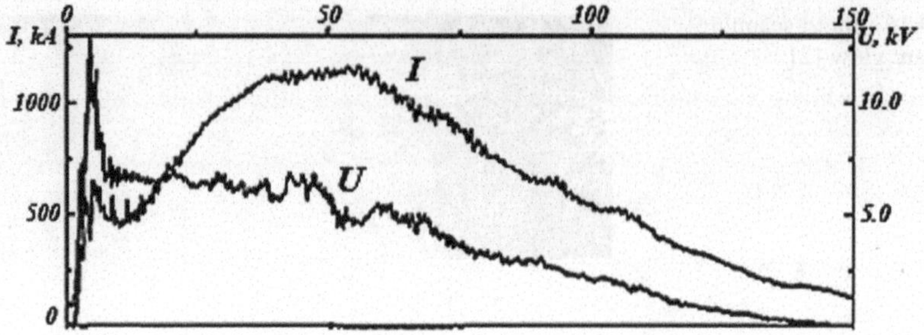

Fig. 10.19 Oscillograms of the current (*I*) and voltage drop across the arc (*U*), time (μs) [2]

To the instannt in time of most arc contraction corresponds the feature on the current oscillogram (*I*) and the peak on the oscillogram of the voltage drop across the arc (*U*) (Fig. 10.19).

The characteristic feature of such arcs is the decrease of the arc glow at arc contraction, which distinguishes them from arcs in vacuum, where the intensity of radiation in the visible range of the spectrum increases at the contraction of the Z-pinch in the vapors of the metal plasma.

The brightness temperature of the arc in hydrogen at wavelength of 550 nm is less than that at wavelength of 694 nm, and its maximum is occurred later (Fig. 10.20).

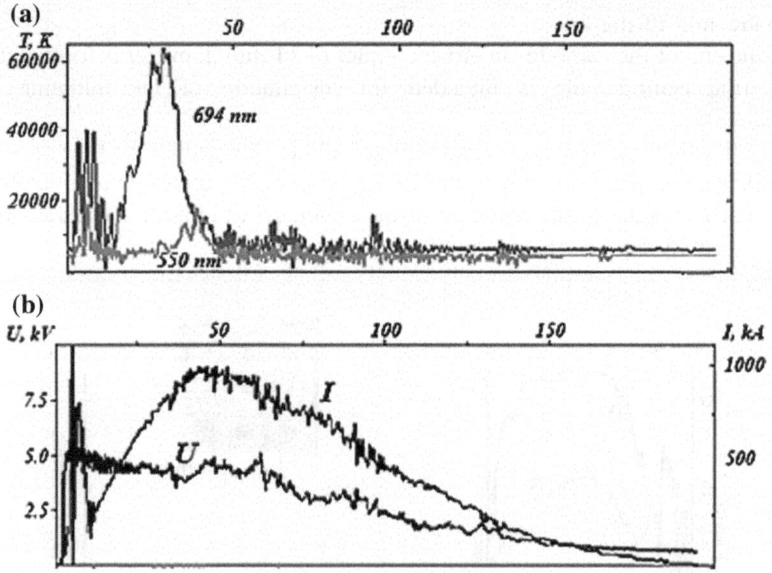

Fig. 10.20 Brightness temperature of the arc in the hydrogen (**a**), oscillogram of the current and voltage drop across the arc (**b**), time (μs) [2]

Fig. 10.21 Photo-scanning of the arc glow in helium [2]

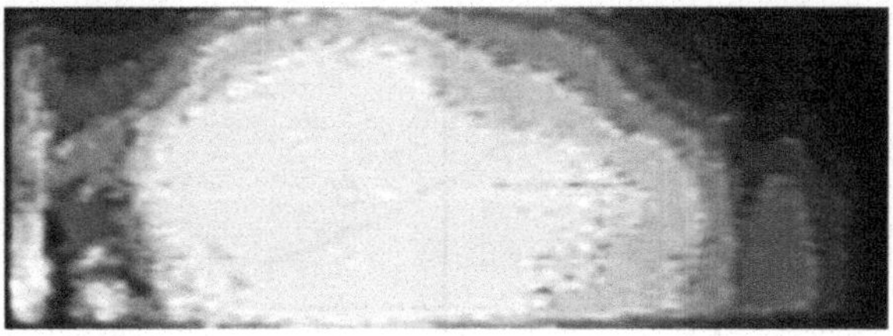

The arc in helium is differed from the arc in hydrogen by the higher brightness temperature.

Figure 10.21 shows the brightness of the arc glow in helium.

At the arcing in helium, as at arcing in hydrogen, the brightness temperature at wavelength of 550 nm is less than at wavelength of 694 nm (Fig. 10.22).

At arcing in helium, the moment of maximum arc contraction corresponds to the feature on the current oscillogram, at this voltage peak practically absent, and arc contraction is occurred by the decrease of the brightness temperature (Fig. 10.23).

Figure 10.24 shows the graphs of the brightness temperature of the arc at wavelength of 694 nm for various energies inputted into the arc. At increasing of the inputted energy, the brightness temperature of the arc grows (the curve 1 corresponds to the greater energy, than curve 2).

The rise rate of the current at these experiments ranged (from 3 to 5) $\times$ 10^{10} A/s. For the arcs with a lower current rise rate, arc contraction, which on time is close to the brightness temperature maximum, occurs earlier than for the arcs with higher current rise rate and lower current. The maximal brightness temperature of such arcs is higher than that of arcs with a higher current rise rate, which is due to the different density of metal vapor in the arc.

To estimate the parameters (T, I, and E) of the arc, which is metallic plasma of multiply-charged ions surrounded by hydrogen or helium at high pressure, the system of equations was used

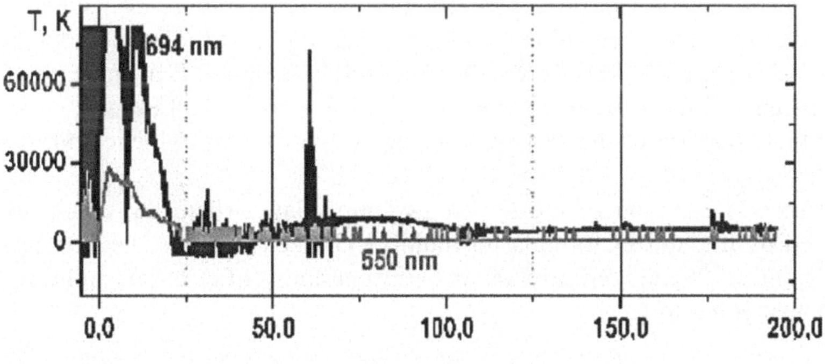

Fig. 10.22 Brightness temperature of the arc in helium [2]

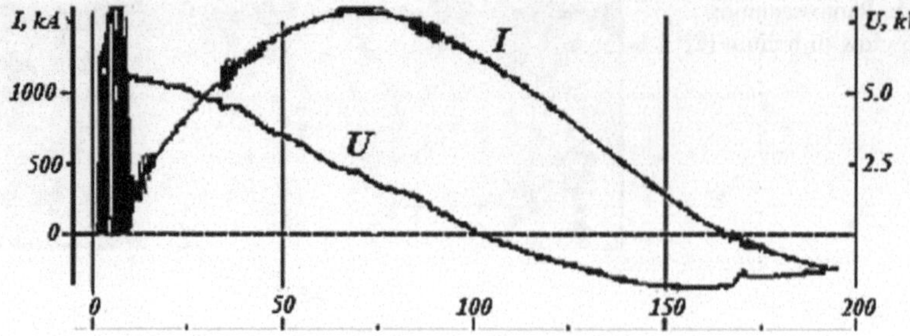

Fig. 10.23 Discharge current (*I*) and arc voltage drop (*U*), time (μs) [2]

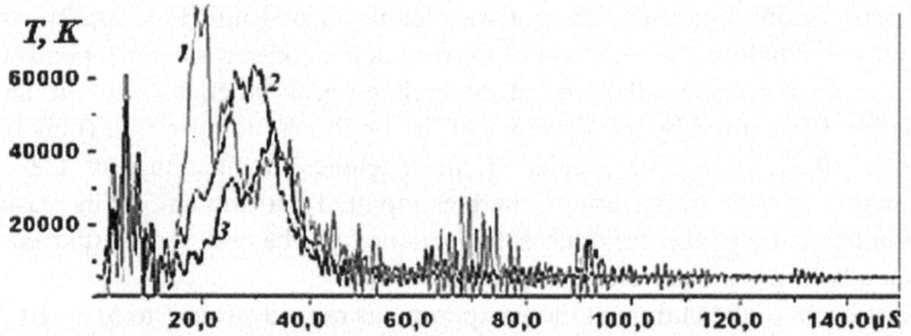

Fig. 10.24 Brightness temperature of the arc at wavelength of 694 nm [2]

$$\begin{cases} (1+z)n_i kT = 1.6\times 10^{-4}\frac{I_2}{r^2} \\ \frac{3/2\times 10^{-4}T^{3/2}}{z\ln\Lambda} = \frac{I}{\pi E r^2} \\ IE = \pi q r^2 f(T) \end{cases}. \quad (10.17)$$

where k is the Boltzmann constant, I is discharge current, E is electric field intensity in the arc, T is arc temperature; r is radius of the arc, n_i is concentration of the metal ions of the wire and electrodes, q is radiation power from unit arc volume, taking into account the recombination-bremsstrahlung and line radiation of multiply-charged metal ions, $\ln\Lambda$ is the Coulomb logarithm, z is average ion charge, which is approximated by the expression $z = 1.6 \times 10^{-2}T^{1/2}$ [26].

The first equation of the system is the equality of the gas-kinetic and magnetic pressures.

The second equation of the system is Ohm's law, where the conductivity is expressed by the Spitzer formula for multiply-charged plasma. The Coulomb logarithm estimated by the temperature and concentrations of electrons and ions in the arc is taken equal to 6.

The third equation of the system is the equality of the electric energy inputted in the arc and energy radiated from the arc. In this case, the arc plasma is assumed to be transparent.

Taking into account the energy yield with radiation, as in the work [27] it was assumed that the average energy of the emission quanta hv is (1–2) kT, and in this case it is assumed to be ~1.5 kT. According to the data of the work [28] $q = 1.09 \times 10^{-26}\, n_e\, n_i$ (W/cm^2), where $n_e = z\, n_i = 1.6 \times 10^{-2}\, T^{1/2}$.

The function $f(T)$ characterizes the energy transfer by radiation passing through the transition layer

$$f(T) = \exp(-\delta n l). \tag{10.18}$$

where n is concentration of molecules and/or atoms of the gas, $l = 10^{-2}$ cm according to the data of the work [29], $\delta = 8.3 \times 10^{-18}(hv/1.5\, kT)^3$ for hv_1—$hv \sim hv_1$, $hv_1 = 13.6$ and 24.6 eV for hydrogen and helium, respectively [4].

After the substitutions, the system of equations is

$$\begin{cases} 1.4 \times 10^{-15} n_i T^{3/2} = \frac{I^2}{r^2} \\ 4.9 \times 10^{-3} T = \frac{I}{Er^2} \\ IE = 5 \times 10^{-28} T^{1/2} r^2 n_i^2 f(T) \end{cases} \tag{10.19}$$

It follows that the electric energy $Q_e = IE$ inputted in the arc can be expressed as $Q_e = 2.9 \times 10^{-13}\, n_i T^{1/2}$.

On the right side of the third equation, there is an expression for the energy radiated by the arc and passing through the transition layer $Q_r = 5 \times 10^{-28}\, T^{1/2} r^2 n_i^2 f(T)$ then $Q_r/Q_e = 1.8 \times 10^{-15} n_i r^2 f(T)$.

If $Q_r > Q_e$, then the arc loses energy and contracts. Therefore, the condition for the beginning of arc contraction is the fulfillment of the relation $r^2 n_i^2 f(T) \geq 5.7 \times 10^{14}$.

The value $r^2 n_i$ is proportional to the total number of ions per unit arc length. Since this value is not known in advance, it is varied at the solution of the system. The value of r was determined from the diameter of the erosion crater, which is close to the diameter of the arc before beginning of contraction (Fig. 9.2).

From the relation $r^2 n_i^2 f(T) = 5.7 \times 10^{14}$, the temperature T was determined, from which the critical current I_{cr} was found by the first equation of the system (10.19).

Table 10.2 Calculated critical current I_{cr}

Test	Gas	Initial pressure (MPa)	$r^2 n_i$ (cm^{-1})	T (eV)	I_{cr} (kA)	I_{exp} (kA)
1	H_2	30	10^{18}	40	659	775 (after 10 μs)
2	H_2	30	3×10^{18}	39	1120	1200 (after 10 μs)
3	H_2	6	3×10^{18}	23	750	700
4	He	21	3×10^{18}	62	1025	1050

The results of estimations of the critical current I_{cr} are presented in Table 10.2, where its value is compared with the experimental value I_{exp} at which arc contraction begins.

The decrease of the arc glow begins by 10–15 μs earlier than its contraction. During this time, the arc temperature rises to $\sim 4 \times 10^5$ K, and the brightness temperature at wavelength of 694 nm decreases to $\sim 2 \times 10^4$ K. After this, due to contraction of the arc, a rapid increase in its temperature begins. Since at this time the third equation of the system (10.19) is no longer satisfied, since $\pi r^2 qf(T) > IE$, so the results of the work [8] were used to estimate the temperature of the arc T. In this case, the arc temperature in section 1 (Fig. 10.18) at the time 8 μs after the arc contraction estimated (time) based on its conductivity and pressure is 3×10^6 K, $n_i \sim 10^{19}$ cm^{-3}, $r = 0.16$ cm, and $n_i r^2 = 2.6 \times 10^{17}$ cm^{-1}. According to the data of Table 8.2, to the beginning of arc contraction is corresponded $n_i r^2 \sim 3 \times 10^{18}$ cm^{-1}.

The decrease of the particles number can be due to the outflow of plasma in the longitudinal direction from the contraction zone. A similar phenomenon is described in the work [30]. The speed of sound calculated by the formula (10.20) is 9.4×10^4 m/s

$$c = \sqrt{\gamma \frac{P}{\rho}} \tag{10.20}$$

where γ is adiabatic index (1.1), P is pressure (7.6 GPa), and ρ is density (9.4×10^{-4} g/cm^3).

At this speed of sound during few microseconds, a part of the metal can escape from the contraction zone, whose length does not exceed the interelectrode distance, i.e., 2 cm. In this case, additional heating of the plasma is possible near the ends of the electrodes, caused by the collision of electrode jets with metal vapor jets moving longitudinally from the arc contraction zone [31, 32]. It is possible that the zones of intense X-ray radiation near the electrodes are due to this phenomenon [33, 34].

At estimation of the conductivity, the electric field intensity E in the arc was determined on the difference of the voltage drops across the arc at various arc lengths. The value $E = 1200$ V/cm, taken for calculations, corresponds to the upper limit, and hence to the lowest arc temperature. As will be shown later, this value of temperature is consistent with the temperature of the arc, determined on the decrease of the arc glow in the visible range of the spectrum and from the intensity of the X-ray radiation.

At the experiments (Figs. 10.23 and 10.24) and at the experiments 1 and 2 of Table 10.2, the critical current was determined from the oscillograms at the time of 10 μs after the brightness temperature began to decrease at wavelength of 694 nm. The agreement with the calculated current I_{cr}, which increases at increasing of the arc power, is satisfied at increasing $n_i r^2$ from 10^{18} to 3×10^{18} cm^{-1}. The difference in the start time of arc contraction in these experiments can be explained by that the plasma in the arc contraction zone, as in the micropinch region [35], is formed as a result of evaporation of the electrodes.

Earlier, it was found that at increasing of arc power, the near-electrode voltage drops increase [36]. This is due to an increase of the concentration of metal vapor in the arc and the energy supplied to the electrodes, what causes the highest concentration of metal vapors near the electrodes.

Experiments 2, 3, 4 (Table 10.2) were carried out at the same current rise rate, so the value $n_i r^2$ for them is the same. When estimating I_{cr} for helium (Experiment 4, Table 10.2; Figs. 10.22 and 10.23), it was assumed that the thickness of the transition layer, depending on the mean free path of X-ray quanta, is inversely proportional to their absorption cross section $l \sim l/\delta \sim 1/(h\nu_1)^3$, therefore, in comparison with hydrogen, where $l \sim 10^{-2}$ cm, for helium $l \sim 1.7 \times 10^{-3}$ cm. Figure 10.21 shows two arc contractions, at this the second, faster contraction is more consistent with the estimates.

The data in Table 10.2 show that the I_{cr} increases at increasing of initial gas pressure and at increasing of metal ions number in the arc. For helium the temperature, at which contraction of the arc begins, is higher than for hydrogen.

Decreasing of the arc glow brightness at arc contraction, which is observed in the visible range of the spectrum, is associated with the formation of transitional layer heated by X-rays. Similar phenomena were observed at the front of strong shock waves at increasing temperature in the wave front [4].

The arc glow after its maximum is decreased, and then remains constant, corresponding to the brightness temperature $T_{br} = (1.0\text{–}1.5) \times 10^4$ K (Fig. 10.23). This can mean that the transitional layer glows at heating by X-ray and the radiation from the arc is completely absorbed. In this case, the mean free path of the radiation quantum in the visible range l_λ becomes at first is equal, and then less than the thickness of the transition layer [4]

$$l_\lambda = 1/\chi = \frac{T_{br}^2 x^3}{0.96 \times 10^{-7}\, n \exp[-(x_l - x)]}, \tag{10.21}$$

where $x = h\nu/kT_{br}$; $x_1 = h\nu_1/kT_{br}$; $h\nu = 1.79$ eV; $h\nu_1 = 13.6$ eV, n is concentration of hydrogen molecules.

In its turn, the mean free path of the SXR quantum is equal to the thickness of the transition layer Δx.

According to the data of the work [37], $\delta = 5.42 \times 10^{-17}\ (h\nu_1/h\nu)^{3.5}$ at $h\nu \gg h\nu_1$, for hydrogen $h\nu_1 = 13.6$ eV, then

$$\Delta x = 1/n\sigma = \frac{1}{5.42 \times 10^{-17}\, n\, (13.6/1.5T)3.5}. \tag{10.22}$$

As mentioned earlier, the main share of the radiation energy is transferred by quanta with $h\nu = 1.5\ kT$, where T is average temperature of the arc.

From the ratio $l_\lambda = \Delta x$ for $\lambda = 694$ nm ($h\nu = 1.79$ eV), at the brightness temperature change from 10^4 to 1.5×10^4 K, the arc temperature changes from 370 to

90 eV. For $\lambda = 550$ nm, these temperatures are slightly higher, because the radiation leaves the deeper zones of the arc.

X-ray radiation from the arc was recorded at initial hydrogen pressure of 5 MPa. Preliminary estimations show that the energy of the registered emission quanta is ≥ 370 eV.

Thus, the critical current at which the contraction of the arc containing the metal of the electrodes and the initiating wire begins at initial gas pressure of 5–30 MPa is ~ 1 MA, whereas for the arc in vacuum is ~ 100–200 kA.

This difference is due to a decrease of the energy loss with radiation due to the absorption of radiation in the transition layer. The critical current also increases with increasing number of metal ions in the arc.

The transition layer surrounding the arc makes it possible to effect the parameters of the arc by changing the initial pressure and the type of gas.

The occurrence of a high-temperature zone in the arc in gases at high initial pressure is due to the same reasons as the formation of "hot spots" at arcing in vacuum, i.e., by radiation contraction.

References

1. A.F. Savvateev, A.A. Bogomaz, A.V. Budin, V.A. Kolikov, F.G. Rutberg, High Temp. **41**, 580 (2003)
2. A.A. Bogomaz, Doctoral thesis (St.-Petersburg, 2012) (in Russian)
3. N.N. Kalitkin, L.V. Kuz'mina, V.S. Rogov, *Tables of Thermodynamic Functions and Transport Coefficients of Plasma* (IPM, 1972) (in Russian)
4. Ya.B. Zel'dovich, Yu.P. Raizer, *Physics of Shock Waves and High-Temperature Hydrodynamic Phenomena* (New York, 1966)
5. D. Saumon, H.M. Van Horn, NATO ASI Series B: Phys. **154** (1987)
6. V.N. Tjulin, *Introduction in the Theory of Sound Radiation and Scattering* (Moscow, 1976) (in Russian)
7. N.B. Vargaftik, *Directory on Thermophysical Properties of Gases and Liquids* (Moscow, 1972) (in Russian)
8. B.V. Zamyshlyaev, E.L. Stupitskiy, A.G. Guz', V.G. Zhukov, *Composition and Thermodynamic Functions of Plasma, Handbook* (Moscow, 1984) (in Russian)
9. Ph.G. Rutberg, A.A. Bogomaz, A.V. Budin, V.A. Kolikov, A.G. Kuprin, A.A. Pozubenkov, in 27th AIAA Plasmadynamic and Laser Conference, New Orleans (1996)
10. A.V. Budin, S.Y. Losev, M.E. Pinchuk, PhG Rutberg, A.F. Savvateev, Instr. Exp. Tech. **49**, 549 (2006)
11. PhG Rutberg, A.A. Bogomaz, A.V. Budin, V.A. Kolikov, A.G. Kuprin, Izvestia RAS. Energy **1**, 100 (1998). (in Russian)
12. Ph.G. Rutberg, A.A. Bogomaz, V.A. Kolikov, in *Hypervelocity Impact Symposium*, Santa Fe (1994)
13. Ph.G. Rutberg, A.A. Bogomaz, A.V. Budin, V.A. Kolikov, A.G. Kuprin, Izvestia RAS, Energy, **100** (1998) (in Russian)
14. V.A. Kolikov, Doctoral thesis (St.-Petersburg, 2005) (in Russian)
15. V.P. Kopushev, V.V. Hrustalev, Sov. J. Appl. Mech. Tech. Phys. **1**, 122 (1980). (in Russian)
16. A.M. Shekhtman, *Gasdinamic Functions of Real Gases* (Moscow, 1988) (in Russian)

17. O.V. Koryshev, D.O. Nogotkov, Yu.Yu.Protasov, and V.D. Telekh, *The Thermodynamic, Optical, and Transport Properties of the Working Media of Plasma and Photon Power Plants* (Bauman Moscow State Technology University, 1999) (in Russian)
18. B.H. Armstrong, R.R. Johnston, P.S. Killy, J. Quant. Spectrosc. Radiat. Transfer **5**, 55 (1965)
19. R.V. Mitin, *Properties of low-temperature plasmas and diagnostics methods,* (Novosibirsk, 1977) (in Russian)
20. N.N. Ogurtsova, I.V. Podmyshenskii, V.L. Smirnov, Sov. J. High Temp. **14**, 1 (1976). (in Russian)
21. E.S. Borovik, V.I. Petrenko, R.V. Mitin, V.P. Kantsedal, Ju.P. Knjazev, Sov. J. Tech. Phys. **39**, 1416 (1969) (in Russian)
22. V.A. Burtsev, N.V. Kalinin, A.V. Luzhinskiy, *Electric Explosion of Conductors and Its Application in Electrophysical Installations* (Moscow, 1990) (in Russian)
23. S.Yu. Gus'kov, G.V. Ivanenkov, A.R. Mingaleev, S.A. Pikuz, V.B. Rozanov, T.A. Shelkovenko, V.V. Nikishin, V.F. Tishkin, W. Stepniewskiy, D.A. Hammer, Plasma Phys. Rep., **26**, 745 (2000)
24. V.N. Korobenko, A.D. Rakhel, A.I. Savvatimskiy, V.E. Fortov, Plasma Phys. Rep. **28**, 1008 (2002)
25. G.V. Ivanenkov, W. Stepniewskiy, Plasma Phys. Rep. **28**, 814 (2002)
26. Yu.G. Kalinin, A.S. Kingsep, V.P. Smirnov, Yu.L. Bakshaev, A.V. Bartov, P.I. Blinov, S.A. Dan'ko, L.G. Dubas, A.V. Korel'ski, V.D. Korolev, V.I. Mizhiritskĭ, G.I. Ustroev, A.S. Chernenko, R.V. Chikin, A.Yu. Shashkov, Z. Li, X. Hua, X. Peng, S. Feng, C. Guo, Plasma Phys. Rep., **32**, 656 (2006)
27. E.V. Aglickiy, V.V. Vikhrov, A.V. Gulov, *Multicharge Spectroscopy in Hightemperature Plasma* (Nauka, Moscow, 1991). (in Russian)
28. S.I. Blinnikov, V.S. Imshennik, Sov. J. Plasma Phys. **8**, 110 (1982)
29. A.M. Voronov, V.L. Gorjachev, V.N. Zhuravlev, Rus. J. Tech. Phys. Lett. **19**, 35 (1993). (in Russian)
30. A.N. Dolgov, V.V. Vikhrev, Plasma Phys. Rep. **31**, 259 (2005)
31. PhG Rutberg, A.A. Bogomaz, A.V. Budin, V.A. Kolikov, M.E. Pinchuk, A.A. Pozubenkov, Tech. Phys. **47**, 26 (2002)
32. A.N. Dolgov, GKh Salakhutdinov, Plasma Phys. Rep. **29**, 757 (2003)
33. H. Chuagui, M. Favre, R. Savedra, E.S. Wyndham, P. Choi, C. Dumitrescu-Zoita, L. Soto, R. Aliaga-Rossel, I.H. Mitchell, Phys. Plasmas **4**, 3696 (1997)
34. H. Chuagui, M. Favre, R. Savedra, E.S. Wyndham, L. Soto, P. Choi, C.D. Zoita, I.E.E.E. Trans, Plasma Sci. **26**, 1162 (1998)
35. V.V. Vikhrev, V.V. Ivanov, K.N. Koshelev, YuV Sidelnikov, Sov. J. Plasma Phys. **8**, 688 (1982)
36. A.A. Bogomaz, A.V. Budin, M.E. Pinchuk, Ph.G. Rutberg, A.F. Savvateev, in 13th SHCE (Tomsk, 2004)
37. A.A. Bogomaz, M.E. Pinchuk, A.V. Budin, S.Yu. Losev, A.A. Pozubenkov, Ph.G. Rutberg, in *Thesis of International Conference Intense Energy Flux Action on Substance-XXII*, 189, Elbrus (2007) (in Russian)

Chapter 11
Energy Features of Plasma Generator

Abstract Thirteen pulsed plasma generators to study megaampere pulsed electric arcs and characteristics of the dense gas plasma at high and ultra-high pressures were developed at the IEE RAS. Plasma-forming gases during the operation of generators were hydrogen, helium, nitrogen, argon, and air. The generators were also intended for solving other scientific problems and therefore differed in the type of the electrode system, the design and volume of the discharge chambers, and some other characteristics. In Chap. 3, the types of discharge chambers of plasma generators and the design of the generators themselves in detail are presented; therefore, in this chapter, we shall dwell only on the results of estimating the energy parameters of plasma generators, the efficiency of their operation, determined by the coefficients of electrical energy transfer from the power supply (capacitor bank), into the arc energy, share of which in its turn is converted into the internal energy of the gas. In addition, we consider the main factors and conditions that determine the efficiency of the pulsed plasma generators.

11.1 Energy Transfer Coefficients

The efficiency η of the plasma generator supplying from a capacitive storage is the ratio of the internal energy of the gas ΔW_g to the electrical energy stored in the power supply W_e

$$\eta = \frac{W_g}{W_e}. \tag{11.1}$$

The process of energy conversion in the system "power supply → plasma generator" occurs at two stages.

The first stage is the transfer of the energy W_e stored in the power supply into the energy of the electric arc W_a; its efficiency is characterized by the coefficient η_e and is determined by the ratio

V. Kolikov et al., *Powerful Pulsed Plasma Generators*,
Springer Series on Atomic, Optical, and Plasma Physics 101,
https://doi.org/10.1007/978-3-319-95249-9_11

$$\eta_e = \frac{W_a}{W_e} \tag{11.2}$$

The second stage is transforming of the arc electric energy into the internal energy of the gas. Its efficiency is characterized by the coefficient η_t and is determined by the ratio of the increment of the internal energy of the gas ΔW_g to the energy inputted in the arc W_e

$$\eta_t = \frac{\Delta W_g}{W_e}. \tag{11.3}$$

Let us represent the values of the energy transfer coefficients obtained based on their dependencies on the gas parameters and the characteristics of the power supply.

The coefficient η_e is defined as the ratio of the electrical energy inputted in the arc W_a to the electrical energy stored in the power supply W_e. The energy inputted in the arc W_a during the discharge time T is given by

$$W_a = \int_0^T IU_a \mathrm{d}t. \tag{11.4}$$

where I is current in the arc and U_a is voltage drop across the arc.

In the case when the power supply is the capacitor bank, W_e is defined by formula

$$W_e = \frac{c(U_c - U_r)^2}{2}. \tag{11.5}$$

where c is capacitance of the capacitor bank, U_c is voltage of the capacitor bank and U_r is residual voltage of the capacitor bank.

Since the electric energy is inputted only in the active resistance of the arc, the coefficient η_e can also be defined as

$$\eta_e = 1 - \frac{R_c}{R_c + R_a}. \tag{11.6}$$

where R_c is active resistance of the discharge circuit and R_a is active resistance of the arc.

The results of studies of the characteristics of pulsed generators of dense hydrogen plasma show that the average active resistance of the pulsed arc in hydrogen at initial pressure of $\sim$40 MPa and currents of $\sim$1 MA is $\sim$6 mΩ.

The active resistance of the discharge circuit of capacitive storages with multi MJ energy used to supply of the high-power arc loads can vary widely; for example, this parameter of the super-power wind tunnel power supply of the

McDonnell Douglas is 0.1 mΩ [1, 2]. However, the development of power supplies based on capacitor banks with such resistance of the discharge circuit is very expensive, since requires a lot of expensive materials.

The most experiments on megaampere arc research were performed using power supplies CPS-25 and CPS-10; their wave impedance of the discharge circuit is 0.68 and 1.47 mΩ, respectively.

One of the factors, which are affecting the energy transfer coefficient from the battery to the arc η_e, is the initial gas pressure P_i. The dependence of η_e on P_i for the capacitor bank CPS-10 is shown in Fig. 11.1. The dependence has growing character, with the average values of η_e varying from 0.48 to 0.85 at change of the initial hydrogen pressure from 5 to 42 MPa.

Despite the fact that the most experiments (at exception of several experiments where were used nitrogen, helium, and air) were performed with hydrogen, the main common factors for all cases, which can affect the arc resistance and, consequently, the efficiency of energy transfer from the power supplies to the load, in addition to the initial gas pressure are the interelectrode distance, charge carriers concentration in the arc, material of the electrodes, the gas type, volume of the discharge chamber, etc.

The dependences of the arc resistance on the interelectrode distance and the mass of the eroded metal of the electrodes are shown in Figs. 11.2 and 11.3.

The dependences plotted for various initial hydrogen pressures have a growing character, and at interelectrode distance from 2 to 14 mm, the arc resistance varies from 3 to 7 mΩ.

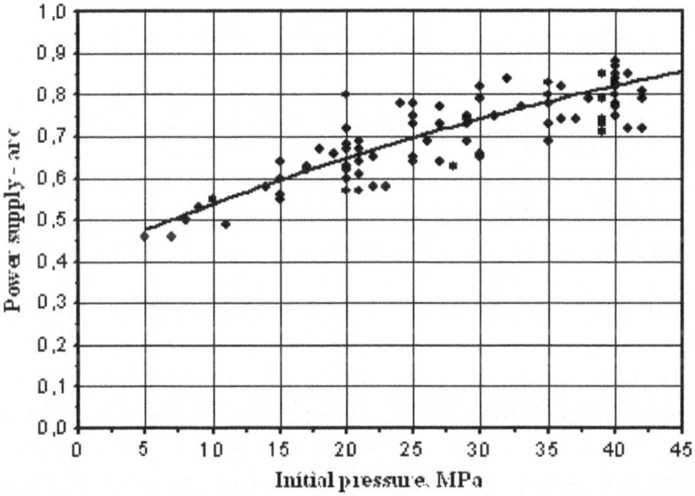

Fig. 11.1 Coefficient of energy transfer from the power supply CPS-10 to the arc versus initial hydrogen pressure [3]

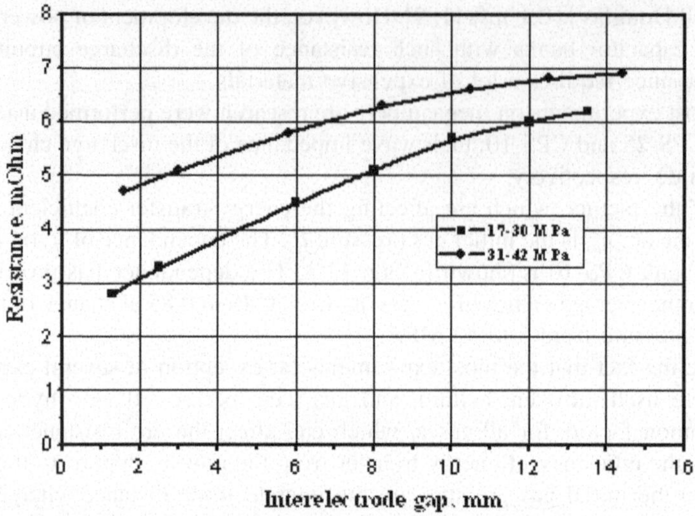

Fig. 11.2 Resistance of the arc versus the interelectrode distance [3]

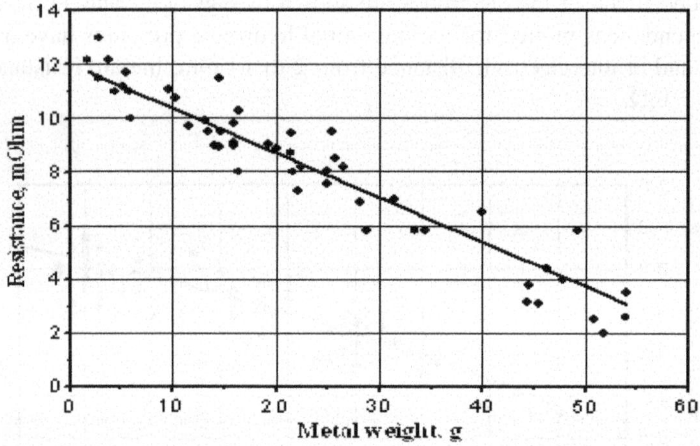

Fig. 11.3 Resistance of the arc versus the weight of the eroded metal [3]

The dependence of the resistance on the weight of the eroded metal has the opposite nature of the dependence presented above. When the weight of eroded metal varies from 2 to 55 g, the arc resistance varies from 11 to 3 mΩ.

The specific resistance of the arc plasma versus the ratio of the weight of hydrogen M_g to the weight of the metal M_m evaporated from the electrodes is shown in Fig. 11.4.

The characteristic feature of the dependence is the prevailing number of points corresponding to the ratio of the weight of hydrogen to the weight of metal

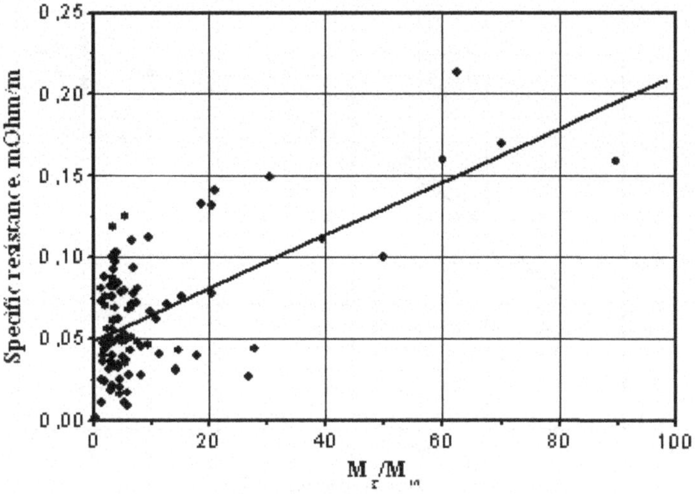

Fig. 11.4 Specific resistance of the arc plasma versus the ratio of the weight of hydrogen M_g to the weight of the metal M_m evaporated from the electrodes [3]

evaporated from the electrodes from 1 to 10, what indicates at significant weight of eroded metal and low values of resistivity in most experiments. This necessitated the minimization of metal erosion to improve the efficiency of energy transfer from the power supply to the arc.

The coefficient η_t is determined by the ratio of the increment of the internal energy of the gas ΔW_g to the energy inputted in the arc W_a

$$\eta_t = \frac{\Delta W_g}{W_e}. \tag{11.7}$$

ΔW_g can be estimated from the formula

$$\Delta W_g = \frac{P(V - \alpha m_g) - P_i V}{\gamma - 1}. \tag{11.8}$$

where P is final gas pressure in the discharge chamber, V is volume of the discharge chamber, m_g is the weight of the gas, P_i is initial gas pressure, γ is the adiabatic index, and α is covolume.

The value of W_e was determined from the oscillograms of the current and voltage. The main parameter affecting the coefficient η_t is the initial hydrogen pressure. Figure 11.5 shows the dependence of the coefficient of the arc electric energy transfer into the internal energy of hydrogen on its initial pressure. Here the average values of the coefficient η_t for initial pressures from 20 to 42 MPa, which are typical for the conditions of the experiments, are 0.7–0.9.

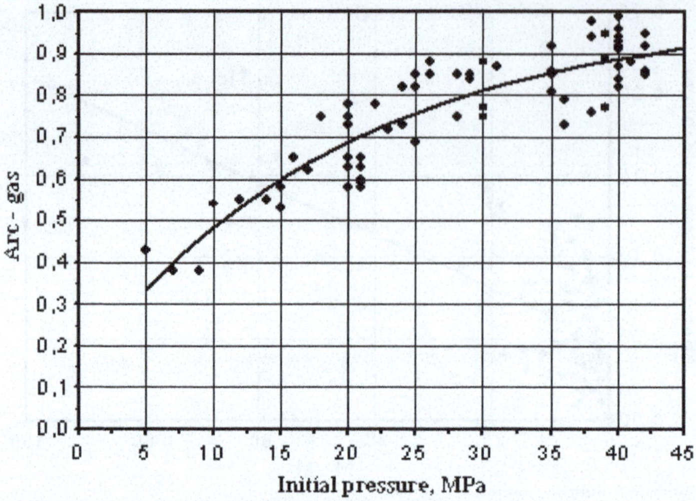

Fig. 11.5 Coefficient of the arc electric energy transfer to the internal energy of hydrogen versus its initial pressure [3]

Figure 11.6 shows the dependence of the coefficient of the arc electric energy transfer to the internal energy of hydrogen on the volume of the discharge chamber, at pressure in the discharge chamber of 20 and 40 MPa.

The dependencies have a growing character, which can be explained both by the increase of the geometric dimensions of the chamber and by the increase of the

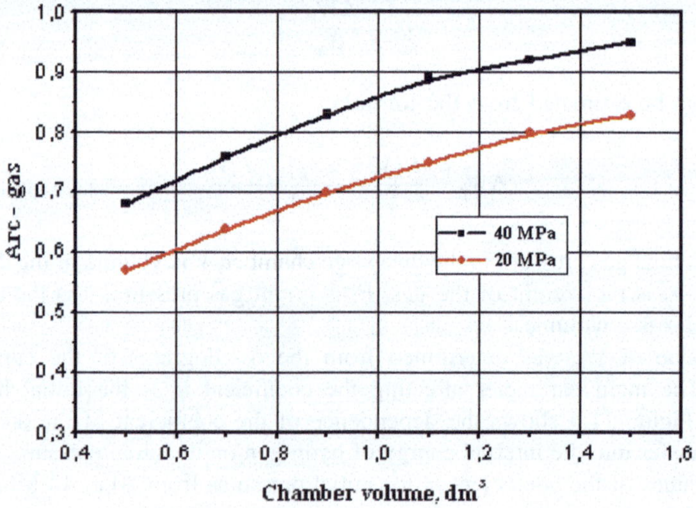

Fig. 11.6 Coefficient of the arc electric energy transfer to the internal energy of hydrogen versus the volume of the discharge chamber [3]

weight of the gas, what should lead to an increase in the degree of dissipation and the efficiency of energy absorption.

11.2 Simulation of Plasma Generators Processes

Some experimental results obtained at the investigation of plasma generators can be explained with the help of simplified mathematical models of the main processes occurring during the operation of the generator.

The operating cycle of the generator can be represented as a combination of two processes:

- Transfer of electrical energy from the power supply to the arc
- Heating by the arc of the operating gas (hydrogen) in the discharge chamber.

The modeling task is greatly simplified if the effect of the operating gas on the arc parameters is not taken into account. In this case, the problem of the discharge simulation is solved independently, and its solution can be written as the dependence of the electric power on time. In this case, the connection between the electric power of the arc and the rate of increment of the internal energy of the gas is given by the empirical relation

$$\eta_{\mathrm{a}} = 1 - \mathrm{e}^{-\frac{16{,}000\rho}{I}}, \tag{11.9}$$

where η_{a} is ratio of the growth rate of the internal energy of hydrogen to the arc power, ρ is initial gas density in the discharge chamber (kg/m^3), and I is average current by process (A).

The arc, being a nonlinear load, with a small error can be considered as a constant counter-emf connected in the discharge circuit [4]. The value of the counter-emf is assumed to be equal to the voltage drop across the arc at the current maximum.

An analysis of the experimental data shows that the average voltage drop across the arc for all experiments is 3.8 kV. The discharge is described by the system of equations

$$\begin{aligned} \frac{\mathrm{d}I}{\mathrm{d}t} &= \frac{IR + U - U_{\mathrm{a}}}{L} \\ \frac{\mathrm{d}I}{\mathrm{d}t} &= \frac{I}{C}, \end{aligned} \tag{11.10}$$

where C is capacitance of the capacitor bank, L is discharge circuit inductance, R is active resistance, U_{a} is voltage drop across the arc, and U is voltage of the power supply.

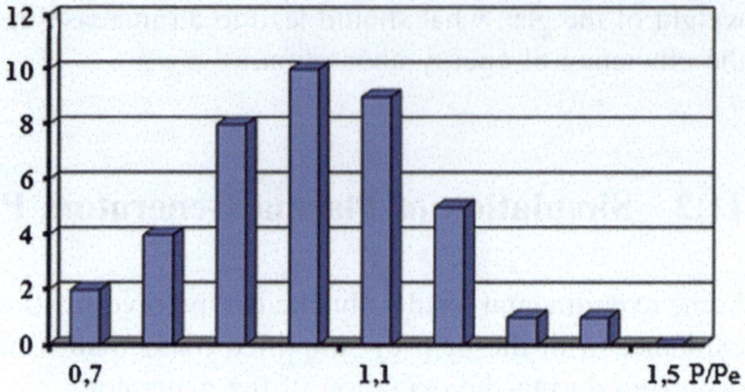

Fig. 11.7 Histograms of the distribution of the ratios of the calculated values of the pressure P to their experimental values P_e [3]

The solution is reduced to calculating of the current discharge parameters until the moment of the current passing through zero, which is equivalent to the condition of the arc extinction.

In this formulation, the possibility of repeated breakdowns is not taken into account, as data analysis shows that in the overwhelming majority of cases, the energy input from them is significantly lower than from the primary pulses.

To estimate the accuracy of this thermodynamic model, the maximal pressure in the discharge chamber was calculated from the data of 40 experiments. The histograms of the distribution of the ratios of the calculated pressures to their experimental values are shown in Fig. 11.7.

The pressure spread is largely due to the spread of the diaphragm rupture pressure, the value of which in the calculations was assumed to be constant and equal to 100 MPa.

References

1. J.F. Bianshetta, K.R. Siver, *Modern Technology of Aerodynamic Research at Hypersonic Speeds* (Mechanical Engineering, Moscow, 1965) (in Russian)
2. R. Rohtert, K. Sivier, *The Technique of Hypersonic Research* (Moscow, 1964) (in Russian)
3. V.A. Kolikov, Doctoral thesis (St. Petersburg, 2005) (in Russian)
4. I.A. Glebov, F.G. Rutberg, *Powerful Plasma Generators* (Moscow, 1985) (in Russian)

Chapter 12
Applications of Pulsed Plasma Generators

Abstract The development of industry is accompanied by the increase of its negative impact on the natural environment and man. One of the main types of such impact is the entry in the environment of toxic chemicals formed at the production stage, in the form of waste and/or in the form of these substances that have become unusable after their use. Until the middle of the twentieth century, most industrial wastes entered the natural environment almost without any processing, in terms of its (nature) adaptive capacity. However, as experience and scientific knowledge on the negative impact of industrial wastes on the biosphere have become evident, as so many toxic substances persist for a long time, migrate and transform, adversely affecting the environment, and significant harm is also caused to a person as at direct exposure to waste and through food chains. This especially applies to organic substances that have no analogues in the natural environment. The number of such substances is steadily increasing with the development of chemistry, chemical technology, and related industries. One of the tasks of researchers is to find ways to safely destroy toxic substances and their hardware design.

12.1 Application of Plasma Generators in Scientific Researches

Super- and hypersonic wind tunnels. These devices can be considered the first, in which powerful pulsed plasma generators were used. Wind tunnels are designed to study super- and hypersonic aerodynamics, simulate the entry of spacecraft into the atmosphere of planets, etc. The results of these studies are presented in the works [1–6]. The required parameters of the heated gas flow in the working part of the wind tunnel determine the parameters of the gas in the chamber of the plasma generator. One of the main requirements is the high temperature of the gas, in order to avoid its condensation at large degree of the gas expansion in the nozzle.

To ensure the stability of the gas flow parameters, it is necessary to maintain the gas parameters in the generator chamber. To meet this requirement, the most suitable is the chamber with rod electrodes. On the other hand, the operating time of

V. Kolikov et al., *Powerful Pulsed Plasma Generators*,
Springer Series on Atomic, Optical, and Plasma Physics 101,
https://doi.org/10.1007/978-3-319-95249-9_12

the wind tunnel (the time of the quasi-stationary gas flow) is usually much longer than the discharge time of the capacitive energy storage, and, therefore, the arcing time. Under these conditions, after rapid heating of the gas in the chamber, its flow through the nozzle occurs at decrease of the parameters of the braking of the flow, but at almost constant Mach number.

In this case, the stability and duration of the arcing process does not play a determining role; therefore, plasma generators with other types of discharge chambers having higher efficiency of heat transfer can be used as a gas heater.

Among the wind tunnels developed based on powerful plasma generators, the above-mentioned hypersonic pulsed wind tunnel of the aviation company McDonnell (Fig. 12.1) occupies a special place [1].

The generator is powered by the capacitor bank with energy of 7 MJ and voltage of 12 kV. The discharge chamber of the plasma generator is designed for a maximum pressure of 700 MPa. The parameters of the gas flow in the working part of the tunnel are next: speed of gas flow from 1800 to 5400 m/s, range of Mach numbers from 10 to 27 at gas density corresponding to an altitude of more than 30 km.

The wind tunnel has a simple conical nozzle. There is two operating parts: one of 1270 mm diameter for large Mach numbers and another of 762 mm diameter to simulate flight conditions at lower altitudes with large Re numbers and small Mach numbers.

The vacuum system is capable to produce a vacuum in the operating part of the tunnel up to 10–6 mm Hg. The design of the discharge chamber provides the possibility of replacing all elements of the electrode system as a separate unit.

The Institute of Theoretical and Applied Mechanics of the Siberian Branch of the Russian Academy of Sciences developed the pulsed hypersonic wind tunnel IT-301 (Fig. 12.2) [6].

In this installation, the coaxial discharge chamber was used [7]. The installation was powered by a capacitive storage with energy of 0.5 MJ and voltage of 5 kV. Nitrogen and air were used as the operating gases. Operating parameters are next:

Fig. 12.1 Plasma generator of the McDonnell's pulsed wind tunnel [1]

Fig. 12.2 Pulse wind tunnel IT-301 [6]

- Initial gas pressure in the discharge chamber 2.5–15.0 MPa
- Maximal pressure in the discharge chamber 200 MPa
- Gas temperature 1000–4000 K
- Mach numbers 8–16
- Reynolds numbers 10^5–10^8
- Diameter of the nozzle is 140 mm at M is 8–12, and 250 mm at 10–16.

Electric rocket engines. The using of plasma generators as a basis for low-thrust rocket engines, for example, spacecraft orientation engines is considered [8]. The main advantage of a plasma generator, in this case, is the ability to design an engine with a very high specific pulse, i.e., with a high gas flow speed. For a plasma generator, this parameter can be realized when hydrogen or helium is used as an operating gas.

Since reliability of operation and stability of output parameters are of great importance for such devices, the generators with a coaxial-type discharge chamber at relatively low initial gas pressure are the most suitable.

Technology and test benches. One of the possible applications of powerful plasma generators is the developing of test benches for testing various materials under extreme conditions. Such facility has the design similar with a hypersonic wind tunnel, but with a wider range of output parameters. Such possibilities of plasma generators as the producing of high-speed and high-temperature gas flows, saturation of the gas flows by micro- and macroparticles of the various materials, and achieving the pressures of several hundred MPa in the chamber, can be used. Obviously, depending on the specific task, the appropriate type of the discharge chamber should be selected.

Another promising use of plasma generators is a developing of facilities for the destruction of high toxic chemical compounds, for example, military poisonous substances. The working gas temperature of several thousand degrees, which is typical for plasma generators, is sufficient for the complete decomposition of any chemical compounds. At the same time, rapid cooling of the decomposed products at flowing through the supersonic nozzle ("hardening") prevents the formation of new toxic compounds.

A somewhat smaller spreading of plasma generators in such areas of science and technology as:

- Sources for gas-discharge, gas-dynamic, and gas-dynamic lasers with combined pumping, which make it possible to increase their efficiency by additional vibrational nonequilibrium
- Simulation of energy removal from a thermonuclear reactor
- Modeling the impact of some factors of nuclear explosion on structural materials
- Plasma chemical technologies at supercritical parameters.

12.2 Application of Plasma Generators for Toxic Substances Destruction

Since the beginning of the 1940s and 1950s, scientific and technical developments and industrial technologies for the processing of toxic waste have appeared with a view to neutralizing them, which in turn has led to a certain decrease of their negative impact on the environment and human beings.

Among the various methods of processing toxic waste, the most universal methods are high-temperature mineralization (HTM). The essence of the HTM method is in the thermal impact on waste, as a result of which their gasification or reduction takes place with the formation of inorganic substances. Typically, this reduces the toxicity of the waste or performed their complete detoxification. If necessary, gaseous products of the HTM can be subjected to additional purification.

The most common methods of the HTM implementation are incinerators. There, combustible wastes having various aggregate states burn in the air stream at temperatures of 1200–1700 K, depending on the chemical nature of the wastes. In those cases where the internal energy of the waste is not sufficient to maintain the required temperature, additional fuels are supplied in the combustion furnace.

At present, dozens of different designs and various technological schemes of the HTM are used in the world industrial practice, the design and instrumental features of which are determined by the parameters of waste, the characteristics of the main production facilities, and regional requirements for such devices and protection of the environment.

The idea of using plasma generators for the destruction of various high toxic substances and the decontamination of industrial and other wastes has become particularly relevant due to the fact that, as is known, the chemical binding energy of the molecules of some of these substances is so high that it is impossible to break these bonds using traditional methods [9]. Taking into account that the arc temperature in pulsed plasma generators is $\geq 25{,}000$ K, and the internal energy of the working gas reaches the multi-MJ level, under these conditions, depending on the chemical composition of the waste and the plasma-forming gas, the stable two–three atomic compounds can be formed: oxides, hydrides, and halides. The exhaust gases formed in the plasma chemical process, containing secondary toxic substances, undergo an additional purification, usually chemo-absorption.

Compared to incinerators, the plasma chemical installations for neutralizing waste have the following advantages:

- Possibility to adjust the temperature of the plasma-forming gas in the reactor from 1000 to 10,000 K
- Short reaction times in the reactor
- Much smaller mass and size characteristics of the reactor and the installation as a whole
- Possibility of full automation of the technological process
- Short time to put into the operating mode.

At the same time, plasma chemical installations have two main shortcomings:

- Limited operating resource of plasma generators
- Increased power consumption.

Comparison of incinerators with plasma chemical installations shows that the use of the former is advisable at neutralizing large quantities of complex mixed waste containing a large proportion of combustible waste and low-toxic substances. The use of plasma chemical installations is advisable at neutralizing relatively small amounts of concentrated high toxic substances. It is particularly advisable to use them at the site of formation or storage of high toxic waste. At destroying of high toxic substances, the special attention is paid to the problem of localizing accidents and eliminating their consequences that may arise in the technological process. Small-sized plasma chemical installations can be placed in hermetical rooms with remote automatic control of operation, both in normal mode and during localization of accidents and elimination of their consequences.

References

1. R. Rotert, K.P. Sivjer, *Technics of Hypersonic Researches* (Moscow, 1964) (in Russian)
2. D. Kaizer, *Technics of Hypersonic Researches* (Moscow, 1964) (in Russian)
3. D. Lukashevich, U. Garris, R. Jekson, D. van-der-Blick, R. Miller, *Technics of Hypersonic Researches* (Moscow, 1964) (in Russian)
4. J.A. Blick, *Modern Technics of Aerodynamic Researches at Hypersonic Speeds* (Moscow, 1965) (in Russian)
5. J.G. Kelley, M.A. Levine, A.L. Besse, A. Tatarian, Review Sci. Instr. **38**, 641 (1967)
6. A.S. Korolev, B.V. Boshenjatov, I.G. Druker, V.V. Zatoloka, *Pulse Tubes in Aerodynamic Researches* (Novosibirsk, 1978) (in Russian)
7. A.S. Antonov, B.V. Boshenjatov, *Aerophysics Researches* (Novosibirsk, 1972) (in Russian)
8. Ju.S. Gil'zin, *Electric Interplanetary Spacecrafts* (Moscow, 1971) (in Russian)
9. V.A. Kolikov, A.A. Bogomaz, A.V. Budin, A.F. Savvateev, Ph.G. Rutberg, in *Fifth European Conference on Thermal Plasma Processes*, St. Petersburg (1998)

Index

V. Kolikov et al., *Powerful Pulsed Plasma Generators*,
Springer Series on Atomic, Optical, and Plasma Physics 101,
https://doi.org/10.1007/978-3-319-95249-9

The manufacturer's authorised representative in the EU is Springer Nature Customer Service Centre GmbH, Europaplatz 3, 69115 Heidelberg, Germany. If you have any concerns regarding our products, please contact ProductSafety@springernature.com

Printed and bound by CPI Group (UK) Ltd, Croydon, CR0 4YY
05/10/2026
02228138-0010